11th International Conference on Energy Materials and Environment Engineering (ICEMEE 2025)

E3S Web of Conferences Volume 639 (2025)

Qingdao, China
13-15 June 2025

ISBN: 979-8-3313-2357-8

Printed from e-media with permission by:

Curran Associates, Inc.
57 Morehouse Lane
Red Hook, NY 12571

Some format issues inherent in the e-media version may also appear in this print version.

For additional information, please contact EDP Sciences – Web of Conferences
at the address below.

EDP Sciences – Web of Conferences
17, Avenue du Hoggar
Parc d'Activité de Courtabœuf
BP 112
F-91944 Les Ulis Cedex A
France

Phone: +33 (0) 1 69 18 75 75
Fax: +33 (0) 1 69 28 84 91

contact-edps@webofconferences.org

Additional copies of this publication are available from:

Curran Associates, Inc.
57 Morehouse Lane
Red Hook, NY 12571 USA
Phone: 845-758-0400
Fax: 845-758-2633
Email: curran@proceedings.com
Web: www.proceedings.com

TABLE OF CONTENTS

ENVIRONMENTAL ENGINEERING AND APPLICATIONS OF NEW MATERIALS

Preface

The 2025 11th International Conference on Energy Materials and Environmental Engineering (ICEMEE 2025) was successfully held in Qingdao, China on June 13, 2025. As the global scientific community continues to face complex challenges in achieving energy transition, environmental sustainability, and material innovation, ICEMEE 2025 provided a timely platform for interdisciplinary exchange among experts in energy systems, environmental technologies, and functional materials.

The conference attracted over 80 participants from academia, industry, and research institutes across multiple countries and regions, reflecting its growing international influence. A total of 77 full-paper submissions were received, among which 25 were selected through rigorous peer review for inclusion in this volume.

ICEMEE 2025 featured a distinguished lineup of keynote speakers who addressed critical issues at the intersection of climate science, materials engineering, and energy technologies. Prof. Hongquan Song (Henan University, China) shared insights on the modeling and ecological implications of climate change and air pollution. Prof. Yu Liu and Assoc. Prof. Dongdong Zhang (Northeast Electric Power University, China) presented frontier research on laser cladding coatings and particle-modulated alloy design, respectively—both contributing to the advancement of energy materials. Assoc. Prof. Md. Hasanuzaman (Universiti Malaya, Malaysia) explored the challenges and future prospects of energy storage materials for renewable power integration, highlighting a key enabler of low-carbon energy systems.

The technical discussions throughout the conference spanned a broad spectrum of topics, reflecting the complexity and interconnectedness of modern sustainability science. Research contributions covered clean and renewable energy systems, electrochemical storage and conversion, hydrogen and fuel cell technologies, nuclear safety, and multiphase heat transfer. Environmental themes included climate-pollutant interactions, air and water pollution control, solid waste toxicity, and ecological risk management. In parallel, materials-related sessions explored advances in degradable polymers, nanomaterials, sustainable catalysts, and lightweight composites—all with an emphasis on energy efficiency and environmental resilience.

By bridging fundamental research with applied innovation, ICEMEE 2025 reinforced the strategic role of integrated energy, environmental, and materials engineering in shaping a sustainable technological future. We would like to express our sincere gratitude to all contributing authors, keynote speakers, reviewers, and the organizing committee whose efforts made this edition of ICEMEE both intellectually rich and technically robust. We hope this volume serves as a valuable reference for researchers and professionals committed to advancing low-carbon, eco-compatible, and resource-efficient solutions.

ICEMEE 2025 Conference Committee

Committee Member

General Chairs
Prof. Haoyong Chen, South China University of Technology, China
Prof. Abdeltif Amrane, University of Rennes 1, France

Technical Program Committee Chairs
Prof. Jincheng Wang, Shanghai University of Engineering Science, China
Assoc. Prof. Md. Hasanuzaman, Universiti Malaya, Malaysia
Assoc. Prof. Pengfei Liu, Henan University, China

Organizing Committee Chairs
Prof. Yu Liu, Northeast Electric Power University, China
Assoc. Prof. Yangyang Wang, Henan University, China

Publication Chairs
Prof. Hongquan Song, Henan University, China
Prof. Jeyraj Selvaraj, University Malaya, Malaysia
Assoc. Prof. George N. Zaimes, Democritus University of Thrace, GREECE
Assoc. Prof. Xiaosheng Qin, Nanyang Technological University, Singapore
Assoc. Prof. Rongkui Su, Central South University of Forestry and Technology, China

Committee Members
Prof. Yu Liu, Northeast Electric Power University, China
Prof. Jincheng Wang, Shanghai University of Engineering Science, China
Prof. Adriano Péres, Universidade Federal de Santa Catarina, Brazil
Prof. Bachir Achour, University of Biskra, Angela
Prof. Vladimir Simón Montoya Torres, National University of Central Peru, Peru
Assoc. Prof. Ziyad Abunada, Central Queensland University, Australia
Assoc. Prof. Bingbing Wang, Northeast Electric Power University, China
Assoc. Prof. Lang Wang, Northeast Electric Power University, China
Assoc. Prof. Dongdong Zhang, Northeast Electric Power University, China
Assoc. Prof. Yali Gao, Northeast Electric Power University, China
Assoc. Prof. Atallah Ouai, Laghouat University, Algeria
Dr. Onn Chiu Chuen, Universiti Malaya, Malaysia
Dr. Xian Liu, Henan University, China

Statement of Peer review

In submitting conference proceedings to *Web of Conferences*, the editors of the proceedings certify to the Publisher that

1. They adhere to its **Policy on Publishing Integrity** in order to safeguard good scientific practice in publishing.
2. All articles have been subjected to peer review administered by the proceedings editors.
3. Reviews have been conducted by expert referees, who have been requested to provide unbiased and constructive comments aimed, whenever possible, at improving the work.
4. Proceedings editors have taken all reasonable steps to ensure the quality of the materials they publish and their decision to accept or reject a paper for publication has been based only on the merits of the work and the relevance to the journal.

Title, date and place of the conference

2025 11th International Conference on Energy Materials and Environmental Engineering (ICEMEE 2025)

June 13-15, 2025
Qingdao, China

Proceedings editor(s):

Prof. Hongquan Song, Henan University, China
Prof. Jeyraj Selvaraj, University Malaya, Malaysia
Assoc. Prof. George N. Zaimes, Democritus University of Thrace, GREECE
Assoc. Prof. Xiaosheng Qin, Nanyang Technological University, Singapore
Assoc. Prof. Rongkui Su, Central South University of Forestry and Technology, China

Date and editor's signature

Hongquan Song

Jeyraj Selvaraj

George N. Zaimes

Xiaosheng Qin

Rongkui Su

E3S Web of Conferences **639**, 01001 (2025) https://doi.org/10.1051/e3sconf/202563901001
ICEMEE 2025

A Discussion of Supply-side Solutions to Carbon Reduction in the Energy Sector

Qi Zhan*

London School of Economics and Political Science, London, UK

Abstract: Climate change is evolving into a very concern that has to be taken into account, and the energy sector is observed to be to blame for the excessive emissions of greenhouse gas (GHG), mainly carbon dioxide, among all kinds of human activities. While great efforts are made to cover the gap between carbon emission and absorption through energy consumption, limited attention has been drawn to the supply-side. Therefore, the study aims at exploring supply-side solutions to emission reduction in the energy sector by drawing a blueprint for low-carbon energy supply and providing suggestions for solutions based on the logic behind each strategy. The article divides the supply-side solutions into industrial solutions and auxiliary solutions in accordance with their connection with the energy industry. Further, it discusses sub-types of solutions along economic and technological feasibilities of renewable energy technologies (RETs). The research concludes that the low-carbon transition of energy supply is in urgent need for systematic supply-side solutions. Admittedly, the alignment with the climate goal of carbon reduction is the key component of supply-side solutions; nevertheless, the solutions should go beyond and give consideration to promoting the transition as a whole.

1. Introduction

With increasing concerns about climate change and other natural disasters along, the latest evidence points out, however, the acceleration of global warming and the significant risks it poses. In the Sixth Assessment Report (2023), the Intergovernmental Panel on Climate Change (IPCC) confirms that the average global surface temperature for 2011–2020 is 1.1°C above the pre-industrial baseline of 1850–1900.[1] The rise leads to a series of unpredictable consequences, such as frequent extreme weather and biodiversity loss – particularly, some of the impacts already go beyond adaptation. The report estimates that anthropogenic factors, GHG emissions in particular, are responsible for the warming. Specifically, among all kinds of human activities, the energy sector is often seen to contribute to emissions the most. In accordance to the United Nations (UN) (2023), energy generation and supply (mainly power and heat supply) brings around 35% of total emissions. [2] In addition to that, the International Energy Agency (IEA) (2024) finds that the source even accounts for 43.8% of global carbon emissions from fuel combustion and that carbon emissions from the power industry increased the most among all sectors between 1970 and 2022.[3] However, when including transport, manufacturing, buildings, and other industries related to energy consumption, the energy sector represented even as much as 73% of global emissions in 2017 in the estimate by the World Economic Forum (WEF) (2020). Therefore, reducing GHG emissions from the energy sector is a critical move in alignment with climate goals of 1.5°C and requirements for sustainable development as demonstrated by sustainable development goals[1].[4] As show in fig 1 and 2.

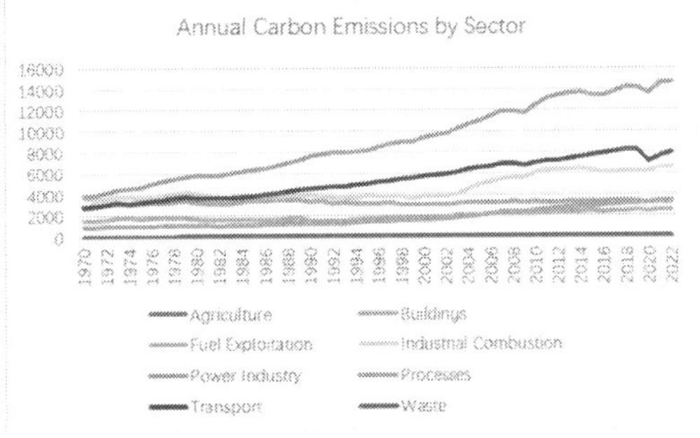

Fig. 1. Annual Carbon Emissions by Sector – IEA (2024).

[1]The climate goal, first put forward in the Paris Agreement, is an international and intergovernmental agreement to limit global temperature within 1.5°C (and well below 2°C) above the pre-industrial levels. The sustainable development goals (also SDGs) are a set of goals adopted by members of the UN, highlighting the inner-connections between different aspects of sustainable development and reflecting a general care for the peace and prosperity of all human beings.

*Zhanqi670@163.com

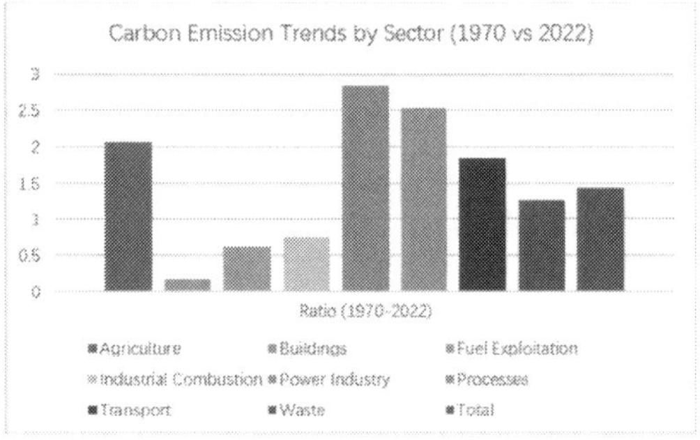

Fig. 2. Carbon emission trends by sector (1970 vs 2022) – IEA (2024).

Generally, GHG emissions from the energy sector are dominated by two forces: energy consumption and energy supply, which also spark a debate between demand-side and supply-side solutions to GHG reduction in the sector[2]. Supporters for demand-side solutions argue that "the higher the energy demand…, the slower the transition to carbon-free energy production." They contend that renewable energy supply must first meet the increasing demand driven by energy consumption growth before replacing and reducing fossil fuel production. Therefore, demand-side solutions are thought to be not merely necessary but flexible, practical, and resistant to environmental risks (Creutzig et al., 2018; Mayer et al., 2024).[5],[6] In contrast, the opposite, like Sinn (2008), refute that demand reductions are ineffective without corresponding responses from suppliers.[7] They emphasise the ineffectiveness in addressing market failures, such as the green paradox. However, it is surprising that supply-side policies have been far less widespread for a long time, and the gap between the two sets of solutions further limit the potential of supply-side solutions in practice. In response, this article aims to draw attention back to the supply-side solutions to carbon reduction in the energy sector by investigating the difficulties confronted by low-carbon energy supply and the solutions accordingly, thus providing a reference for policymaking.

To analyse the low-carbon transition in energy supply, this article is structured into five main sections. The first part is a brief introduction to the context and the structure of the research. The second part reviews previous studies on key definitions and recent theories regarding supply-side solutions for low-carbon energy supply and summarises the research gap in accordance. The third and fourth sections examine how the energy sector promotes low-carbon energy supply and how financial, fiscal, legal, and intelligent support complement the supply-side solutions. The final part includes a conclusion that finds that the industrial solutions and auxiliary solutions collectively constitute the implication of the supply-side solutions to carbon reduction in the energy sector, where both types of solutions promote the low-carbon transition of energy supply quickly and smoothly.

2. Literature review

2.1. Energy sector and supply-side solutions to carbon reduction

Before further discussion, the research needs to first qualify two definitions – "energy sector" and "supply-side solutions". Typically, the energy sector is a collective term from the production to the consumption of energy, or the totality of all the industries involved in the process. In practice, however, the categorisation of the energy sector may vary. A common approach is based on different types of energy, which often involves traditional energy, such as coal, oil, gas, and electricity of different types, like thermal power, wind power, water power, and nuclear power. Particularly, some studies tend to place renewable energy under the category of electricity with increasing practices of green electricity. Another widely adopted approach categorises the energy sector based on usage scenarios, such as electricity and heat generation, as well as sectors like buildings, agriculture, transportation, and chemicals. Though the two methods have different focuses, they both reflect the development of energy as a whole, indicating the interdependence and interaction between supply and demand. Therefore, the analysis below will take both definitions of the energy sector into consideration.

In contrast, the categorisation of supply-side solutions to carbon reduction in the academic literature is expansive and developing. Generally, the solutions refer to policies/strategies that "target a range of sectors, infrastructures, and products to achieve sustainability aims" or aim at mitigating GHG emissions by limiting the supply of carbon-intensive energy and coordinating decrease in fossil fuels consumption with demand policies (Trencher et al., 2023).[8] Early thinking concentrates on the economic dynamics of specific fossil fuels with production, pollution, and investment patterns, while recent studies explore the relationship more inclusively (Bohm, 1993).[9] For instance, some researchers divide the supply-side policies into restrictive (e.g., subsidy reduction, tax, and production quotas) and supportive ones (e.g., R&D subsidies and feed-in tariffs) (Green and Denniss, 2018).[10] Le Billion and Kristofferson (2020) make the interaction with fossil fuel consumption a primary concern.[11] Meanwhile, Lazarus et al. (2015) and Scott et al. (2022) attempt to expand the scope of research to include technology solutions, regulatory approaches, and information programs in the framework.[12],[13] In short, discussions on supply-side solutions have

[2] In a general context, the supply-side solutions to adapting and mitigating climate change refer to a range of measures to promote low-carbon/carbon-free energy supply or/and to reduce fossil energy supply, such as developing renewable energy technologies, improving energy efficiency, keeping fossil fuels underground and so on. Sometimes, the supply-side solutions may be generalized as the supply-side technology solutions centered on renewable energy technologies. In contrast, the demand-side solutions imply strategies that target consumer behavior, technology choices, lifestyle, and so on. However, there are also many measures covered by both sides, such as reducing energy waste and promoting carbon tax.

transitioned from industrial policies and fiscal and financial policies to a larger framework of technological and legal policies. In response to the evolution, the article attempts to develop a new framework to integrate the solutions based on Scott et al. and Lazarus et al. by dividing supply-side solutions into industrial solutions and auxiliary solutions, including financial, fiscal, legal, and intelligent policies, that support the implementation of industrial policies.

2.2. Criticisms on low-carbon supply-side solutions in the energy sector

Admittedly, there remains a controversy on the supply-side solution to carbon reduction in the energy sector. One group of scholars expresses concerns about the inefficiency and uncertainty of the supply-side solutions for reducing carbon emissions from the energy sector. Quiggin (2023) attributes the limitations to the need for scaling up and investment in the renewable energy sector, as some supply-side options for emission reduction are "still at the R&D phase." [14] Zhao, Li, and Zhang (2024), however, concentrate on the uncertainty of renewable energy supply due to its weather dependence.[15] Mashayekhi, Read, and Lindhult (2016) argue that neither gains in energy efficiency nor renewable supplants are efficient and effective in reducing energy demand. Some researchers express these concerns more mildly. Pellegrini and Arsel (2022), for instance, suppose supply-side policies should be "effective and complementary to existing policies" despite operational uncertainties. Wiedmann et al. (2020) think that the low-carbon energy technologies do play a positive role in controlling past emissions but contribute little to absolute emissions reductions at present. Several studies acknowledge the potential of low-carbon energy supply solutions but highlight the great barriers to implementing the policies, like Rayner (2021), who emphasises the inadequate institutional responses as an impediment to restricting international fossil fuel supply.

In contrast, another group of scholars considers supply-side solutions to be essential, especially in their relation to demand-side solutions. The demand-side solutions refer to a sum of policies and practices that aim at reducing energy demand and consumption by motivating consumers, including reducing energy waste, improving energy efficiency, and so on. In comparison, though also categorised according to the policy purpose, the supply-side solutions concentrate on improving low-carbon energy supply and energy efficiency from the supply side. On one hand, the reduction requires cooperation and coordination between demand-side and supply-side solutions. For instance, Paul and Moe (2023) agree that the coordination between supply-side and demand-side action is important not only for industrial analysis but also for achieving rapid decarbonisation "in a way that protects health and economic security." Prest (2022) start from carbon leakage and find it able to be effectively mitigated when policies from both sides are implemented "in parallel with similar vigor." On the other hand, solutions for low-carbon energy supply are thought to have distinct advantages over demand-side solutions. Sinn (2008) believes that effective mitigation measures must "succeed in flattening the carbon supply path" in global energy markets. Similarly, Dion (2019) attributes their contribution to the prevention of the "green paradox" and lock-in and stranded assets. Furthermore, Lazarus et al. (2015) identify the cost advantage of supply-side solutions, which allows for "greater emission reductions at the same (or lower) cost than demand-side policies alone." From the perspective of regulation, the cost-effectiveness is also reflected by cost-efficiency and easy enforcement of supply-side treaties compared to demand-side treaties (Stankovic et al., 2024).

In summary, most criticisms of low-carbon supply-side solutions in the energy sector focus on the technological and financial uncertainties, as well as potential inefficiencies in echoing changes in energy demand. Supporters, however, emphasise the coordination benefits of supply-side solutions and their cost-effectiveness in promoting the transition. One reason for this divergence is whether the supply-side solutions to carbon reduction in the energy sector are treated at the same level as demand-side solutions rather than, for example, a combination of technological solutions and financial solutions, which will be discussed in the following part.

2.3. A Return to supply-side solutions to carbon reduction in the energy sector?

Although the role and effectiveness of supply-side solutions to carbon reduction in the energy sector remain debated, the demand side increasingly dominates the discourse. In comparison, supply-side solutions are often seen as, according to Creutzig et al. (2018), a set of technology solutions or financial solutions. There appear to be different explanations for the mismatch between solutions of both sides. Lazarus et al. (2015) attribute the underestimation of supply-side solutions to resistance from political economies, consumption-orientated national accounting standards, and public perceptions. According to Santos et al. (2022), however, low compatibility with mainstream economics, choice of discount rate, and national heterogeneity in the rates of GHG emission and growth are to blame. In other words, the underperformance of supply-side solutions to carbon reduction in the energy sector is largely resulted from the constraints of established frameworks and outdating thinking patterns. To empower supply-side solutions, it is necessary to have systematic integration and enrichment of supply-side solutions. Newell and Daley (2024) believe supply-side climate policies are supposed to move beyond addressing specific fossil fuels and policies targeting certain stages in the fossil fuel value chain. Creutzig et al. (2017) agree on the significance of technological learning and support in promoting renewable energy deployment but emphasise the importance of effective yet various financing instruments and careful yet considerate management of system integration, like market stability reserve, public-private partnerships (PPPs), and other new forms of carbon credit like "pro set, for future success.

Additionally, Hille, Althammer, and Diederich (2020) complement how the provision of legislative and regulatory support promotes supply-side solutions.

In conclusion, the reduction in emissions from the energy sector is confronted with inevitable uncertainties generated by renewable energy substitutes and inefficiencies arising from research, development (thereof R&D), and deployment at the early stage, which underscores how technological and economic feasibility shape supply-side solutions. However, the challenges are more than that. The policy implications of supply-side solutions are largely simplified and distorted—technological and financial constraints have been overstated in previous literature. Another weakness revealed by past studies is the lack of systematic integration between different supply-side solutions. Consequently, excessive focus has been placed on isolated supply-side initiatives, like promoting energy efficiency or finding renewable energy supplants. In contrast, the interconnections among various supply-side solutions are ignored. As an equal party to demand-side solutions, supply-side solutions are not only important but necessary. Therefore, as pointed out, more efforts are required to develop supply-side solutions to carbon reduction in the energy sector.

3. Supply-side solutions to carbon reduction in the energy sector

To reduce carbon intensity from the energy supply, it is essential to phase out fossil fuels and switch to low-carbon or carbon-free energy sources. For this sake, both industrial and auxiliary solutions are needed. In detail, the low-carbon transition of energy supply involves examining how low-carbon energy supply displaces traditional fossil fuel supply and how the transition is motivated. Following the issue, this chapter will first investigate the industrial transition to low-carbon energy supply by considering technological and economic feasibility and then examine how financial, fiscal, legal, and intelligent supports and provisions align with reducing carbon intensity of energy supply.

3.1. Industrial solutions to carbon reduction in the energy sector

As mentioned above, there are mainly two factors determining the replacement of low-carbon/carbon-free energy for fossil fuels: technological feasibility and economic feasibility—the former refers to whether renewable energy can replace fossil fuels for specific uses, and the latter one indicates whether the renewable energy alternative can be scaled up. Accordingly, this section first analyses two types of industrial solutions: promoting the replacement of renewable energy and reducing carbon emissions from fossil fuels, based on technological feasibility, and then develops supply-side policy implications for each type of solution by combining economic feasibility and specific energy application scenarios.

3.1.1. Promoting the replacement of renewable energy

When RETs allow for the replacement of fossil fuels, there are mainly two sets of measures in promoting the replacement according to economic feasibility and attraction: the electrification of renewable energy (or green electricity/power) and non-electrical use of renewable energy. In detail, compared with thermal power, green power promises a low price with the help of strong grid systems and versatility of electricity. For example, since the National Development and Reform Commission promoted the development and deployment of wind power in 2005, the cost of power in China has decreased dramatically, even lower than that of thermal energy by 2021 (CICC, 2019). Meanwhile, similar practices can also be found across the world. However, green electricity may struggle to cover all the application scenarios of fossil fuels, like long-distance transportation and the chemical industry (Ficca, Marulo, and Sollo, 2023). Therefore, though less economically attractive and feasible, non-electrical use of renewable energy remains necessary.

During replacement for thermal power, the electrification of renewable energy is challenged by two issues: the instability of green electricity and the adaptability of the grid system. Firstly, green electricity is less stable than thermal power due to dependence on weather, geography, season, and so on, leading to variance in the quality of green electricity and challenges with grid connection (Zhao, Li, and Zhang, 2024). Secondly, apart from dependence on environmental issues, instability is reflected by inflexibility. Some renewable energy generation struggles to respond quickly to changes in grid load compared to thermal power, and consequently, additional costs for peak shaving also challenge green electricity (Sun et al. 2024). Thirdly, the introduction of green electricity also brings challenges to the preset grid capacity and grid-connectedness, as green electricity is taken little consideration during the design of the grid system (Mureddu et al., 2015). In response to the concerns, the electrification of renewable energy can be promoted in the following ways. On one hand, improving the stability of green electricity through energy storage is crucial—such technologies can both reduce pressures from peak shaving and effectively mitigate local grid overloads brought by green electricity access (Islam et al., 2024). For example, UK energy providers, like British Gas, promote solar panels and solar batteries for smart energy management. On the other hand, grid expansion and implementation of smart grids are expected to improve the adaptability of the grid by accommodating more intermittent renewable generation and increasing the capacity of grid-connected renewable energy (Hu et al., 2014). In addition to that, there is also evidence suggesting that the combination of multiple green powers is potentially able to compensate for the fluctuation of a single green energy, which improves stability and flexibility at the same time (Liu et al., 2020). In the case of King Island, Australia, the use of a mix of wind and solar power produces around 65% of its electricity with great stability.

Undeniably, the electrification of renewable energy offers a promising path for fossil fuels. However, due to the nature of electricity, the electrification does not suffice to meet all the replacement scenarios. Firstly, for long-distance transportation, like aviation, which is highly dependent on endurance and safety, green electricity, along with battery technologies, is challenged by low energy density, storage difficulty, and high-temperature sensitivity (Hızarcı and Arifoğlu, 2023). Secondly, for the metal smelting industry, like iron and aluminium, the redox reaction involved in the industrial process inevitably leads to the use of fossil fuels as well. Though the appearance of electric arc furnace technology has reduced coal usage to some extent, completely eliminating fossil fuels remains challenging (Reyes-Bozo et al., 2024). Thirdly, for some chemical industries, such as plastic, FF remains the ideal raw material (Levi and Cullen, 2018). Therefore, the non-electrical use of renewable energy serves as a necessary complement to electrification solutions. Hydrogen energy is one of the key alternatives to green electricity, particularly in the transportation and chemical industries. On one hand, hydrogen energy, along with hydrogen fuel cells, has high energy density and fast charging speed, which will match the need for long-distance transportation. On the other hand, hydrogen is also expected to replace fossil fuels as raw material in chemical industries, such as synthetic ammonia and methanol, or as a reductant in the smelting industry (Islam et al., 2024). Additionally, biofuel is considered a promising alternative energy source, though it remains under further investigation.

In the case of hydrogen-powered aircraft, for example, as the aircraft becomes technologically feasible, it appears to be an attractive alternative to the kerosene-powered aircraft in the long term. It is worth noting that fuel cells contribute little to reducing in-flight emissions unless they become lighter, more powerful, and more durable to ensure large, fuel cell-powered transport aircraft reliable in long-distance transportation. Therefore, lightweight liquid hydrogen tanks along with the integration into the airframe become crucial to the hydrogen-powered aircraft. With regard to the energy performance of liquid hydrogen, Gallagher, Stuart, and Spence (2024) compare liquid hydrogen aircraft with three different sustainable aviation fuel aircrafts and find liquid hydrogen aircraft yield the lowest energy consumption in almost all cases. In terms of environmental effectiveness, in the case of green hydrogen from electrolysis with wind energy, the total carbon emissions can decrease to 80% of that of fossil fuels according to Donateo et al. (2024).

3.1.2. Reducing carbon emissions from fossil fuels

However, RETs do not always guarantee the complete replacement for fossil fuels nor the provision of carbon-free energy supply. For instance, the production of natural-gas-based hydrogen (blue hydrogen) inevitably generates GHG emissions. Meanwhile, some RETs have difficulty breaking through within the short term (Ohene, 2023). As a result, the supply-side solutions to carbon reduction should not only focus on renewable energy suppliers but control carbon emissions where renewable energy is not able to be involved. In this sense, the clean use of fossil fuels could be promoted through inter-industry integration and inner-industry compensatory measures.

From the inter-industry level, industrial integration and cooperation can improve energy efficiency and help reduce systematic carbon emissions. Based on current modes of industrial collaboration, there are two ideas for industrial solutions to carbon reduction from fossil fuels. For industries with possibilities of material circulation or product circulation, it is important to establish business ecosystems and, hence, a circular economy (Marques-McEwan et al., 2023). Depending on the potential for circulation, two approaches to establishment can be considered. Specifically, when industrial symbiosis is possible, the by-product of one industry becomes the resource for another within the ecosystem, and the promotion of key technologies and the change of paradigm are necessary. When, however, the recycling of by-products does not suffice to provide resources for another industry, it is necessary to increase the proportion of recyclable products in the industrial ecosystems as much as possible, especially those involving carbon emissions. For example, the carbon emitted by chemical production can be captured and used as material for chemical production again (Budzianowski, 2017). Industries without opportunities for circulation can cluster together to enhance the energy efficiency of fossil fuels and infrastructures. Specifically, carbon emissions can be mitigated through collective investments in energy-saving equipment, resource sharing, or centralised carbon emission treatment.

Here is an instance of developing crcular economy within chemical industry: The surfactants of laundry detergent capsules, typically occupying 15–50% of the product weight, generally come from fossil carbon. In an innovative alternative practice, Unilever develops a new equipment to capture and liquify carbon emitted within the steel manufacturing plant. The liquified carbon is transformed into ethanol through a fermentation process and finally into surfactants for laundry detergent capsules. After the trial in Aisa, similar pilots are esablished in Europe and South Africa. Besides, once the detergent capsule is consumed, it will dissolve in water, and then be processed at individual water treatment plants worldwide. The following image by Marques-McEwan et al. demonstrates the industrial processes. As show in fig 3.

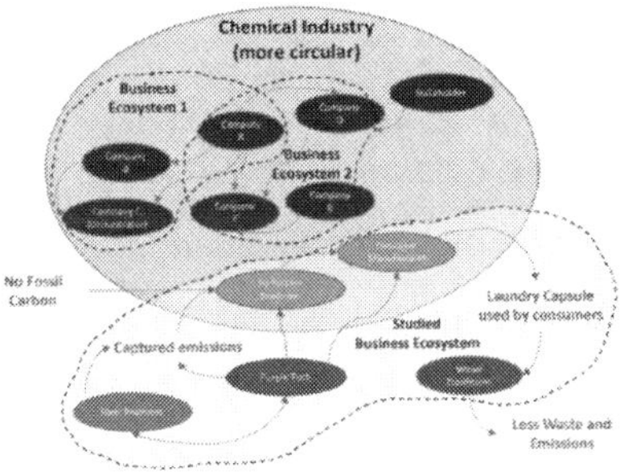

Fig. 3. Circular chemical industry system by Marques-McEwan et al.

From the inner-industry level, compensatory measures include both the introduction of carbon capture, utilisation, and storage (CCUS) technologies and the optimisation of industrial processes. Admittedly, apart from renewable energy supplants and inter-industry integration, there are still cases where neither technological nor economic feasibility is high enough to reduce the pressure from carbon emissions due to the limitations of RETs. Therefore, CCUS technologies are needed for carbon removal in industrial links involving carbon emissions, such as in the metal smelting, chemical, and cement industries but potentially do not suffice to be the main measure of carbon reduction. For example, the captured carbon emissions account for less than 0.5% of total emissions in the US, according to Mon, Tansuchat, and Yamaka (2024). Meanwhile, as mentioned above, the captured carbon can be used as a material as well, which further reduces the waste of energy. Similarly, optimising industrial processes can further reduce carbon emissions. The first approach involves introducing green equipment, such as switching to electric vehicles for material transportation, to reduce the reliance on fossil fuels. The second is to improve the efficiency of energy supply with improvements in the supply chain—for example, using the Internet of Things (IOT), big data, and blockchain to establish a smart supply chain by monitoring the input of fossil fuels or updating real-time carbon emission data. According to Arana-Landín et al. (2023), the four technology groups (AI, Big Data, IoT, and Robotics) will potentially improve energy efficiency by approximately 25% in processes where they are integrated globally. The third approach is to clean fossil fuels, such as replacing coal with natural gas and washing coal to improve combustion efficiency. As show in fig 4.

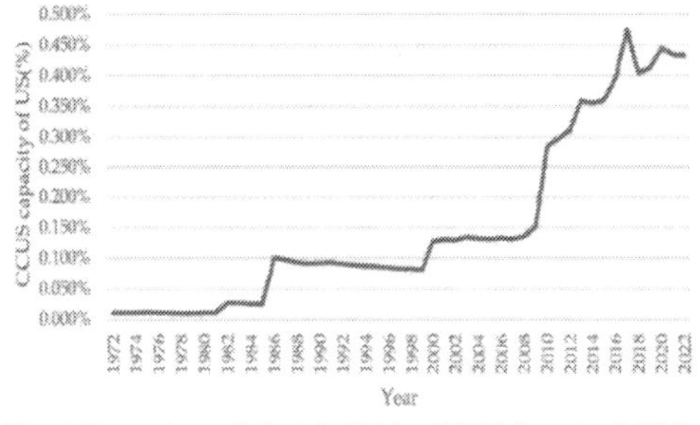

Fig. 4. Percentage of abated CO2 by CCUS from total CO2 emissions of US by Mon, Tansuchat, and Yamaka.

To sum up, according to the difference in technological and economic feasibility of renewable energy, there are two paths for the energy industry to achieve carbon reduction from the supply side. One approach is to promote the displacement for fossil fuels with renewable energy, and the other is to reduce carbon intensity of fossil energy. In the case of high technological feasibility, carbon reduction can be promoted by the electrification of renewable energy. Still, the non-electrical use of renewable energy, though likely less economically attractive, is necessary to complement the application of green electricity. In the absence of key technological breakthroughs in renewable energy, carbon emission intensity can be reduced at both inter-industry and inner-industry levels by improving energy efficiency and minimising energy waste, thereby smoothing the transition. In a word, in the process of carbon reduction, the energy sector needs to take both technological and economic feasibilities of renewable energy into consideration to achieve the relative advantage of renewable energy over fossil fuels and thus integrate the energy sector to align with climate change.

3.2. Auxiliary solutions to carbon reduction in the energy sector

In addition to industrial solutions of carbon reduction, the supply-side solutions also include how the low-carbon transition of the energy supply is guided and supported from the supply side. Therefore, auxiliary solutions to carbon reduction are crucial for aligning the energy sector with climate mitigation goals. During the process, a series of challenges emerge, and hence, there is a need for financing, fiscal, legal, and intelligent support to promote the transition. In this regard, this section examines auxiliary solutions from two aspects: promoting a targeted supply of funds and improving support services.

3.2.1. Promoting targeted supply of funds

As a key component of the low-carbon transition in the energy sector, the renewable energy industry is characterised by great uncertainty and requires substantial, long-term investment. Consequently, there is a great need for financing. However, the need for funds and investments is hardly met with current financing and fiscal solutions. On one hand, challenges such as insufficient green financing, difficulty in fund access, and mismatch of financing terms restrict the performance of low-carbon technologies. On the other hand, the lack of fiscal support makes it difficult for the renewable energy industry at the current stage to confront the challenges brought by the fossil fuel industry and market failure. Therefore, to improve the provision of low-carbon energy, it is necessary to promote the effective supply of funds with financing solutions and macro-guidance of funds with fiscal solutions.

Admittedly, whether for the R&D or deployment of renewable energy, there is a great need for funds "to get the projects off the ground and running." In contrast, the financing of renewable energy faces a series of challenges (Walsh, 2014). Firstly, renewable energy projects are subject to high upfront costs and high sensitivity to interest rates but often lack good collateral, which makes the projects less attractive to investors. Secondly, the traditional financial market, with its limited range of financial instruments, is unable to address the complex and diverse financial needs of the renewable energy industry (Eti et al., 2024). Thirdly, the renewable energy industry lacks effective measures to respond to financial and economic risks, further increasing its risk exposure (Wang et al., 2022). Therefore, financing solutions for low-carbon energy supply need financing channel expansion and innovation of financing mechanisms. On one hand, the motivation of public and institutional investments is important for green financing expansion. Particularly, these investments can help advance renewable energy technologies (RETs) from concept to commercialisation, as they are better equipped to address risks associated with R&D and information asymmetry. In 2022, for example, the world invested $154 billion in nature-based solutions, with public finance contributing 83% of the total (UNEP, 2022). On the other hand, similarly, innovative financing mechanisms are supposed to promote long-term, low-cost, stable financing by, for example, developing green mortgages and green insurance to reduce credit risks, establishing community-based trust funds to provide flexible and customised service, and initiating syndication to further split risks. In the case of the US, the Inflation Reduction Act of 2022 includes energy price insurance mechanisms with substantial tax credits and subsidies to reduce the costs and risks of renewable energy production.

Meanwhile, financing for the renewable energy sector is also confronted with challenges of market failures, such as strong market power of fossil energy groups, externalised environmental costs, limited market incentives, and high cost of capital for renewable energy, so fiscal solutions will aim at optimising the allocation of resources to overcome the additional costs brought by the transition and provide guidance for funds. Subsidies and taxes are effective approaches to internalising the price of carbon emissions and reallocating capital resources by subsidising the renewable energy industry and increasing taxes on the fossil energy industry (Meng and Yu, 2023). It should be noted, however, that to maintain a balance between tax and subsidy, mere reliance on tax or subsidy may still lead to investment inefficiencies. In addition to that, fiscal support should also involve the improvement of the identification of renewable energy and green technology industries to reduce the distortion of capital supply from greenwashing[3]. Specifically, identification such as ecolabelling, the green certificate, and ESG data exposure provides tools for tracking the carbon footprint of energy industries and keeping continuous monitoring, which would be helpful to provide a reference for subsidy and tax by promoting the targeted supply of funds. For instance, the Danish Ministry of Environment appoints Ecolabelling Denmark, responsible for EU Ecolabel and the Nordic Swan Ecolabel, to certify products and services that meet environmental requirements.

3.2.2. Improving supporting services

Beyond the provision of targeted funds, the external environment is also indispensable for empowering the transition and increasing technological and economic feasibility. Particularly, the supply of supporting services—such as legal and intelligent services—not only fosters a favourable institutional environment by promoting market access to renewable energy, protecting renewable energy intellectual property (thereof IP), and improving regulatory compliance but increases the technological feasibility of low-carbon/carbon-free technologies and hence flattens the path to the transition through preparation for innovation and expansion of the talent pool.

The need for legal services is mainly divided into two categories: legislative and regulatory support. For the former, the need comes from the update of legal measures compatible with renewable energy, protection of RETs intellectual properties, and compatibility between renewable energy and existing energy systems (Oduro, Uzougbo, and Ugwu, 2024). For the latter, the challenges are brought by access to renewable energy projects, constraints on investment, and uncertainty in environmental performance (Grove and Clouse, 2021). Accordingly, the legislative solutions should first simplify legal procedures (especially for licensing), increase the consistency of law when crossing jurisdictional areas, and establish a clear legal framework to improve the stability and reliability of legal support. Apart from that, intellectual property protection should be strengthened but balanced. Undeniably, IP protection will guarantee the returns of RETs as a general-purpose technology, but the protection can potentially hinder the adoption of renewable energy by increasing the cost. For regulatory support, similarly, regulation should make a trade-off between motivating renewable options and improving compatibility with environmental sustainability, which might need a closer combination with transparency improvement and data exposure.

For example, UK enacted The Energy Act 2023, a major new legal framework that establishes modified rules for energy production, security and sectoral regulation. The Act introduces new frameworks and mechanisms to support green technologies such as low-carbon hydrogen projects and seeks to facilitate speedier deployment of or projects by, for example, making changes to environmental regulations for offshore wind. Likewise, on 8th November 2024, the Energy Law of the People's Republic of China is passed and will be brought into force on 1st January 2025. The law sets out requirements in a number of areas, ranging from energy structure to green energy consumption and is expected to promote the development of the renewable energy industry against climate goals of carbon peak and carbon neutrality.

Finally, intelligent support for the low-carbon transition encompasses not only the accumulation and innovation of technologies but also the development of human capital. From the perspective of technological innovation, on one hand, the disconnection between basic research and applied R&D makes it difficult for knowledge tanks to support technological breakthroughs; on the other hand, there is a "valley of death" from the conceptualisation of RETs to industrialisation and marketisation (Raihan et al., 2024).In terms of human capital, there is a significant demand for green human capital to master clean technologies and be familiar with

[3] Greenwashing refers to that a company makes misleading or exaggerated claims about the environmental impact of their products or services.

environmental policies and standards. However, a significant gap remains between education and industrial needs. In this sense, the connection between education, innovation, and industrialisation can be strengthened with the industry-university-research (IUR) cooperation, where universities serve as research centres supplying technologies to the renewable energy industry, by which education and innovation make a good match. Take the University of Edinburgh, for example, the university invests over £3 million of quality research funding to drive the development of innovative energy technologies in cooperation with the Tyseley Energy Park, an energy innovation zone focused on energy storage, clean transport fuels, and related advancements. Concerning the gap between basic research and technological application, a possible solution is to promote cooperation between upstream and downstream industries with complementary advantages and resource sharing and to bridge the technology supply with demands.

In summary, improving the technological and economic feasibility of the industrial solutions mentioned above, auxiliary solutions are expected to complement the former in terms of financing, fiscal, legal, and intelligent support. On one hand, the supply of low-carbon energy comes with a large capital demand but high uncertainty. Therefore, it is necessary not only to expand the financing channels of public and institutional investment but also to provide long-term, stable, and reliable funds compatible with the characteristics of the renewable energy industry. On the other hand, the effective supply of capital as a driver of low-carbon transition requires fiscal guidance to overcome the distortion of financing, especially from market failure and additional costs. In terms of supporting services, improvements in legislation and regulation help create a favourable institutional environment where the transition is promoted smoothly. Particularly, the stability and consistency of laws are essential to empower the development of a clean energy supply. At the same time, IP protection and regulation, though equally important, are supposed to be carefully balanced between different needs. Finally, in terms of intelligent support, the university-industry partnerships contribute to bridging training and research and filling the talent gap, while supply chain would be a good opportunity to accelerate the process from conceptualisation to application. In short, the auxiliary solutions are, as indicated, a complement to industrial solutions—together, they will leverage the huge potential from the supply side and promote the low-carbon transition of the energy sector quickly and smoothly.

4. Concluding remark

The study attempts to explore the supply-side solutions to carbon emission reduction in the energy sector by drawing a blueprint for low-carbon energy supply solutions and identifying the logic behind each strategy and the solutions as a whole. The article divides the supply-side solutions into industrial solutions and auxiliary solutions in accordance with their connection with the energy industry. Based on the economic and technological feasibility of renewable energy technology, industrial solutions can be further divided into electrical and non-electrical uses of renewable energy, industrial integration and cooperation, and cleaner use of fossil fuels. Concerning the support for improving the feasibility, auxiliary solutions include financial, fiscal, legal, and intellectual support. The article finds that the low-carbon transition in the energy sector displays a huge and systematic need for supply-side solutions. While aligning the energy sector and the goal of carbon reduction is a key component of the supply-side solutions. Still, the solutions are supposed to move beyond and pay attention to the promotion of the process as a whole. Meanwhile, even though industrial solutions have rich policy implications and a basic logic for implementation, the technological and economic feasibility of renewable energy remains pivotal. In addition to that, the study also concludes that there are complex dynamics in the low-carbon transition of the energy sector, indicating that the promotion of carbon reduction is supposed to be carried out in a quick but smooth manner.

The study has potential limitations. Despite an attempt to reconceptualise the supply-side solutions to carbon reduction in the energy sector, it remains primarily focused on the intersection of technological and economic solutions. On one hand, the study starts from the supply of low-carbon energy motivated by renewable energy technologies, presuming technological and economic feasibility as the rationale for selecting solutions. On the other hand, the study fails to include other potential topics, e.g., culture and entrepreneurship, into the discussion nor contribute to filling the gap of why the supply-side solutions have long been underestimated, such as a shift to supply-orientated national accounting standard. Meanwhile, the study pays limited attention to the interaction and synergy between supply-side and demand-side solutions in promoting carbon reduction, despite the collective influence of energy supply and consumption dynamics on the process. However, the article still contributes to drawing a blueprint of current practices to provide a policy toolkit accordingly and illustrating the rationale for choosing different strategies of the supply-side solutions to bridge the gap between achieving climate goals and harmonising the transition.

Moving forward, the further discussion may concentrate on the following topics. Firstly, how effective are supply-side and demand-side solutions in reducing carbon emissions, and what contributes to the difference in effectiveness, For instance, technological breakthroughs like renewable energy technology potentially promote the alignment with climate goals in the long term but require a long period of R&D and investment beforehand and thus are less likely to be effective in the short term. Secondly, how does the political economy reshape the discussion of the supply-side solutions versus demand-side solutions in the energy sector and dominate the discourse, which would be helpful to complement current research of productive factors and contribute to the discussion of a just and sustainable transition, Thirdly, how do the supply-side solutions to carbon reduction in the energy sector respond to multiple development goals, such as economic growth, equality, and climate goals, in a national case study, which would be important for identifying the dynamics between different development goals and promoting a better trade-off wherever there is a conflict.

References

1. Aldunce, P., Blanco, G., Calvin, K., Cheung, W., Dasgupta, D., Denton, F., Diongue-Niang, A., Dodman, D., Garschagen, M., Geden, O., Hayward, B., Jones, C., Jotzo, F., Krinner, G., Lasco, R., Lee, J., Meinshausen, M., Mukherji, A., Otto, F., Revi, A., Roy, J., Slangen, A., Sörensson, A. A., Thorne, P., Trisos, C., Van Vuuren, D., Wei, Y. M., Winkler, H., Zommers, Z. (2023) *The Sixth Assessment Report of the Intergovernmental Panel on Climate Change*. Available at: https://www.ipcc.ch/assessment-report/ar6/ (Accessed: 10 November 2024).
2. United Nations–Facts and Figures (https://www.un.org/en/actnow/facts-and-figures#:~:text=Home%20energy,-Electricity%20drives%20our&text=The%20energy%20supply%20sector%20(electricity,are%20supplied%20from%20fossil%20fuels).
3. Greenhouse Gas Emissions from Energy Data Explorer (https://www.iea.org/data-and-statistics/data-tools/greenhouse-gas-emissions-from-energy-data-explorer).
4. 3 Charts that Show Countries and Sectors with the Highest Greenhouse Gas Emissions (https://www.weforum.org/stories/2020/12/climate-change-greenhouse-gas-emissions-environment-paris-agreement/#:~:text=2.,to%20net%20zero%20by%202050).
5. Creutzig, F., Roy, J., Lamb, W.F., Azevedo, I.M., Bruine de Bruin, W., Dalkmann, H., Edelenbosch, O.Y., Geels, F.W., Grubler, A., Hepburn, C. and Hertwich, E.G. (2018) 'Towards Demand-Side Solutions for Mitigating Climate Change', Nature Climate Change, 8(4), pp.260-263.
6. Mayer, P., Heer, M., Shu, D.Y., Zielonka, N., Leenders, L., Baader, F.J. and Bardow, A. (2024) 'Flexibility from Industrial Demand-Side Management in Net-Zero Sector-Coupled National Energy Systems'. Frontiers in Energy Research, 12, pp.1443506.
7. Sinn, H.W. (2008) 'Public Policies Against Global Warming: A Supply Side Approach', International Tax and Public Finance, 15, pp.360-394.
8. Trencher, G., Rinscheid, A., Rosenbloom, D., Koppenborg, F., Truong, N. and Temocin, P. (2023) 'The Evolution of "Phase-Out" as a Bridging Concept for Sustainability: From Pollution to Climate Change', One Earth, 6(7), pp.854-871.
9. Bohm, P. (1993) 'Incomplete international cooperation to reduce CO2 emissions: alternative policies', Journal of Environmental Economics and Management, 24(3), pp.258-271.
10. Green, F. and Denniss, R. (2018) 'Cutting with both arms of the scissors: the economic and political case for restrictive supply-side climate policies', Climatic Change, 150(1), pp.73-87.
11. Le Billon, P. and Kristoffersen, B. (2020) 'Just cuts for fossil fuels? Supply-side carbon constraints and energy transition', Environment and Planning A: Economy and Space, 52(6), pp.1072-1092.
12. Lazarus, M., Erickson, P. and Tempest, K. (2022) *Supply-side climate policy: the road less taken*. Stockholm Environment Institute.
13. Scott, K., Smith, C.J., Lowe, J.A. and Garcia-Carreras, L. (2022) 'Demand vs supply-side approaches to mitigation: What final energy demand assumptions are made to meet 1.5 and 2° C targets?', Global Environmental Change, 72, pp.102448.
14. Quiggin, D. (2023) 'Net zero and the role of the aviation industry'.
15. Zhao, J., Li, F. and Zhang, Q. (2024) 'Impacts of renewable energy resources on the weather vulnerability of power systems', Nature Energy, pp.1-8.

Design of a Smart Street Light-Integrated Electromec-hanical System for Willow Catkin Control

Ting Zhou [1,a], Ye Yuan [2,b], Xue Yin [1,c*], Yuqian Liu [1,d*], Junxiu Sun [3,e], Fenghua Qi [1,f*]

[1]School of Electronic Engineering, Nanjing Xiaozhuang College, Nanjing, China
[2]School of Mechanical Engineering, Nanjing Engineering College, Nanjing, China
[3]School of Information Engineering, Nanjing Xiaozhuang College, Nanjing, China

Abstract. Aiming at the lack of development of extended functions for environmental management and failure to fully utilize the potential of streetlights as urban infrastructure, this study proposes a modular intelligent streetlight system integrating electrostatic adsorption and resource recycling. By integrating electrostatic spinning technology (electric field strength of 15 kV/m), photovoltaic power generation (conversion efficiency ≥21%) and bionic water-collecting device (water-collecting efficiency of 85%), an urban infrastructure with environmental purification function is constructed. The experimental data confirmed that under the working conditions of 30%-90% relative humidity and 0-15 m/s wind speed, the system's willow capture efficiency reaches (85±3)% (n=5), of which (80±2.1)% is still maintained at 90% humidity ($p<0.01$, ANOVA), and the comprehensive energy efficiency is improved by 42.7% compared with that of the traditional street light (CI: 39.5%-45.8%). After 0.5 MPa pulse spray treatment, the willow residue was significantly reduced to (5.2±0.7)% (Δ=75.3%)". This study provides a scalable electrostatic adsorption technological solution for urban willow wadding pollution treatment.

1. Introduction

With the acceleration of urbanization, willow floc pollution has become a seasonal environmental management challenge in the northern regions of China. Willow flakes carry allergens and particulate matter (e.g., PM10), and their seasonal dispersion leads to deterioration of urban air quality. Studies have shown that PM10 concentrations can increase by 30%-50% during periods of high willow flocculation, and the rate of allergic disease visits rises by about 25%. Willow flocculation not only affects air quality, but is also prone to respiratory allergies, fire hazards and urban cleaning burden [1] . Traditional means of control such as manual sweeping, water spraying and flocculation suppression are inefficient (less than 40% coverage), and mechanical sweeping is prone to secondary dust, so there is an urgent need for an intelligent and sustainable solution. At the same time, the demand for multi-functional public facilities in the construction of smart cities is growing, and streetlights, as the most widely distributed infrastructure in the city, are gradually evolving from a single lighting function to an integrated and intelligent direction [2] . How to utilize the street light carrier to integrate environmental governance functions has become an important issue to enhance the effectiveness of urban management.

In recent years, electrostatic spinning technology has demonstrated significant advantages in the fields of air purification and particle capture due to its high efficiency, low energy consumption and adaptability. Studies have shown that electrostatic adsorption devices can effectively adsorb micron-sized particles (e.g., PM2.5, pollen, etc.) through high-voltage electrostatic fields [3] , as well as there are electrostatic adsorption devices[4] that face the problem of humidity sensitivity and energy consumption in outdoor (efficiency decreases by 50% at humidity > 70%), and they have been successfully applied in scenarios such as robotic grasping and ship exhaust treatment[5]. Streetlights, as urban distributed nodes, have been integrated with environmental monitoring, data transmission, and other functions (e.g., Beijing Smart Streetlight Demonstration Project)[6][7][8]. However, existing technologies are mostly limited to closed or specific scenarios and have not yet been deeply integrated with urban public facilities. Nonetheless, the existing smart streetlight functions are still dominated by lighting and communication, and there is a lack of development of extended functions for environmental governance [9][10][11], which fails to give full play to the potential of streetlights as urban infrastructures.

In summary, this study innovatively combines the electrostatic spinning technology[12]with urban street lamps, and realizes: ① high-voltage electrostatic field (15

[a]e-mail: 1257356926@qq.com
[b]e-mail: 2992442971@qq.com
[c*]e-mail: 2458128019@qq.com
[d*]e-mail: 1551726194@qq.com
[e]e-mail: 3412924558@qq.com
[f*]e-mail: Qifenghua.hi@163.com

kV/m) directional adsorption; ② nano-fiber membrane (φ80-200 nm) depth filtration; ③ bionic water collection-photovoltaic power supply synergistic mechanism by the design of multi-physical field coupling, which not only solves the limitations of the traditional willow wadding management but also optimizes the utilization of the urban space resources through multi-functional integration in line with the concept of smart city construction and sustainable development. It also optimizes the utilization rate of urban space resources through multi-functional integration, which is in line with the sustainable development concept of smart city construction [13] .

2. System design and implementation

2.1. Mechatronics architecture

The multifunctional street light system proposed in this study adopts a modular design concept [14][15] , and the main structure contains:

Composite functional light pole: the main structure made of Q345B low carbon alloy steel has excellent mechanical properties (tensile strength ≥ 470 MPa, yield strength ≥ 345 MPa) and weather resistance characteristics. The pole adopts main and vice pole segmented design (main pole Φ200×8 mm, vice pole Φ150×6 mm), and realizes reliable connection through the high-strength flange in accordance with GB/T 3098.1-2010 standard. Polycrystalline silicon photovoltaic module (conversion efficiency ≥21%) is integrated on the top of the main pole, combined with MPPT intelligent controller to maximize the photoelectric conversion efficiency. Circumferentially distributed LED lighting modules (color temperature 5000K, luminous efficacy ≥ 120 lm/W) realize intelligent power supply management through IP65 protection grade distribution box.

Multi-stage pollutant adsorption system: based on electrostatic spinning technology to build a two-stage filtration system: the top layer is equipped with a high-voltage electrostatic field module with an electric field strength of 15 kV/m, which can effectively capture willow particles with a particle size of >5 μm; the second layer adopts a nano-fiber membrane (fiber diameter of 80-200 nm) to build a depth filtration barrier, with a retention efficiency of 98.7% for PM2.5 particles. The bottom is equipped with a double-chamber water storage device (PP material, volume 80 L), through the solenoid valve-controlled spray system (flow rate 0.8 m^3/h) to realize the pollutant automatic rinsing, supporting the willow collection barrel built-in 316L stainless steel filter (pore size 100 μm) to achieve the solid-liquid separation.

Umbrella Intelligent Water Collector: It is made of 210T polyester tarpaulin[16] and aluminum alloy skeleton, which is lightweight and durable, with a water collection efficiency of up to 85%, and a maximum water capacity of 120L/h, which improves the efficiency of water resources utilization. The umbrella-shaped design can effectively expand the water collection area and adapt to a variety of weather conditions, especially outstanding in rainwater collection, and can also provide shade from the sun and rain. The aluminum alloy skeleton ensures structural stability and wind resistance, suitable for long-term outdoor use.

Intelligent Surveillance Module: Integrated Hikvision DS-2CD3T47EWD-L 4-megapixel CMOS sensor with 1/1.8" oversized target surface realizes HD imaging (2560×1440@30fps) under 0.001 Lux low-light environment. Deep learning algorithms to achieve face recognition (accuracy ≥ 99.2%), behavior analysis and other intelligent functions, support ONVIF protocol for multi-platform access. Protection level up to IP68, operating temperature range -40°C to +70°C.

Intelligent Drainage System: The three-stage drainage system built based on PLC includes: differential pressure sensor (range 0-10 kPa) for real-time monitoring of filter status; PID-controlled pulse spray system (pressure 0.2-0.6 MPa adjustable); and IoT transmission module (NB-IoT, transmission interval ≤15 min) for remote status monitoring. The system is equipped with a double safety protection mechanism, which automatically triggers the emergency drainage program when the water level sensor detects that the water storage capacity is ≥90%.

The overall structure of the system is shown in Fig. 1, and its innovativeness is reflected in:①the functional intensification of municipal facilities through mechatronic design; ② the umbrella-like water collection structure based on the principle of bionics optimizes the aerodynamic performance of the traditional water collection device; and ③ the development of an intelligent operation and maintenance system based on edge computing to realize the real-time monitoring of equipment status and predictive maintenance. Experimental tests show that the system's comprehensive energy efficiency in typical application scenarios is 42.7% higher than that of traditional street lights, and the PM2.5 removal rate can reach 81.3%, which has significant environmental benefits and engineering application value.

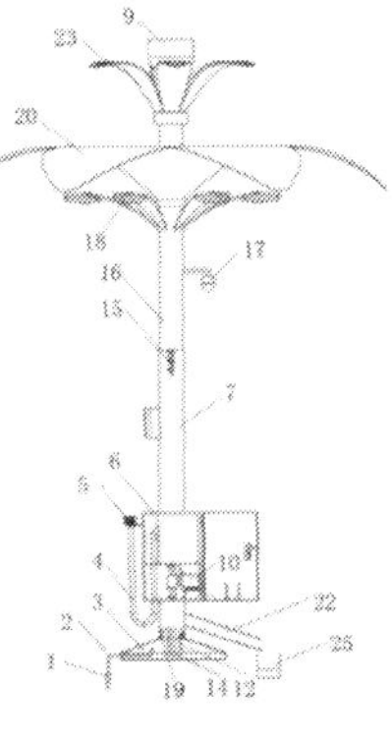

Figure 1. Schematic diagram of the overall street light structure of the multi-function street light system.

2.2. Power system design

The multifunctional street light system proposed in this study mainly controls the switching operation of each structural component through the control chip (STM32G474RET6) as shown in Figure 2, which is connected to the first water pump, the second water pump, the motor, the photovoltaic panel, the static generator, the lighting lamp, the monitor, and the water level sensor, respectively, by means of different pins, to realize the intelligent management of each power-using equipment, distribute the power according to the demand, and ensure that the system is Efficient operation. Its technical advantages include: ① 12-bit ADC realizes maximum power point tracking of PV array (tracking error <1.5%), ② hardware CRC checksum ensures NB-IoT communication data integrity, and ③ dynamic power management reduces the standby current to 8.3 μA, etc., which is suitable for use in the street light retrofit project.

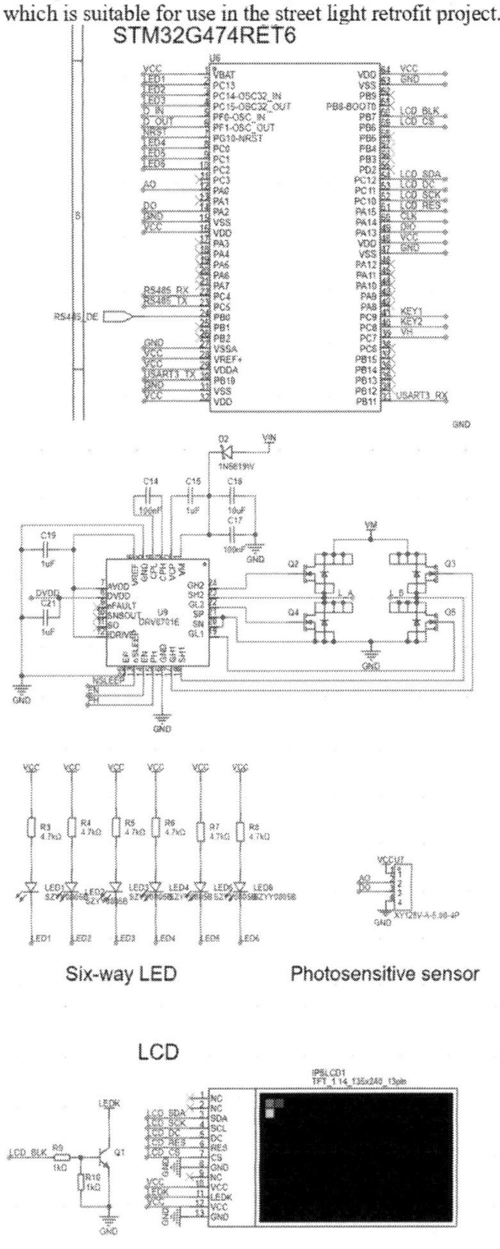

Figure 2. Local schematic of the microcontroller STM32G474RET6.

3. Core technological innovations

3.1. Model of electrostatic spinning fusion street lighting system

Electrostatic spinning utilizes a high-voltage electrostatic field to charge and deform the polymer solution, forming an overhanging cone-shaped droplet at the end of the nozzle. When the droplet surface charge repulsion exceeds the surface tension, tiny jets are ejected from the top of the Taylor cone. The jets are stretched at high speeds by electric field forces, and the solvent evaporates and solidifies to form nanofibers. The final fibers are deposited on the receiving device to obtain nanofiber material with small diameter, high specific surface area and high porosity[17].

One of the core driving force of electrostatic spinning of the applied electric field, electric field distribution based on Maxwell's equations formula is as follows, where E represents the electric field strength (V/m), V is the applied voltage (V), d is the spinneret and acceptor plate spacing (m), is the electric potential (V), is the space charge density (C/m3):

$$E=\frac{V}{d} \tag{1}$$

$$\nabla.(\varepsilon\nabla\varphi) = -\frac{\rho}{\varepsilon_0} \tag{2}$$

When the electric field force and the liquid surface tension reaches equilibrium, the formation of the Taylor cone formula is as follows, where is the solution surface tension (N/m), is the vacuum dielectric constant, and r is the radius of the nozzle (m):

$$E_{critical}=\sqrt{\frac{4\gamma}{\varepsilon_0 r}} \tag{3}$$

$$\gamma = \frac{rE^2\varepsilon_0}{2} \tag{4}$$

The motion of the jet is governed by electric field forces, viscoelastic forces and surface tension. The Ferndon fluid eigenequations are formulated as follows. T is the stress tensor (Pa). λis the relaxation time (s), η is the solution viscosity (Pa.s), and D/Dt is the convective conductance:

$$T+\lambda\frac{DT}{Dt} =\eta(\nabla v + (\nabla v)^T) \tag{5}$$

This nanofiber material applied to streetlights can effectively adsorb tiny particles in the air, such as PM2.5, pollen, willow flakes, etc., and purify the air around the streetlights[18]. And with photocatalytic self-cleaning function of electrostatically spun material, can be covered with a layer of electrostatically spun nanofiber film on the surface of the streetlight lampshade, to improve the service life and safety of the streetlight, or the top of the lamp post to install an electrostatically spun fiber filter such as the label 24 shown in Figure 3 (electrostatic generator). The former use the larger surface area of the lampshade, so that more air and fiber film contact, improve the adsorption efficiency, reduce dust into the street lamp internal circuit and parts of the erosion; the latter through the high-voltage electrostatic field will be introduced into the top of the street lamps, after the fiber filter purification and then discharged, direct as a small air purifiers, continuing to around the air, "Detox! ".

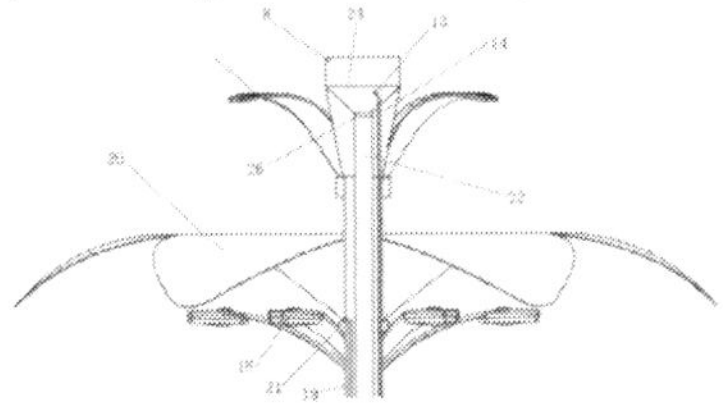

Figure 3. Localized view of street light.

Since the electrostatic spinning technology may face problems such as aging and breakage of the fiber membrane in complex outdoor environments, this study uses weather-resistant polymer materials [19] to enhance the fiber membrane's resistance to ultraviolet rays and erosion by wind and rain; the electrostatic spinning device on the streetlight is regularly inspected and maintained, and the damaged parts are replaced in a timely manner. In addition to this, this study installs drying rods inside the street light near the circuit and key components, such as in the electrical box at the bottom of the pole, or at the connection between the lampshade and the pole. Drying rods are usually made of highly absorbent materials such as silica gel and diatomaceous earth. Its function is to absorb moisture in the air around the street light and reduce humidity. In a humid environment, moisture may lead to short-circuiting of internal circuits and rusting of metal parts, affecting the service life of streetlights. The drying rod can effectively alleviate these problems and ensure the stable operation of streetlights. For example, in the rainy season in the south, the air humidity is high, and the drying rod can absorb the excess moisture and prevent the street light from being damaged due to moisture. In order to play a better role, according to the size of the street lamp and the humidity of the environment, the number and distribution of drying rods can be reasonably arranged to ensure that the entire interior space of the street lamp can be effectively dehumidified.

3.2. Efficient cleaning and water recycling mechanisms

There is a willow wadding collection chamber inside the top of the street lamp, located below the electrostatic spun fiber filter, the design of which gives full consideration to the collection and cleaning needs of willow wadding. The lower end of the collection cavity is connected to the sewage pipe in the body of the pole through the solenoid valve, when a certain amount of willow flakes are collected, the solenoid valve can be controlled to open, the willow flakes will be discharged into the sewage pipe. The diameter of the sewage pipe is not less than 10cm, and in this embodiment it is 20cm, which effectively avoids the situation of the willow flakes sticking and blocking the sewage pipe[20] , and ensures the smooth discharge of the willow flakes.

In order to ensure the continuous operation of the willow wadding system adsorbed by the electrostatic spinning technology, this study adds a cleaning component including a nozzle, a water pipe and a water storage chamber, and the nozzle is set in the willow wadding collection chamber and is connected to the water storage chamber through the water pipe. When it is necessary to clean the willow wadding in the willow wadding collection chamber, a motor is activated to drive the water pump, and the water in the water storage chamber is extracted to the nozzle, which sprays out the water at a certain pressure and angle to flush the willow wadding. The flushed willow flakes will be discharged to the willow flakes collection barrel outside the pole through the sewage pipe under the action of water flow.

In terms of water recycling, natural precipitation and external water sources are fully utilized. Rainwater collected by the arc umbrella is stored in the water storage chamber as the main water source supplement. When the water level inside the water storage chamber reaches a certain height, interaction with external water sources is realized through water level sensors and water pipes to ensure that the water storage chamber always maintains the right amount of water to meet the cleaning needs. At the same time, the willow collection bucket provided at the end of the sewage pipe can be equipped with a filter mesh as shown in FIG. 1 with the label 25, to realize the separation of the willow and the water flow, and the separated water flow can be further recycled, which further improves the utilization efficiency of the water resources, and embodies the environmental protection and sustainability of the system [21] .

4. System testing and evaluation

4.1. Experimental tests

In order to comprehensively evaluate the performance of this multifunctional street lamp in willow adsorption and environmental management, as well as to approach the real application scenarios, the experiments are modeled after representative urban areas that cover different environmental conditions, such as traffic flow, distribution of green vegetation, sparsity of people, and meteorological factors, etc., so as to ensure the universality and reliability of the results of the experiments [22] . The temperature in this laboratory is accurately controlled at about 23 degrees, the relative humidity is maintained at 40%-60%, and it is equipped with high-precision environmental testing equipment, which can simulate different wind speeds and humidity environments for pre-laboratory simulation tests. There are also professional analytical testing instruments, such as electron microscopes and particle counters, to analyze the collected willow samples.

With the help of vibration testing equipment to simulate the vibration of street lamps in severe weather such as strong winds, the stability of the street lamp structure is checked. With the help of electrical safety testing equipment, the safety performance indexes such as grounding resistance and leakage protection of street lamps are strictly tested. Using artificially prepared

 https://doi.org/10.1051/e3sconf/202563901002

willow floss, a multifunctional street lamp model installed with electrostatic spinning materials and drying rods is placed in an experimental chamber simulating different Nanjing spring environmental conditions (e.g. different wind speeds and humidity). After a certain period of operation, the initial and residual concentrations of willow flakes in the chamber were measured by a high-precision particle counter to calculate the collection efficiency; at the same time, the willow flakes on the streetlights were collected and weighed.

High-precision pressure sensors were used to monitor the pressure changes of the high-pressure flushing system[23] in real time, to ensure that the pressure remained stable at different treatment stages; to analyze the influence of pressure fluctuations on the treatment effect of willow wadding, filter, dry and weigh the treated willow wadding, and calculate the amount of residue (%) as shown in Eq. 6, to assess the removal efficiency of the high-pressure flushing technology on willow wadding; and to compare the data on the amount of residue under different treatment parameters.

$$R=R0e^{-k.tP}+C \tag{6}$$

In the above equation, R is the amount of willow residue after cleaning (%), R0 is the initial amount of willow (g/m3), k is the cleaning efficiency coefficient, P is the spray pressure (MPa), t is the cleaning time (s), and C is the intrinsic residual constant of the system.

4.2. Data processing and analysis

The core function of the system was tested, of which the test data are shown in the following figure 4, the system in the humidity 30%-90%, wind speed 0-15 m/s conditions, the average capture rate of willow floss reaches 85%, humidity 90% still maintains the efficiency of 80%, compared with the traditional electrostatic device to improve 32%; and with the change of the wind speed, the capture rate of the drying rod enabled is relatively stable, with small fluctuations; the capture rate of the drying rod is not enabled at the wind speed of 8m/s around began to decline. The capture rate without the drying rod starts to decrease at wind speed around 8m/s. This shows that the humidity control device is helpful for the capture rate. This shows that the humidity control device can help to improve the capture rate and maintain a better effect under different wind speed environments.

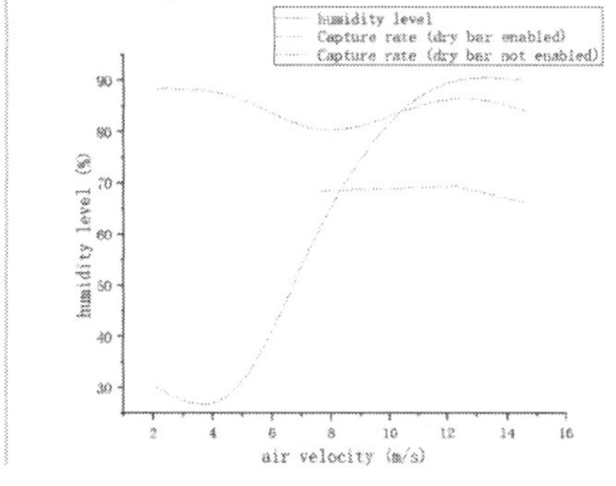

Figure 4. Capture rate graph.

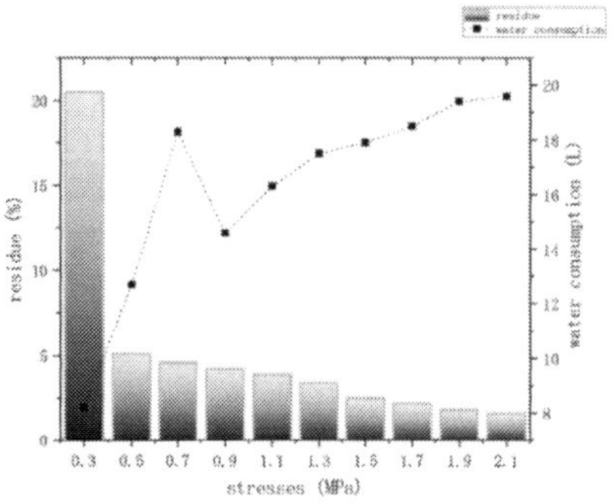

Figure 5. Pressure versus willow residue and water consumption.

As shown in Figure 5, the system can achieve willow residue <5% at 0.6 MPa spray pressure (Eq. 5), with a 37% reduction in water consumption compared to the conventional method. The humidity adaptability test showed that the capture efficiency under high humidity environment (RH90%) was improved by 32% after the addition of humidity regulation device ($p<0.01$, ANOVA). Long-term operation data showed that the nanofiber membrane maintained 78% of the initial efficiency after 3000 h of service, as shown in Table 1 to validate the system reliability.

Table 1. Key performance indicators.

norm	this system	traditional approach	promotion rate
Willow Capture Efficiency	85%	53%	60.4%
unit energy consumption	0.8kWh/kg	1.7kWh/kg	52.9%
O&M costs	¥0.23/m^3	¥0.51/m^3	54.9%

5. Conclusions and outlook

Aiming at the deterioration of air quality, public health risks and urban management problems caused by willow flocculation in northern cities of China, this study proposes a multifunctional streetlight system with integrated environmental management functions, which mainly embodies the electromechanical integration design to realize the functional intensification of municipal facilities and the electrostatic spinning-photocatalytic synergy to enhance the pollutant removal efficiency. An electrostatically spun fiber filter is installed on the top of the pole, and an electrostatically spun nanofiber membrane is covered on the surface of the streetlight lampshade to improve the capture rate of willow flakes and the service life of the streetlight; in addition, to prevent the streetlight from being damaged by moisture and to enhance the capture rate of willow flakes, a humidity control device is integrated in the interior of the streetlight close to the electric circuit and the key components, and in order to reduce the loss of energy and the waste of resources, solar energy collected by photovoltaic panels is converted into electric energy and stored by using intelligent photocatalysts.

In order to reduce energy loss and resource waste, solar energy collected through photovoltaic panels is converted into electricity and stored, and the collected willow flakes are flushed into willow flakes collection buckets by constructing a pulsed high-pressure spray system using an intelligent umbrella-shaped water collector and an intelligent regulation system for water level, which realizes self-sufficiency of energy and recycling of water resources, and effectively optimizes the utilization rate of urban spatial resources. Future research will focus on optimizing the following directions: ① developing deep learning-driven dynamic energy allocation algorithms; ② exploring the application of 5G communication technology in pollutant monitoring data transmission; and ③ enhancing the photocatalytic self-cleaning performance of nanofiber membranes to extend the service life.

Acknowledgments

Supported by the National Natural Science Foundation of China under the Youth Fund (Grant No. 12104232); and the Great Innovation Program of Jiangsu Province (202311460036Z, 202411460027Z).

References

1. DANG Shuai, WANG Qing, LU Xushan, et al. Research on electrostatic adsorption gripping technology based on blended plain fabric[J]. Light Industry Machinery, 2024, 42(1): 13-20.
2. LIU Li. Design and application of intelligent street light[J]. Building Materials and Decoration,2023,19(3):153-155.
3. JIANG Xudong, MO Zhiyuan, WANG Yuanshu, et al. Technological application of active regenerative electrostatic adsorption in ship engine exhaust PM treatment industry[J]. China Storage and Transportation, 2022(6): 84-85.
4. Shenzhen Junchuang Technology Co. An adsorption device based on electrostatic adsorption technology: CN202120804258.2 [P]. 2021-11-16.
5. D. Akbas, T. Gurbuz and G. I. Alptekin, "Smart Street Lighting Systems for Sustainable Smart Cities: A Conceptual Framework," 2024 IEEE Sustainable Smart Lighting World Conference & Expo (LS24), Eindhoven, Netherlands, 2024, pp.1-4
6. Agramelal, F.; Sadik, M.; Moubarak, Y.; Abouzahir, S. Smart Street Light Control: A Review on Methods, Innovations, and Extended Applications. Energies 2023, 16, 7415.
7. Deepaisarn, S.; Yiwsiw, P.; Chaisawat, S.; Lerttomolsakul, T.; Cheewakriengkrai, L.; Tantiwattanapaibul, C.; Buaruk, S.; Sornlertlamvanich, V. Automated Street Light Adjustment System on Campus with AI-Assisted Data Analytics. Sensors 2023, 23, 1853.
8. S. S, S. M, C. V. Josphine, M. T. Kingslin, S. Sivarajan and C. L. K R, "IoT based on Smart Traffic Lights and Streetlight System," 2023 2nd International Conference on Edge Computing and Applications (ICECAA), Namakkal, India, 2023, pp. 1311-1316, Dizon, E., Pranggono, B. Smart streetlights in Smart City: a case study of Sheffield. J Ambient Intell Human Comput 13, 2045–2060 (2022).
9. Wang Hui. Design and realization of intelligent street light control system [J]. Encyclopedia Forum Electronic Magazine, 2020 (14):1781.
10. DANG Shuai, WANG Qing, LU Xushan, et al. Research on electrostatic adsorption gripping technology based on blended plain fabric[J]. Light Industry Machinery, 2024, 42(1): 13-20.
11. WANG Yu, LIU Fuhua, DANG Hongtao, et al. Design of multifunctional street light based on microcontroller[J]. Wireless Interconnection Technology,2022,19(5):67-68.
12. Tvilight intelligent street light[J]. Industrial Design, 2021 (8):18.
13. CHEN Yingfei, ZHANG Zhenpeng. AI-based real-time decision optimization of intelligent street light system[J]. China Lighting Electric Appliance.2024(11):114-116.
14. CHEN Zhonglin, ZHANG Shijing, ZHAN Yuchuan, et al. Intelligent street light control system based on ZigBee[J]. New Industrialization, 2021, 11(7):15-18.
15. ZHOU Minghui, ZHANG Suting, ZHANG Pingcai, et al. Design and research of analog intelligent street light[J]. North-South Bridge.2021(4):111,103.
16. Jiangsu Pule Electric Co. A switchgear anti-interference device based on electrostatic adsorption technology: CN202220475638.0 [P]. 2022-08-23.
17. Shenzhen Xindian Technology Co. A primary filter structure applied to electrostatic adsorption technology: CN202121781245.4[P]. 2022-02-01.
18. JIA Zhenzhen, LI Yiming, ZHENG Zhihong, et al. Preparation of flexible high specific surface electrostatically spun carbon nanofibers and their performance in adsorption of VOCs[J]. Materials Guide, 2024, 38(18):239-246.
19. XIE Li-Cheng, JIANG Yan, ZHANG Zhi-Jun, et al. Preparation of PAN/CNTs composite fibers by electrostatic spinning and their VOCs adsorption performance[J]. Plastics Science and Technology,2024,52(2):44-48.
20. Que Fanbo. Research on intelligent street light system based on LORA and NB-IOT technology[J]. Western Transportation Science and Technology, 2023(9): 164-166,214.
21. JING Kaisheng, PEI Weiwei, MU Xiaoxue, et al. Modification and functionalization of electrostatically spun PAN nanofibers for filtration and adsorption materials[J]. Advances in Textile Science and Technology, 2024, 46(12): 1-4,12.
22. Wu Wenjun. Design of PI-based electrostatic adsorption device[D]. Hubei: Huazhong University of Science and Technology, 2022.
23. LIU Fu-You. Theoretical and experimental study on dynamic electrostatic adsorption and its performance optimization[D]. Chongqing: Chongqing University, 2022.

Study on Microstructure Control and Mechanism of Al-60wt%Sn Alloy Particles Fabricated by Pulsated Orifice Ejection Method

Fumin Xu[a], Qirui Ma[b], Wei Dong[*]

School of Materials Science and Engineering Dalian University of Technology Dalian, China

Abstract. Advanced electronic packaging imposes high demands on core-shell solder balls, creating an urgent need for new preparation technologies. The pulsated orifice ejection method (POEM) offers particles with uniform size, high sphericity, and consistent thermal history, enabling flexible control of the microstructure through process parameters. In this study, POEM was used to prepare Al-60wt%Sn alloy particles by varying the melt superheat and particle size. By combining SEM, EDS, and DSC analysis with calculations of the cooling rate and internal temperature gradient of the droplets, a window diagram correlating melt superheat, droplet size, and particle microstructure was established. The results indicate that melt superheat and droplet size influence the liquid-liquid phase separation state and the migration behavior of the secondary-phase droplets by affecting the cooling rate and internal temperature gradient of the droplets. When the particle size is 285 and 320 μm, particles with uniform size, centered cores, and a perfect core-shell structure are obtained, meeting the requirements for advanced electronic packaging. This study provides valuable insights into the preparation of monotectic alloy particles with different structures and offers reliable results for understanding the formation mechanisms of core-shell structures in monotectic alloys.

1. Introduction

Monotectic alloys exhibit a dome-shaped miscibility gap in their phase diagrams. When the alloy melt enters this region, liquid-liquid phase separation occurs. Even in a microgravity environment in space, this separation phenomenon persists and may lead to melt stratification [1]. Leveraging this characteristic allows for the fabrication of materials with unique properties, attracting considerable attention from researchers. When a binary alloy has a large positive enthalpy of mixing, its equilibrium phase diagram typically exhibits a flat and uniformly inclined liquidus line. Under rapid cooling of the melt, a metastable miscibility gap may exist. Depending on the processing technique, novel microstructures can be obtained, such as a dispersed distribution of one phase within the matrix of the other or the formation of core-shell particles [2].

At present, conventional droplet preparation methods primarily include gas atomization and jet breakup techniques, both of which are containerless solidification technologies. Compared with traditional methods, these techniques effectively prevent heterogeneous nucleation induced by container walls, allowing the melt to solidify under high cooling rates and deep undercooling conditions, thereby optimizing material properties. By leveraging the liquid-liquid phase separation characteristics of monotectic alloys, the containerless solidification method can be used to fabricate core-shell structured Cu@(Sn-Bi) solder balls [3]. In this structure, the high-melting-point Cu core endows the solder ball with excellent electrical conductivity and mechanical strength, while the low-melting-point Sn-Bi soft solder shell ensures good wettability and solderability. This type of solder ball not only exhibits outstanding mechanical properties, electrical conductivity, and thermal conductivity but also shows great potential for application in next-generation electronic packaging.

However, due to the uncontrollable particle size and flight trajectory of droplets produced by these conventional methods, the thermal history of their solidification varies significantly. Although particles with various microstructures, such as second-phase dispersion or core-shell structures, can be obtained, it is challenging to produce uniform particles with a single microstructure. In particular, for core-shell solder balls, not only must the particles exhibit a highly spherical core-shell structure, but the core and shell must also be strictly concentric, with all core-shell particles maintaining identical core and shell dimensions. Only such precisely engineered and perfectly structured core-shell solder balls can meet the high-performance requirements of advanced electronic packaging.

Pulsated Orifice Ejection Method (POEM) is an innovative technique for producing metal particles. Its working principle, as shown in Fig. 1, involves using piezoelectric actuator to drive the a transmission rod movement [4].The periodic motion of the transmission rod disrupts the force equilibrium at the orifice, causing a controlled volume of molten alloy to be ejected from the

[a]email: fuminxu@dlut.edu.cn
[b]email: mqr062907@163.com
[*] Corresponding author's email: w-dong@dlut.edu.cn

orifice during each cycle, forming droplets. These droplets then undergo containerless solidification in a drop tube, ultimately generating particles. In addition, noble gases can not only serve as cooling gases for melts but also facilitate the preparation of particles.

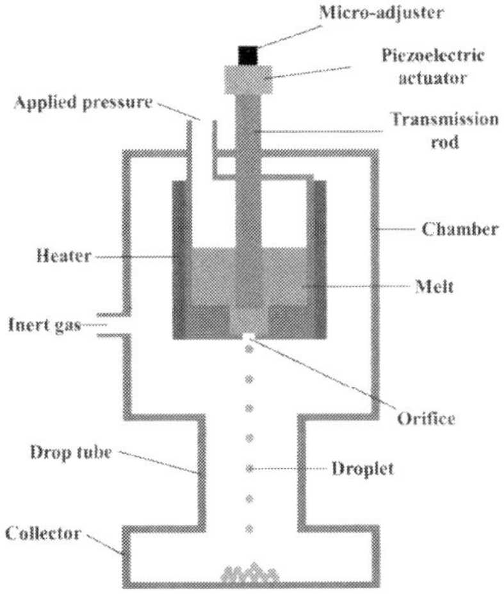

Fig. 1. Schematic diagram of POEM device.

Compared with traditional methods, POEM produces particles with excellent sphericity, uniform and controllable particle size, and highly consistent thermal histories. More importantly, this method allows independent control over key parameters such as droplet size and cooling atmosphere, enabling precise regulation of the cooling rate. Additionally, due to the regular shape and consistent flight trajectory of POEM-produced particles, theoretical modeling can be conducted with reduced discrepancies between predictions and experimental results. Therefore, utilizing POEM to precisely control the solidification process of monotectic alloys, combined with theoretical modeling, is of significant scientific and practical value in studying the morphology and formation mechanisms of alloy particles.

Al-Sn alloys exhibit a high positive mixing enthalpy and possess a metastable miscibility gap below the liquidus line, making them widely applicable across various fields [5]. By suppressing liquid-liquid phase separation, Al-Sn alloy particles with a dendritic structure can be obtained. Due to their uniform internal composition, these particles demonstrate excellent homogeneity in powder metallurgy processes, making them highly promising for use in self-lubricating and wear-resistant materials.

On the other hand, utilizing the liquid-liquid phase separation characteristics enables the fabrication of Al-Sn core-shell structured particles. Since the core layer has a relatively high melting point, while the shell is composed of low-melting-point solder, this structure meets the performance requirements of composite solder balls. Additionally, compared to Cu, Al has slightly lower electrical and thermal conductivity but is more cost-effective and has a lower density. The Al-rich phase in the core layer can enhance the electrical and thermal properties of the solder ball. As a result, Al@Sn core-shell structured particles have the potential to replace Cu-based solder balls, offering promising applications in the field of electronic packaging.

This study aims to utilize the Pulsated Orifice Ejection Method (POEM) to fabricate Al-60wt%Sn alloy particles by adjusting process parameters. The obtained samples will be systematically analyzed in terms of sphericity, particle size distribution, microstructure, and composition. Combined with simulation calculations, the formation mechanisms of different microstructures will be explored to optimize processing conditions. The ultimate goal is to achieve highly spherical, concentric, and perfectly structured core-shell particles that meet the high-performance requirements of core-shell solder balls in advanced electronic packaging.

2. Materials and methods

In this study, Al-60wt%Sn alloy was selected as the research material, with high-purity Al and Sn metals (99.99%, supplied by Beijing Zhongnuo Xincai) as raw materials. Al and Sn were weighed at a 4:6 mass ratio for melting. It is important to note that due to the lower melting point and higher density of Sn compared to Al, a specific loading sequence was adopted to prevent Al from floating on the Sn surface during melting. Specifically, Sn was first placed at the bottom of a crucible, followed by the addition of Al on top. To ensure a stable experimental environment, a mechanical pump and a molecular pump were used to achieve low and high vacuum conditions, respectively. Once the chamber pressure was reduced to 4×10^{-3} Pa, Ar gas was introduced until the pressure stabilized at 1 atm, and the heating process commenced. When the melt reached the target temperature, it was held at a constant temperature for 15 minutes to ensure uniform composition. After the holding stage, the piezoelectric actuator was activated to initiate the formation of particles. During the experiment, the driving voltage and pulse waveform of the drive rod were adjusted, and combined with the micro-adjuster, the descent distance of the drive rod was continuously changed to ensure uniform spraying of the melt. The overheating of the melt was controlled by adjusting the size of the induction current and the heating time under different currents. At the same time, by replacing the crucible with an orifice of different sizes at the bottom, the particle size of the resulting particles could be precisely controlled, as it is typically approximately equal to the orifice size.

The obtained particles were analyzed using multiple characterization techniques. An optical microscope (LV100NPOL, Nikon, Japan) was used to capture particle morphology, and image analysis software (WinROOF 2018, Mitani Corporation, Japan) was employed to measure and statistically analyze particle size distribution. After embedding and polishing, the microstructure of the particles was examined using a field-emission scanning electron microscope (SEM, SU5000, Japan), and compositional analysis of specific regions was performed with an energy-dispersive spectrometer (EDS). Additionally, the thermal properties of the particles were evaluated using a differential scanning calorimeter (DSC,

DSC3, Switzerland) at a heating rate of 20°C/min over a temperature range of 200–680°C.

3. Results and discussion

3.1. Morphological analysis

Fig. 2 presents the optical micrographs of Al-60wt%Sn alloy particles with different sizes, prepared using various micropore diameters, along with the corresponding particle size distribution analyzed using WinROOF 2018 software. It is evident that all four groups of particles exhibit excellent sphericity, with an average sphericity exceeding 0.9. Additionally, the particle size uniformity is remarkable, with the smallest standard deviation recorded at 1.73 μm and the largest at only 3.67 μm.

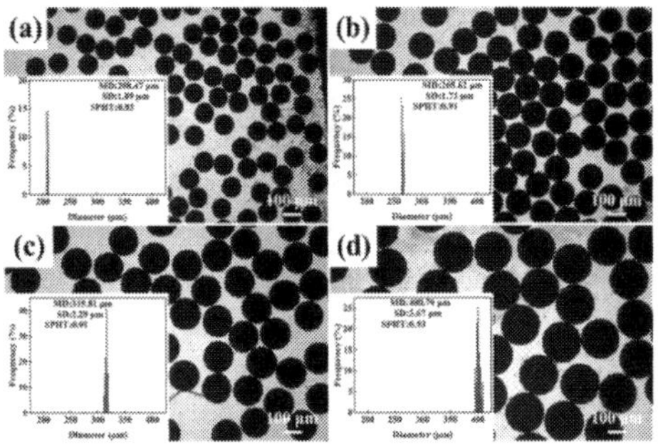

Fig. 2. Optical micrographs and particle size distribution diagrams of Al-60wt%Sn alloy particles with different particle sizes.

To observe the internal microstructure of the particles, resin embedding and polishing were performed. Fig. 3 presents several typical microstructural features observed in Al-60wt%Sn alloy particles. Specifically, Fig. 3a displays a dendritic structure, Fig. 3b shows a mixed structure where dendrites coexist with a secondary phase formed by liquid-liquid phase separation, and Fig. 3c illustrates all particles with uniform diameter and core size, where the core and shell layers are concentric, forming a perfect core-shell structure. Regardless of the specific microstructure, all particles exhibit excellent size uniformity and morphological consistency. This is primarily attributed to the Pulsated Orifice Ejection Method (POEM), which introduces pulsed perturbations into the compositionally homogeneous melt through a driving rod at the bottom of the crucible, precisely controlling droplet volume and ensuring consistent thermal history across all particles. Consequently, any particle obtained under the same processing conditions is highly representative, making it suitable for microstructural formation mechanism analysis, without interference from variations caused by differences in thermal history.

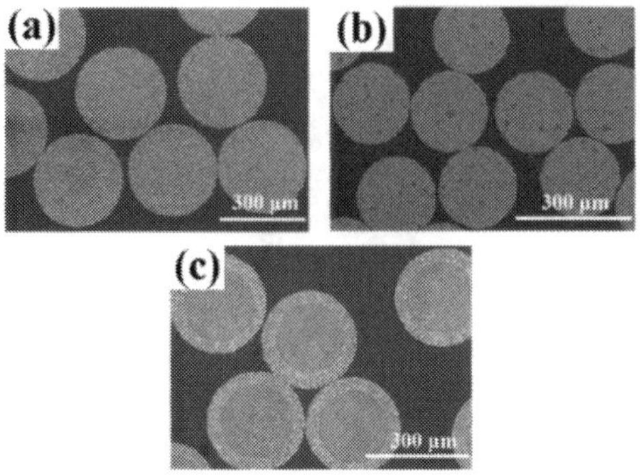

Fig. 3. Several typical morphologies in Al-60wt%Sn alloy particles.

3.2. Component analysis and melting characteristics

To determine the composition of various microstructures in particle electron microscopy images, EDS was used to perform surface analysis on dendritic structures and core-shell structure particles. The results are shown in Fig. 4. From the surface analysis and the analysis of points P_1 and P_2, it is evident that the dark regions in the dendritic structure are Al-rich phases, while the light regions are Sn-rich phases. For the core-shell structure, combined with the analysis of points P_3 and P_4, it is determined that the core layer consists of an Al-rich matrix with dispersed Sn-rich phases, whereas the shell layer consists of an Sn-rich matrix with dispersed Al-rich phases. Additionally, DSC was performed on both the dendritic structure and core-shell structured particles, with the results shown in Fig. 5. In the DSC curve of the dendritic structure, only a single endothermic peak appears at approximately 235°C, corresponding to the melting of the Sn-rich phase. However, for the core-shell structured particles, due to the compositional differences between the core and shell layers and the significant melting point difference between Al and Sn, a two-stage melting behavior is observed. Specifically, the DSC curve exhibits two endothermic peaks: the first peak appears at approximately 236°C, corresponding to the melting of the Sn-rich matrix, while the second peak appears at approximately 595°C, corresponding to the melting of the Al-rich matrix.

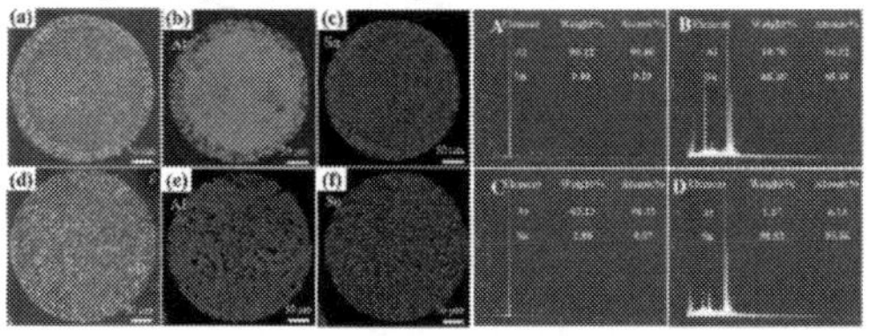

Fig. 4. The elemental results of dendritic structure and core-shell structure.

https://doi.org/10.1051/e3sconf/202563901003

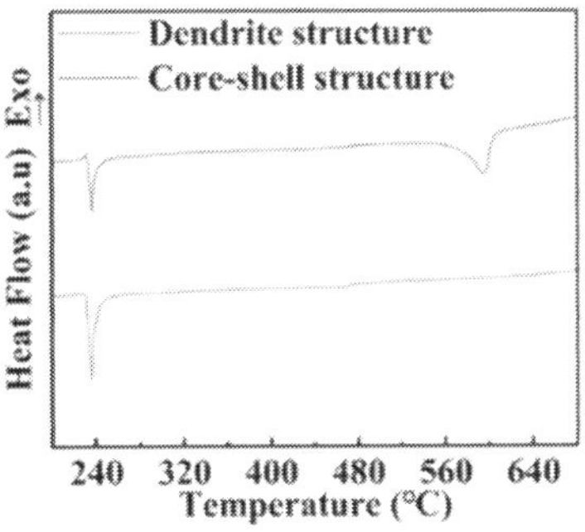

Fig. 5. The DSC curve of dendritic structure and core-shell structure.

3.3. Microstructure of Al-60wt%Sn under different process parameters

3.3.1. The influence of superheating

Fig. 6 presents the microstructures of Al-60wt%Sn alloy particles with a relatively small particle size of approximately 283 μm under different degrees of superheating. As the degree of superheating increases, the microstructure undergoes a sequential transformation, progressing through dendritic structures, a mixed structure of dendrites and phase-separated secondary phases, imperfect core-shell structures, perfect core-shell structures, and finally, core-shell structures with aggregated secondary phases.

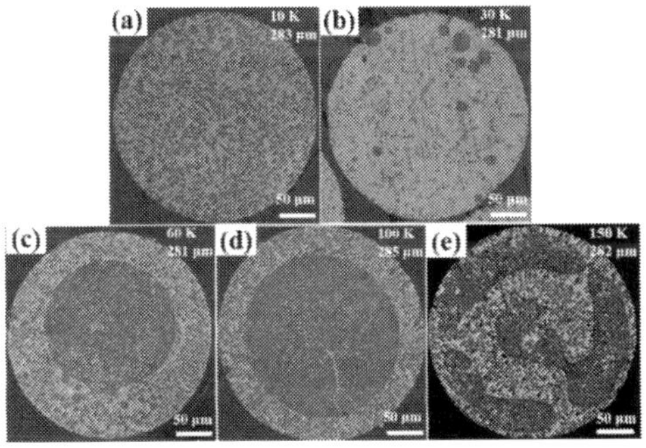

Fig. 6. Microstructure of Al-60wt%Sn alloy particles with smaller particle sizes under different superheating conditions.

Further investigation was conducted on the effect of superheating on the microstructure of particles with a larger size of approximately 320 μm. As the degree of superheating increased, the microstructure evolved into a perfect core-shell structure, a core-shell structure with aggregated secondary phases, and finally, a dendritic structure, as shown in Fig. 7.

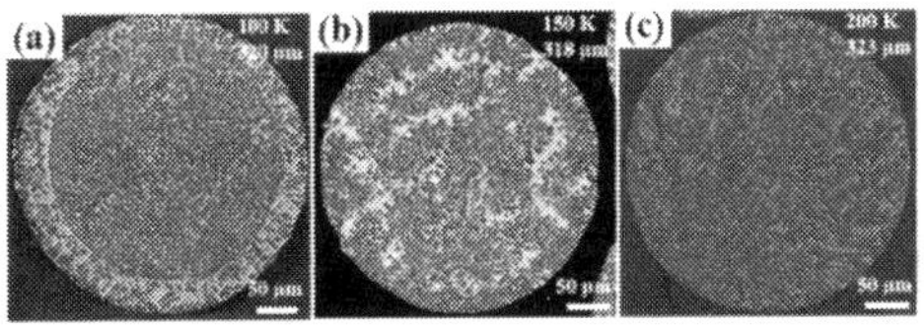

Fig. 7. Microstructure of Al-60wt%Sn alloy particles with larger particle sizes under different superheating conditions.

3.3.2. The influence of particle size

Fig. 8 presents the microstructures of Al-60wt%Sn particles under different levels of superheating and particle sizes. As shown in the Fig, at a degree of superheating of 10 K, 221 μm and 283 μm alloy particles exhibit a fine dendritic morphology, with smaller particles displaying even finer dendrites. Under high superheating of 200 K, both particle sizes develop coarser and longer dendrites, with narrow secondary dendrite arm spacing. Additionally, in larger 373 μm particles, the dendrites appear significantly longer.

At a degree of superheating of 100 K, when the particle size is 272 μm, the microstructure exhibits an imperfect core-shell structure. As the size increases to 285 μm and 320 μm, the microstructure transitions into a perfect core-shell structure. However, with a further increase in size to 360 μm, the microstructure reverts to an imperfect core-shell structure, similar to that observed at 272 μm. Moreover, in large particles of 360 μm, the core-shell structure deviates from a standard concentric circular shape, with the core layer exhibiting a noticeable shift in position.

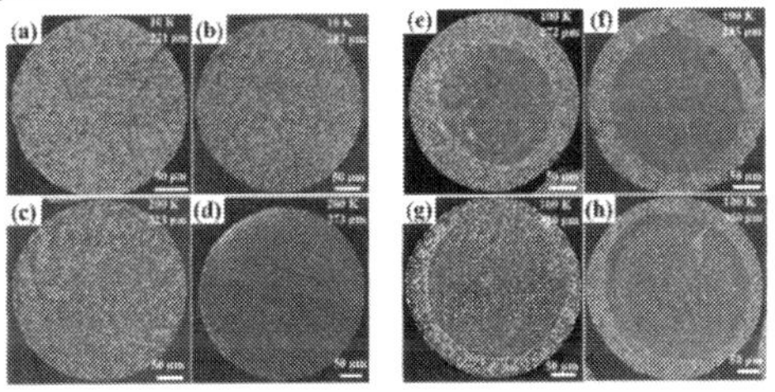

Fig. 8. Microstructure of Al-60wt%Sn particles under different superheating conditions and particle sizes.

Based on the above results, a microstructure window diagram was constructed to summarize the relationship between particle size, degree of superheating, and microstructure, as shown in Fig. 9. This diagram provides valuable insights into the processing conditions required to achieve specific microstructural features in Al-60wt%Sn particles. In particular, it serves as an important guideline for producing solder balls with a perfect core-shell structure, where all particles exhibit identical core and shell sizes with concentric alignment, as shown in Fig. 3c. Such a structure is highly desirable for advanced electronic packaging applications.

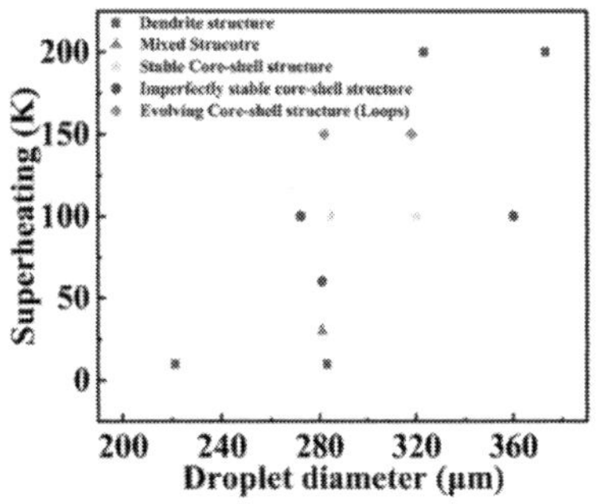

Fig. 9. Microstructure window determined by superheating and droplet diameter.

4. The formation mechanisms of different structures

4.1. Calculation of cooling rate and temperature gradient

For monotectic alloys, under containerless solidification conditions, after the melt undergoes liquid-liquid phase separation, the secondary phase exists as small liquid droplets within the matrix. Due to the temperature difference between the surface and the center of the droplets, a temperature gradient is generated, leading to thermally induced Marangoni migration of the secondary phase globules, as well as Stokes motion caused by gravity. These phenomena are the primary causes of macroscopic segregation in monotectic alloys. The Marangoni migration velocity V_m and the Stokes motion velocity V_s of the secondary phase globules are given by the following equations [6]:

$$V_m=\frac{-2r}{3(3\mu_d+2\mu_m)}\cdot\frac{\partial\sigma}{\partial T}\cdot\frac{\partial T}{\partial x} \tag{1}$$

$$V_s=\frac{2g\Delta\rho r^2}{3\mu_m}\cdot\frac{\mu_d+\mu_m}{3\mu_d+2\mu_m} \tag{2}$$

From the above equations, it can be seen that the Marangoni migration velocity V_m is directly proportional to the temperature gradient $\frac{\partial T}{\partial x}$ and the globules radius r, while the Stokes motion velocity V_s is proportional to r^2. During the early stages of liquid-liquid phase separation, since r is small and $\frac{\partial T}{\partial x}$ is large, V_m far exceeds V_s. Additionally, because the surface of the droplet cools faster than the center, there is a temperature difference between the surface and the center of the droplet. Given the small size of the droplet, a significant tension gradient caused by the temperature gradient exists within the droplet. Ratke and others have shown that when a second-phase liquid exists in the matrix and there is a tension gradient between the second phase and the matrix, Marangoni migration occurs to reduce surface energy [7]. For falling droplets, the secondary-phase globules will migrate towards the hotter center of the droplet and collide. With the collision and aggregation of these small globules, the radius of the secondary-phase globules increases. As r increases, the V_s increase in with radius is much faster than the increase in V_m, and the impact of gravity becomes more significant. If the secondary-phase globules have sufficient time to move before solidification, the gravitational segregation will lead to the formation of an off-center core-shell structure.

Thus, it can be concluded that the cooling rate (which affects liquid-liquid phase separation and solidification time) and the temperature gradient (which affects V_m) play key roles in the evolution of the particle's microstructure.

The cooling rate of monotectic alloy droplets has a significant impact on the final solidification microstructure. However, it is difficult to directly measure the cooling rate of droplets. Typically, an approximate cooling rate is obtained through simulation calculations. In contrast, particles produced by atomization or jet breakup methods often have significant variations in size and poor thermal history consistency, causing discrepancies between the calculated results and the actual situation. In this study, however, the alloy particles prepared using the POEM method exhibit high uniformity in particle size and consistent thermal history. As a result, the simulated results are more representative of the actual conditions. This study uses a particle heat transfer model based on Newton's Law of Cooling to calculate the temperature-time variation curves of Al-60wt%Sn alloy droplets under different particle size conditions [8]. The difference between the center temperature and the surface temperature of droplets with different particle sizes is calculated. Using these results, the average cooling rate of the droplets and the average radial temperature gradient, when the surface temperature drops to 863 K, are computed, as shown in Fig. 10, 11, and 12. The thermal property parameters used in the calculations are listed in Table 1.

Table 1. Thermo-physical parameters of materials used for cooling rate calculation.

Parameter	Value
Thermal conductivity of Ar, λ/W·m^{-1}·K^{-1}	1.763×10^{-2}
Density of Ar, ρ_g/kg·m^{-3}	1.624
Viscosity coefficient of Ar, μ/ Pa·s	22.7×10^{-6}
Specific heat of Ar, C_g/J·g^{-1}·K^{-1}	0.5215
Liquidus temperature, T_L/K	863
Droplet initial velocity, u /m·s^{-1}	2.0
Density of droplet, ρ/kg·m^{-3}	4040
Specific heat of, C_p/J·kg^{-1}·K^{-1}	769.9

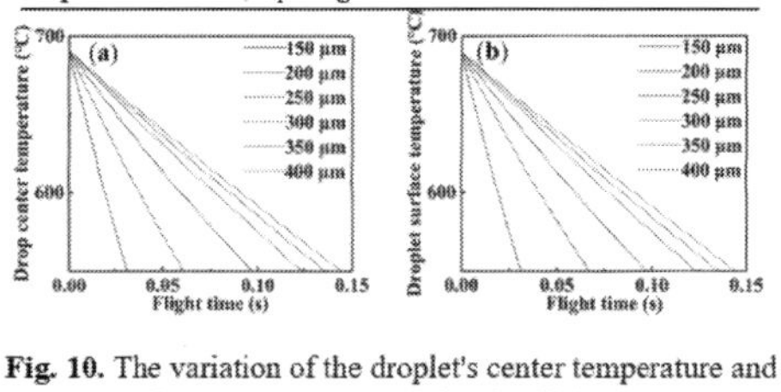

Fig. 10. The variation of the droplet's center temperature and surface temperature with droplet flight time.

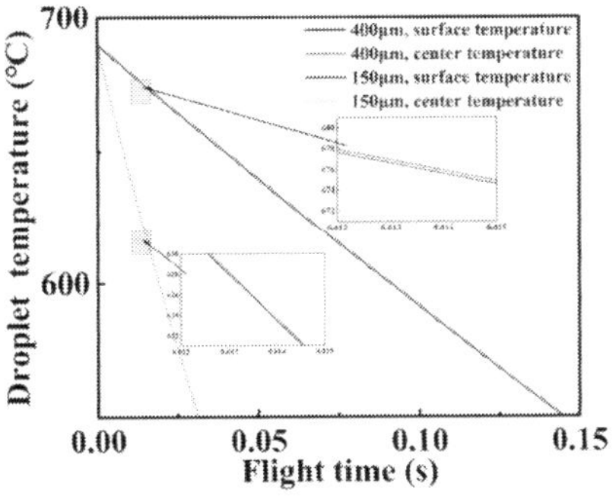

Fig. 11. The difference between the center temperature and surface temperature of droplets with different particle sizes.

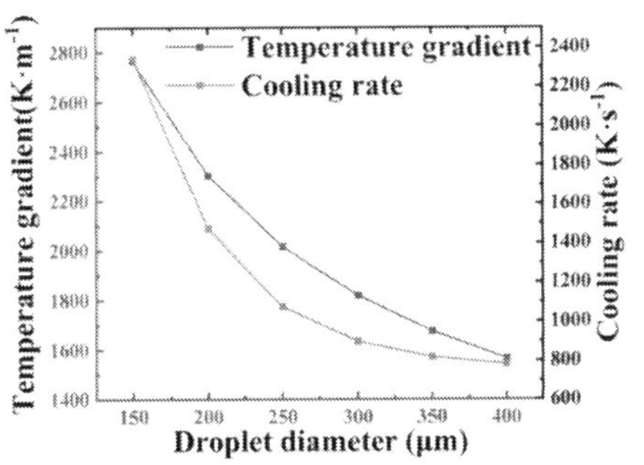

Fig. 12. The cooling rates of droplets with different particle sizes and the average radial temperature gradients when the surface temperature drops to 863 K.

4.2. The mechanism of the influence of superheating on the microstructure of particles

For a particle size of approximately 283 μm, when the degree of superheating is only 10 K, the undercooling of the melt is minimal, insufficient to cool the melt into the metastable monotectic gap. As a result, no liquid-liquid phase separation occurs within the droplet, and the alloy particles form dendritic structures through nucleation and growth mechanisms. When the degree of superheating increases to 30 K, liquid-liquid phase separation occurs, and many small Al-rich globules form. Driven by the temperature gradient, these globules migrate toward the center of the droplet, but before they can fully migrate and collide to grow, the entire alloy droplet has already solidified, resulting in the formation of many small nuclei. When the degree of superheating increases to 60 K, the small second-phase globules have sufficient time to migrate, collide, and grow, forming an imperfect core-shell structure. When the degree of superheating increases to 100 K, the undercooling of the melt increases further, and the initial cooling rate of the droplet is higher. As the surface of the droplet rapidly cools, the radial temperature gradient within the droplet increases, enhancing the Marangoni movement of the Al-rich small globules. Therefore, compared to the 60 K degree of superheating, at 100 K, the core-shell structure becomes more complete, and the size of the second-phase small globules decreases slightly. When the degree of superheating increases to 150 K, although the Marangoni driving force is stronger and the movement speed of the second-phase globules increases, the excessive undercooling caused by the high degree of superheating leads to a dramatic reduction in nucleation points, severely delaying the liquid-liquid phase separation process [9]. This effect is much stronger than the enhanced movement of the second-phase globules caused by the increased superheating, leading to the formation of an agglomerated core-shell structure inside the particle.

For the larger particle size of 320 μm, as the degree of superheating increases from 100 K to 150 K and then to 200 K, the effect of excessive undercooling on delaying liquid-liquid phase separation becomes more pronounced. Therefore, the internal microstructure transitions from a perfect core-shell structure → an agglomerated core-shell structure → dendritic structure. At a degree of superheating of 200 K, due to the rapid cooling of the droplet, a large temperature gradient develops within the droplet, promoting dendrite growth. As a result, the dendritic structure becomes coarse with a small arm spacing.

4.3. The mechanism of the influence of particle size on the microstructure of particles

As shown in Fig. 12, both cooling rate and temperature gradient are negatively correlated with particle size. Therefore, under the same degree of superheating, larger particles solidify later than smaller particles. This means that the second-phase structure within larger particles generally has a longer structural evolution time, while the Marangoni migragation of the second-phase small globules in smaller droplets is more intense. Under the same superheating conditions, the final solidification structure of the particles is influenced by both factors. When the degree of superheating is 10 K and 200 K, the particle internal structure is dendritic in both cases. Due to the lower cooling rate of larger droplets, dendrites in larger droplets have a relatively longer growth time compared to smaller droplets, resulting in coarser dendritic structures for larger particles at both of these degrees of superheating. At 100 K degree of superheating, when the particle size is 272 μm, the Al-rich globules have not had enough time to fully aggregate to the core, and the entire droplet has solidified, forming an imperfect core-shell structure. When the particle size increases to 285 μm and 320 μm, the cooling rate of the droplet decreases, giving the Al-rich globules relatively more time to migrate, resulting in a perfect core-shell structure. However, when the particle size increases further to 360 μm, although the cooling rate decreases even more, the excessive particle size leads to a further reduction in the temperature gradient, which decreases the Marangoni movement rate of the second phase and reduces the collisions between second-phase small globules. This results in the formation of an imperfect, stable core-shell structure. Due to the slower cooling rate at 360 μm, the solidification time is relatively long, and in the later stage of solidification, the influence of Stokes motion gradually increases, causing a deviation in the position of the entire core-shell structure.

5. Conclusions

Using POEM, Al-60wt%Sn alloy particles with uniform particle size, high sphericity, and consistent thermal history were fabricated by adjusting the melt superheating and droplet size, resulting in particles with various microstructures. Based on the results and discussions presented above, the following conclusions were obtained:

1. For droplets of the same size, as the superheating increased from 10 K to 150 K, the degree of liquid-liquid phase separation and the migration speed of the secondary-phase globules changed, leading to a transformation of the particle microstructure from dendritic structures to mixed structures with a secondary phase of varying diameters, then to imperfect stable core-shell structures, perfect stable core-shell structures, and finally agglomerate-like core-shell structures.
2. At both 10 K and 200 K superheating, alloy particles of different sizes all exhibit a dendritic structure. At a superheat of 100 K, particles with sizes of 285 and 320 μm exhibited a perfect core-shell structure, whereas particles with sizes of 272 μm and 360 μm displayed an imperfectly evolved core-shell structure. Additionally, in 360 μm particles, the core layer exhibited a noticeable positional deviation.
3. The variation in melt superheating and droplet size affected the cooling rate and internal temperature gradient of the droplets, thereby influencing the degree of liquid-liquid phase separation and the Marangoni and Stokes motion of the secondary-phase globules, ultimately impacting the particle microstructure. A microstructure window diagram was developed based on different particle sizes and superheats, which provides insights into the process conditions corresponding to specific microstructures of Al-60wt%Sn particles. This diagram serves as an important guideline for obtaining particles with uniform size and a concentric, perfect core-shell structure that meets the requirements of advanced electronic packaging. Moreover, it offers reliable data for investigating the formation mechanism of core-shell structures in monotectic alloys.

References

1. J. Legros, A. Sanfeld, M. Velarde, Fluid sciences and materials science in space. Springer-Verlag, Berlin, 1987.
2. T. Nagase, M. Todai, S. Ichikawa, A. Matsugaki, T. Nakano, "Alloy design and solidification microstructure of Ti-Zr-Hf-Ag-V multi-component alloys with a dual bcc structure." Mater. Trans. Vol.65, pp. 1041-1048, 2024.
3. C. P. Wang, X. J. Liu, I. Ohnuma, R. Kainuma, K. Ishida, "Formation of monotectic alloy powders with egg-type microstructure." Science vol. 297, pp. 990-993, 2002.
4. W. Dong, Y. M. Yuan, C. G. Li, F. M. Xu, Y. Zhao, "Core-shell structure with reversible composition of Al-Bi monotectic alloy particles prepared by pulsated orifice ejection method." Mater. Lett. Vol. 354, p. 135331, 2024.
5. T. A. Zhang, X. M. Chen, Y. Lu, J. L. Hang, D. P. Tao, "Estimation of activity and molar excess Gibbs energy of binary Liquid alloys Pb-Sn, Al-Sn and In-Zn from the partial radial distribution function simulated by ab initio molecular dynamics." Metals, vol. 14, p. 102, 2024.
6. D. L. Geng, S. Y. Wang, N. S. Hou, B. Wei, "Synergic effects of time dependence and thermodynamic driving on metastable phase separation of liquid Fe50Cu50 alloy." Appl. Phys. A, vol. 130, p. 954, 2024.
7. L. Ratke, "Coarsening of liquid Al-Pb dispersions under microgravity-a eureka experiment.", Eureca Sci. Results, vol. 16, pp. 95-99, 1995.
8. W. Dong, Z. R. Jia, X. D. Wang, F. M. Xu, Y. Zhao, "Size-on-demand preparation of SAC305 solder balls based on pulsated orifice ejection method. " Vacuum, vol. 217, p. 112522, 2023.
9. J. He, J. Zhao, H. Li, H. L. Li, X. F. Zhang, Q. X. Zhang, "Directional solidification and microstructural refinement of monotectic alloys." Metall. Mater. Trans. A, vol. 39A, pp. 1174-1182, 2008.

Anion/Cation Co-(de)Intercalation in $Ti_3C_2T_x$ Modified with Cobalt Phthalocyanine for Lithium-Ion Batteries

Yuchen Duan[1], Jinwen Qin[1,2a*], Minhua Cao[1b*]

[1]Key Laboratory of Cluster Science, Ministry of Education of China, Beijing Key Laboratory of Photoelectronic/Electrophotonic Conversion Materials, School of Chemistry and Chemical Engineering, Beijing Institute of Technology, Beijing 100081, P. R. China
[2]Shandong Provincial Key Laboratory of Monocrystalline Silicon Semiconductor Materials and Technology, Dezhou University, Dezhou 253023, P. R. China

Abstract: Due to the shortage of lithium resources and the requirement for high-energy density in lithium-ion batteries (LIBs), anion/cation co-(de)intercalation based on the incorporation of both anions and cations from the electrolyte is expected to be a viable solution. In this work, through cobalt phthalocyanine (CoPc) modification, $Ti_3C_2T_x$ MXene utilizes cationic/anionic carriers for charge storage. Specifically, during the charging process, a reversible intercalation of PF_6^- anions occurs subsequent to the removal of Li ions. By unlocking the capability of anion/cation co-(de)intercalation, the as-designed $Ti_3C_2T_x$-CoPc anode exhibits significant merits in terms of storage capacity (achieving up to nearly 3-fold improvement over the $Ti_3C_2T_x$) and excellent rate performance. Our findings highlight new anion/cation co-(de)intercalation chemistry for MXenes.

1. Introduction

With the rapid development of social economy, the demand for energy has increased significantly to meet daily life and industrial production[1]. CO_2 produced by the extensive use of traditional fossil fuels will lead to global warming and abnormal climate change, and cause serious environmental pollution. Therefore, the development and utilization of renewable energy sources have attracted the attention of researchers.

Even to this day, lithium-ion batteries (LIBs) still play a dominant role the mobile power source market and are extensively used as the most mature electrochemical energy storage (EES) technologies[2]. However, the rapid development of portable devices and new energy power systems has invoked higher demands for the reversible capacity and high rate cycling characteristics of LIBs.

Pseudocapacitive ion storage has been regarded as a promising approach to achieving both improved energy density and high-rate performance. $Ti_3C_2T_x$ MXenes, featuring metallic-like conductivity and transition metal oxide-like surfaces, are considered highly attractive electrode materials for pseudocapacitive applications. These properties enable rapid electron transfer and provide a large number of redox-active sites, which are crucial for efficient pseudocapacitive behaviors. However, the application of $Ti_3C_2T_x$ in energy storage faces inherent limitations due to a cation-only intercalation mechanisms. Specifically, conventional rocking-chair type storage utilizing single cationic species (e.g., Li^+, Na^+, K^+) fail to exploit the synergistic contributions of anion redox chemistry, thereby restricting the theoretical energy density ceiling. Therefore, unlocking the accommodation of large anions into the intrinsic and/or extrinsic storing sites of $Ti_3C_2T_x$ is essential for the enhancement of energy density, mainly originating from the additional anion storing capacity.

Recently, carbon-based electrode materials have emerged as favorable host materials for both cations and anions, primarily achieved through approaches such as nanoporous structure construction, interlayer spacing engineering, and chemical modification[3]. However, their practical implementation faces critical challenges, because anion intercalation brings about much larger volume change of graphitic carbon, compared to cation intercalation, thus leading to the irreversible structural collapse. Impressively, $Ti_3C_2T_x$, with its highly adjustable interlayer spacing and tunable band structure, is poised to facilitate a uniform and efficient flux of both cations and anions to accommodate both of them within its interlayers[4].Up to now, although theoretical and experimental efforts have been devoted to understanding the anion storage of $Ti_3C_2T_x$ electrodes in energy storage devices[5], the realization of reversible cation/anion co-(de)intercalation remains a fundamental challenge.

In this work, we have demonstrated a cobalt phthalocyanine (CoPc) modification strategy on $Ti_3C_2T_x$ to achieve anion/cation co-(de)intercalation (specifically, Li^+ cations and PF_6^- anions) in lithium-ion batteries (LIBs). In the as-designed $Ti_3C_2T_x$-CoPc anode, a reversible extra capacity from PF_6^- intercalation following Li^+ deintercalation during charging processes has been experimentally verified. This unique ion storage

Corresponding author's e-mail: [a*]qjw@bit.edu.cn;
[b*]caomh@bit.edu.cn

outperforms conventional cation-only intercalation, achieving a capacity that exceeds the theoretical limit of $Ti_3C_2T_x$ (413.8 vs. ~320 mAh g^{-1}).

2. Materials and methods

2.1. Preparation of few-layer $Ti_3C_2T_x$.

Typically, LiF (0.998 g) was dissolved in HCl (10 mL, 12 M) under stirring. Then, Ti_3AlC_2 MAX powder (1.0 g) was dispersed slowly into the above mixture. The suspension was stirred for 24 h at 36°C to remove Al layer of Ti_3AlC_2. Afterward, the multilayer $Ti_3C_2T_x$ products were washed with deionized water and centrifuged several times until the pH was above 6. Next, the as-obtained multilayer $Ti_3C_2T_x$ was re-dispersed into deionized water (20 mL) and sonicated for 1 h under N_2. Finally, few-layer $Ti_3C_2T_x$ suspension were collected after centrifugation at 3500 rpm for 1 h. Note that the $Ti_3C_2T_x$ nanosheets were directly obtained via freezing-dry for comparison.

2.2. Preparation of $Ti_3C_2T_x$-CoPc.

Typically, above few-layer $Ti_3C_2T_x$ nanosheets (40 mg) were dispersed into 40 mL DMF after being sonicated for 2 h. Then, CoPc (15 mg, Tokyo Chemical Industry) was dispersed into another DMF solution (40 mL), followed by the addition of 10 mL HCl (1.0 M). Next, the above CoPc solution was added dropwise into the few-layer $Ti_3C_2T_x$ suspension (1 mg mL^{-1}). Under the protection of N_2 atmosphere, the mixture subsequently keeps stirring at 80 °C for 3 h under reflux conditions. After cooling to room temperature, the mixture was filtered, collected and washed with deionized water. Finally, $Ti_3C_2T_x$-CoPc was obtained after freezing-dry.

2.3. Materials characterizations

The powder X-ray diffraction (XRD) analysis was carried out to analyse composition on a Bruker D8 Advance diffractometer with Cu K α radiation (λ=1.54178 Å) at a scan rate of 5° min^{-1} with 40 kV voltage and 50 mA current. The morphology of the samples was characterized by field emission scanning electron microscopy (FE-SEM, HITACHI S-4800) equipped with energy dispersive spectrometer (EDS). Nitrogen adsorption/desorption curves were measured by Brunauer Emmett Teller measurement to analyze the pore size distributions and surface area of the samples under N_2 physisorption at 77 K. Fourier transform infrared spectra (FTIR) were carried out from 400 to 4000 cm^{-1} on a Nicolet 170SX FTIR spectrometer with KBr pellets. X-ray photoelectron spectroscopy (XPS) measurements were carried on Thermo Fisher Scientific-ESCALAB 250Xi. Freeze Drier (Beijing BIOCOOL FD-1C-50) was used to carried out freeze-drying.

2.4. Electrochemical measurements

The electrochemical measurements were carried out using standard CR2025 coin cells in the voltage range of 0.01-3.0 V (vs. Li^+/Li). Active materials, carbon black and polyvinylidene fluoride binder (PVDF) were mixed with a weight ratio of 7:2:1 and pasted on the Cu foil to prepare the working electrode. The mass loading of the active materials was about 1.0-1.3 mg cm^{-2}. Coin-type cells were assembled in an argon-filled glove box, by using the lithium foil (thickness is 0.45 mm) as the counter electrode, a Whatman Celgard 2400 membrane as the separator and the solution of 1 M $LiPF_6$ in ethylene carbonate (EC)/diethyl carbonate (DEC)/ ethyl methyl carbonate (EMC) (1:1:1 vol.%) as the electrolyte. Cyclic voltammetry (CV) curves were measured by a CHI-760E electrochemical workstation at various scan rates. Galvanostatic discharge/charge profiles were conducted on a LAND CT2001A.

3. Results and discussion

3.1. Structure and composition characterizations of $Ti_3C_2T_x$-CoPc

We have successfully modified $Ti_3C_2T_x$ by compositing it with CoPc using few-layer $Ti_3C_2T_x$ in dimethylformamide (DMF) solution. The X-ray diffraction (XRD) patterns of both $Ti_3C_2T_x$-CoPc and $Ti_3C_2T_x$ are presented in **Figure 1a**. After CoPc modification, $Ti_3C_2T_x$-CoPc reveals a (002) peak shift from 6.6 to 6.2°. This shift corresponds to a slight d-spacing expansion from 1.34 to 1.42 nm. Due to a small amount of CoPc functionalization on $Ti_3C_2T_x$ surface, the characteristic diffraction peaks of the CoPc crystalline phase (PDF#14–0948) have not been observed in the XRD pattern of $Ti_3C_2T_x$-CoPc. To further investigate the modification of $Ti_3C_2T_x$ with CoPc, Fourier transform infrared spectroscopy (FTIR) was employed. As depicted in **Figure 1b**, the typical peaks of C=O, -OH, and Ti-O bands from $Ti_3C_2T_x$ remain clearly observable in $Ti_3C_2T_x$-CoPc. This suggests that the chemical structure of

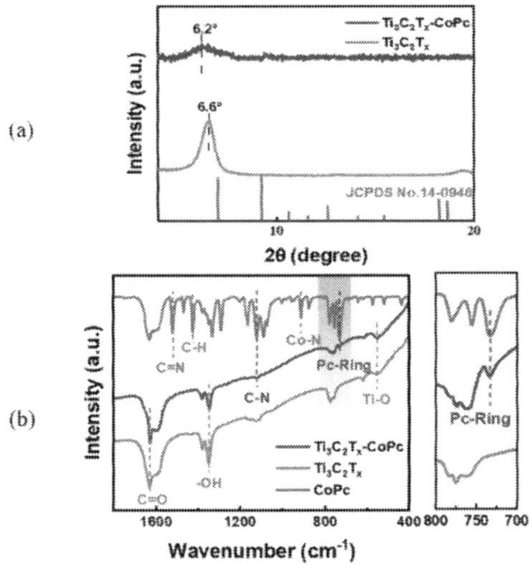

Figure 1. (a) XRD patterns of $Ti_3C_2T_x$-CoPc and $Ti_3C_2T_x$. (b) FTIR spectra of $Ti_3C_2T_x$-CoPc, $Ti_3C_2T_x$ and CoPc.

(a)
(b)
(c)

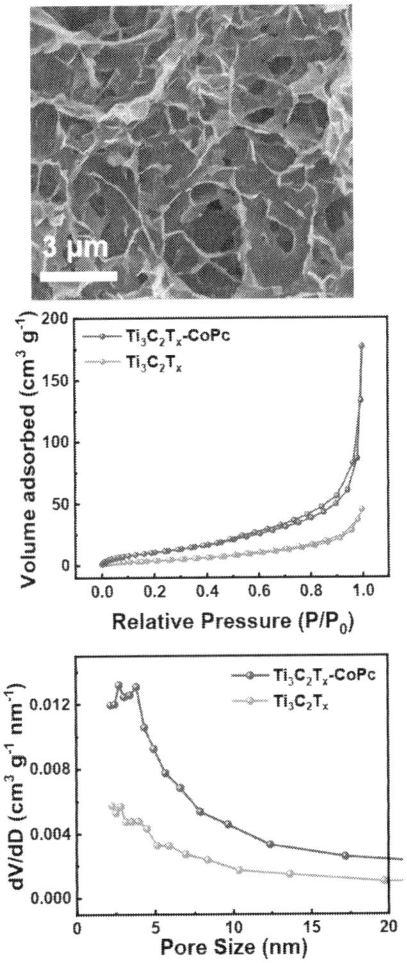

Figure 2. (a) SEM image of $Ti_3C_2T_x$-CoPc. (b) Nitrogen adsorption/desorption isotherms of $Ti_3C_2T_x$-CoPc and $Ti_3C_2T_x$. (c) The corresponding pore size distributions of $Ti_3C_2T_x$-CoPc and $Ti_3C_2T_x$.

(a)
(b)
(c)

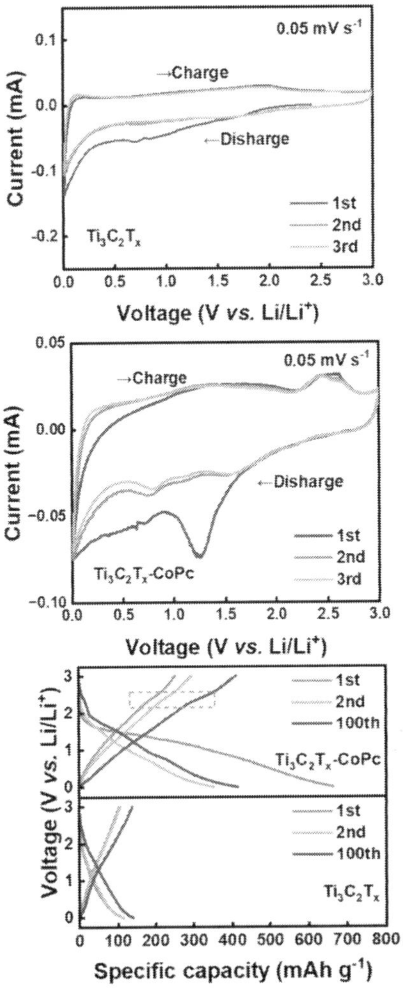

Figure 3. (a) CV curves of $Ti_3C_2T_x$ at 0.05 mV s^{-1}. (b) CV curves of $Ti_3C_2T_x$-CoPc at 0.05 mV s^{-1}. (c) Glvanostatic discharge/charge voltage curves of $Ti_3C_2T_x$ and $Ti_3C_2T_x$-CoPc at 0.1 A g^{-1}.

$Ti_3C_2T_x$ is retained after the modification process. The emergence of a new band at 733 cm^{-1} in $Ti_3C_2T_x$-CoPc, which can be assigned to the skeleton vibration of the phthalocyanine ring (Pc-ring). This provides direct evidence confirming the successful modification of CoPc onto the $Ti_3C_2T_x$ surface, given that the Pc-ring is a key structural motif of CoPc[6-7]. The scanning electron microscopy (SEM) image of $Ti_3C_2T_x$-CoPc in **Figure 2a** shows highly wrinkled nanosheets that are interconnected through a 3D cross-linking structural network. From the nitrogen adsorption/desorption isotherm curves (**Figure 2b, c**), the specific surface areas of $Ti_3C_2T_x$-CoPc and $Ti_3C_2T_x$ are 43.2 and 16.2 m^2 g^{-1}, respectively. The wrinkle structure of $Ti_3C_2T_x$-CoPc provides a larger specific surface area, along with a broader pore size distribution. The modification of $Ti_3C_2T_x$ with CoPc notably enhances its surface area, thereby providing a greater number of active sites for charge storage and consequently resulting in a higher capacity.

3.2. Effects of anion/cation co-(de)intercalation on energy storage

Cyclic voltammetry (CV) was firstly performed to investigate the ion storage process within a potential range of 0.01−3.0 V. As shown in **Figure 3a**, $Ti_3C_2T_x$ displays pseudo-rectagular shaped CV profiles, verifying its capability to accommodate Li^+ ions. The initial CV curve shows an irreversible capacity loss caused by the formation of solid electrolyte interface (SEI) and side reactions between Li^+ with functional groups of $Ti_3C_2T_x$[8]. The reduction peak at about 1.60 V is because Li^+ is captured between the $Ti_3C_2T_x$ nanosheets, and Li^+ will be captured again in the subsequent discharge process, so there is a reversible peak at this position[9]. As for $Ti_3C_2T_x$-CoPc in **Figure 3b**, a new oxidation peak located at 2.2−2.6 V in the charge curves is observed. The appearance of the wide peak associates with the PF_6^-

intercalation[10], highlighting the significant impact of the CoPc modification on the electrochemical properties of $Ti_3C_2T_x$. **Figure 3c** displays the galvanostatic discharge/charge voltage curves of $Ti_3C_2T_x$ and $Ti_3C_2T_x$-CoPc at 0.1 A g^{-1} for the 1st, 2nd, and 100th cycles. The discharge/charge voltage curve of $Ti_3C_2T_x$ shows a sloping feature, whereas $Ti_3C_2T_x$-CoPc displays a distinct plateau within 2.2–2.6 V, consistent with

(a)

(b)

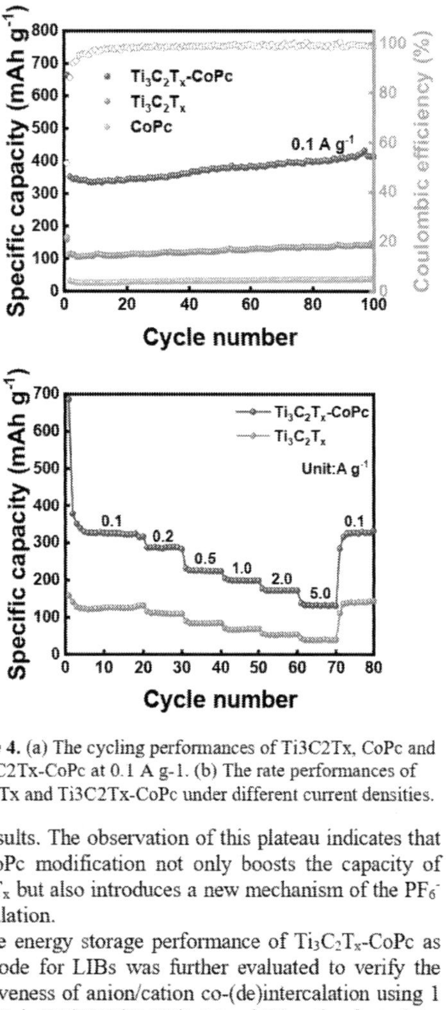

Figure 4. (a) The cycling performances of Ti3C2Tx, CoPc and Ti3C2Tx-CoPc at 0.1 A g-1. (b) The rate performances of Ti3C2Tx and Ti3C2Tx-CoPc under different current densities.

CV results. The observation of this plateau indicates that the CoPc modification not only boosts the capacity of $Ti_3C_2T_x$ but also introduces a new mechanism of the PF_6^- intercalation.

The energy storage performance of $Ti_3C_2T_x$-CoPc as the anode for LIBs was further evaluated to verify the effectiveness of anion/cation co-(de)intercalation using 1 M $LiPF_6$ in EC/DEC/EMC (1:1:1 vol.%) as the electrolyte. The cycling performance of $Ti_3C_2T_x$-CoPc at a current density of 0.1 A g^{-1} is shown in **Figure 4a**,. Clearly, $Ti_3C_2T_x$-CoPc maintains a reversible capacity as high as 413.8 mAh g^{-1} after 100 cycles, about a 3-fold enhancement of that of $Ti_3C_2T_x$ (139.8 mAh g^{-1}) and significantly surpassing the capacity of CoPc (37.1 mAh g^{-1}). There are two factors contributing to the capacity enhancement of $Ti_3C_2T_x$-CoPc. Firstly, the larger surface areas facilitates improved accessibility of active materials to electrolyte, thereby promoting the ion diffusion. Secondly, $Ti_3C_2T_x$-CoPc benefits from additional anion storing capacity derived from the PF_6^- (de)intercalation. In addition, the Initial Coulombic Efficiency (ICE) of $Ti_3C_2T_x$-CoPc, $Ti_3C_2T_x$ and CoPc electrode materials were 51.6%, 48.1% and 18.9%, respectively. The higher ICE of $Ti_3C_2T_x$-CoPc may be due to the inhibition of the adverse side reactions of the functional groups on the surface of $Ti_3C_2T_x$ and the inhibition of the formation of adverse SEI[11]. In **Figure 4b**, the rate performance of $Ti_3C_2T_x$-CoPc is further explored. At various current densities ranging from 0.1 to 5.0 A g^{-1}, $Ti_3C_2T_x$-CoPc delivers discharge capacities of 316.7, 283.4, 224.3, 199.6, 172.4, and 131.8 mAh g^{-1} at 0.1, 0.2, 0.5, 1.0, 2.0, 5.0 A g^{-1}, respectively. Notably, these values are significantly higher than those of $Ti_3C_2T_x$ across the entire range of current densities. It is worth noting that at 5 A g^{-1}, the capacity of $Ti_3C_2T_x$-CoPc. is 3.3 times that of $Ti_3C_2T_x$ (38.9mAh g^{-1}). When the current density returns to 0.1 A g^{-1}, the reversible capacity can be recovered to 332.9 mAh g^{-1}, manifesting the excellent reversibility of $Ti_3C_2T_x$-CoPc.

3.3. The experimental validations of anion/cation co-(de)intercalation

To obtain deeper insight into the anion/cation co-(de)intercalation, we firstly monitored the structural evolution of $Ti_3C_2T_x$-CoPc by voltage-dependent XRD measurements after the 10th cycle. As shown in **Figure 5a,b**, during the charging process, as the potential increases from 0.01 V to 2.0 V, the (002) peak of $Ti_3C_2T_x$-CoPc gradually upshifts from 5.9° to 6.5°. This upshift corresponds to the deintercalation of Li^+ ions from the $Ti_3C_2T_x$-CoPc interlayers. The removal of Li^+ ions leads to a decrease in the d-spacing from 1.48 to 1.34 nm. When the charging continues to 3.0 V, the d-spacing of $Ti_3C_2T_x$-CoPc increases significantly. Specifically, the d-spacing expands from 1.36 nm at 2.0 V to 1.53 nm at 3.0 V, corresponding to a downshift in the (002) peak from 6.5° to 5.7°. The observed increase in d-spacing is consistent with the intercalation of PF_6^- anions into the interlayers of $Ti_3C_2T_x$-CoPc. This is because the change in interlayer spacing induced by PF_6^- anions is comparatively greater than that caused by the smaller Li^+ ions. And as the PF_6^- anions occupy the space previously vacated by deintercalated Li^+ ions, a substantial expansion of the interlayer spacing occurs. In the subsequent discharging process, the (002) peak of $Ti_3C_2T_x$-CoPc exhibits an inverse position change. This observation indicates that the anion/cation co-(de)intercalation processes are reversible. Therefore, this detailed structural analysis through voltage-dependent XRD measurements provides compelling evidence of the anion/cation co-(de)intercalation mechanism in $Ti_3C_2T_x$-CoPc. In contrast, with the continuous progress of the discharge process, the (002) peak of $Ti_3C_2T_x$ gradually decreases until the complete discharge reaches 0.01V, reaching the minimum value, which corresponds to the lithium insertion process. In the subsequent charging process, the peak direction of (002) gradually increases, corresponding to the delithium process, until the full charge returns to 3.0V, and the peak position returns to the initial position[12].

Accordingly, as illustrated in **Figure 5c**, the voltage-dependent Li 1s and P 2p X-ray photoelectron spectroscopy

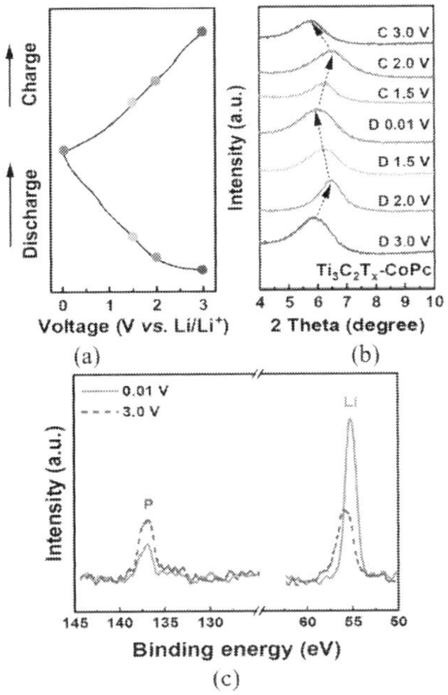

Figure 5. (a) Galvanostatic discharge/charge voltage curves showing the positions for voltage-dependent XRD. (b) voltage-dependent XRD patterns of $Ti_3C_2T_x$-CoPc. (c) Li 1s and P 2p XPS spectra of $Ti_3C_2T_x$-CoPc at fully charged and discharged states.

(XPS) spectra provide further insights into the anion/cation co-(de)intercalation processes in $Ti_3C_2T_x$-CoPc. At the fully charged state, the intensity of P 2p XPS signal is much stronger than that of the fully discharged state, due to the PF_6^- intercalation. Meanwhile, the intensity of the Li 1s XPS signal correspondingly decreases with the deintercaltion of Li^+ ions[13]. This result is consistent with the electrochemical behaviors observed in the voltage-dependent XRD measurements.

4. Conclusion

In summary, we have successfully synthesized $Ti_3C_2T_x$-CoPc through a facile modification strategy, enabling the anion/cation co-(de)intercalation. The experimental investigations of voltage-dependent XRD and XPS have confirmed the anion/cation co-(de)intercalation behavior in $Ti_3C_2T_x$-CoPc. Upon inducing the PF_6^- intercalation at a voltage of ~2.0 V, the as-designed $Ti_3C_2T_x$-CoPc anode in LIBs delivers a remarkable capacity of 413.8 mAh g^{-1}, nearly a 3-fold increase than that of $Ti_3C_2T_x$. Our study provides novel insights into the dual utilization of anions and cations in LIBs, overcoming the limitations associated with traditional cation-only intercalation into MXenes.

Acknowledgment

This work was supported by the National Natural Science Foundation of China (No. 21872008) and the Natural Science Foundation of Beijing, China (No. 2212019). We would like to thank the Analysis & Testing Center of Beijing Institute of Technology for measurements.

References

1. Yan J Y, Yang Y, Campana P E, et al, "City-level analysis of subsidy-free solar photovoltaic electricity price, profits and grid parity in China," Nat. Energy, vol. 4, 2019, pp. 709-717.
2. Du J, Zheng J Q, Liang Y T, et al, "A hybrid deep learning framework for predicting daily natural gas consumption," Energy, vol. 257, 2022, pp. 124689.
3. Yang S, Yu Y, Gao X, et al, "Recent advances in electrocatalysis with phthalocyanines," Chem. Soc. Rev, vol. 50, 2021, pp. 12985-13011 .
4. Jia R, Yang R, Zhen Y, et al, "Anomalous electrochemical aging strengthening behavior of MXene electrodes for synergistic anion-cation storage in dual-Ion batteries," Adv. Funct. Mater, vol. 35, 2024, pp. 2419013.
5. Wang G, Park J M, Kang T, et al, "Anion Storage of MXenes," Small Methods, vol. 7, 2023, pp. 2201440.
6. Li M Z, Yan C L, Ramachandran R, et a, "Non-peripheral octamethyl-substituted cobalt phthalocyanine nanorods supported on N-doped reduced graphene oxide achieve efficient electrocatalytic CO2 reduction to CO," Chem. Eng. J, vol. 430, 2022, pp. 1385-8947.
7. Chen K, Liu K, An P, et al, "Iron phthalocyanine with coordination induced electronic localization to boost oxygen reduction reaction," Nat. Commun., vol. 11, 2020, pp. 4173-4181.6
8. Naguib M, Come J, Dyatkin B, et al, "MXene: A promising transition metal carbide anode for lithium-ion batteries," Electrochem. commun, vol. 16, 2012, pp. 61-64.
9. Zhao S S, Meng X, Zhu K, et al, "Li-ion uptake and increase in interlayer spacing of Nb_4C_3 MXene," Energy Storage Mater, vol. 8, 2017, pp. 42-48.
10. Zou G D, Zhang Z W, Guo J X, et al, "Synthesis of MXene/Ag composites for extraordinary long cycle lifetime lithium storage at high rates," ACS Appl. Mater, vol. 8, 2016, pp. 22280-22286.
11. Oishi M, Yamanaka K, Watanabe I, et al, "Direct observation of reversible oxygen anion redox reaction in Li-rich manganese oxide, Li_2MnO_3, studied by soft X-ray absorption spectroscopy," J. Mater. Chem. A, vol. 4, 2016, pp. 9293-9302.
12. Okubo M, Sugahara A, Kaijiyama S, et al, "MXene as a Charge Storage Host," Acc. Chem. Res, vol. 51, 2018, pp. 591-599.
13. Wang T, Gaugler J A, Li M, et al, "Construction of fluorine and piperazine-engineered covalent triazine frameworks towards enhanced dual-Ion positive electrode performance," Chemsuschem, vol. 16, 2023, pp. e202201219.

FDIA Attack Detection Technique for Smart Grids Based on Graph Reconstruction and Spatio-Temporal Joint Modeling

Yuzhang Gao[1,a], Jing Xia[2*]

[1]Beijing Jiaotong University Beijing China
[2]Goldwind Science &Technology Co., Ltd., Beijing China

Abstract: With the widespread application of smart grids, the false data injection attack (FDIA) has become a major threat to power grid security. Traditional detection methods often have difficulty in effectively identifying such attacks, especially in complex environments. To address this problem, this paper proposes a spatial-temporal joint FDIA detection framework named HSGT-Net, which integrates the Hodge diffusion, Sparge sparsity mechanism, and time series modeling. This framework unifies the branch power and bus power data through a graph reconstruction mechanism, and combines Hodge multi-order diffusion with spatial feature modeling of Sparge-iTransformer, so it can effectively capture the local and global dependencies of the power grid. A GRU module is formulated to model time series features, with improved perception ability of dynamic attack features.

Experimental results show that the HSGT-Net performs better than traditional machine learning methods (such as SVM and KNN) and deep learning models (such as MLP, CNN, and GNN+LSTM) on both IEEE-30 and IEEE-57 standard test systems. Especially in large-scale power system scenarios, it can still maintain high accuracy and robustness. In addition, ablation experiments further verify the effectiveness of each module and prove the key role of joint modeling on spatial and temporal features in FDIA detection. In a word, the HSGT-Net demonstrates strong adaptability and computational efficiency in FDIA attack detection and has good engineering application prospects.

1. Introduction

In recent years, with the deep integration of energy Internet and new generation information as well as communication technology (ICT), the power system is constantly moving towards intelligent and digital transformation [1]. Smart grids improve resource allocation and operational efficiency through distributed energy integration, real-time monitoring, and flexible scheduling. Compared to traditional grids, they enable tight integration of physical infrastructure with communication, information processing, and data analytics, forming an interactive cyber-physical system [2]. However, Smart grids face network security threats, especially in state estimation and data collection. False data injection attacks (FDIAs) can bypass residual-based detection, distort state estimation, and lead to abnormal power flow and imbalance, posing serious risks to grid security [3].

Many FDIA detection methods have been proposed, ranging from early residual analysis and power flow-based approaches to machine learning models like SVM [4]. With the development of deep learning, CNNs have been used for spatial feature extraction [5], and GNN-LSTM models have been introduced for joint spatial–temporal learning [6]. However, most existing methods still focus on static spatial features and lack modeling of dynamic temporal evolution.

Generally, existing FDIA detection studies face several challenges: (1) Inconsistent modeling between branch flow and bus injection leads to weak feature fusion; (2) GNNs struggle to capture long-range or high-order spatial dependencies; (3) Temporal modeling is limited, and spatial-temporal joint learning remains insufficient; (4) Some deep models are computationally intensive, limiting scalability. To address these problems, this paper proposes HSGT-Net, a multi-module framework integrating Hodge diffusion, sparse mechanisms, and time series modeling to enhance detection accuracy and efficiency for large-scale applications.

The main work and innovations of this paper are as follows:

1) A unified graph reconstruction mechanism is proposed to achieve the fusion input of branch power and bus power data, ensuring the integrity and physical rationality of feature expression.

2) A spatial feature modeling method that integrates Hodge multi-order diffusion and Sparge-iTransformer is constructed, which can effectively capture the local and long-range spatial dependencies and reduce the computational complexity with sparse mechanisms.

[a]22120205@bjtu.edu.cn
[*] Corresponding author, xiajingbeijing@126.com

3) The GRU time series are used to achieve dynamic evolution modeling of spatial features, with improved dynamic perception of FDIA attack features.

4) Multiple experimental verifications are performed on the IEEE-30 bus and IEEE-57 bus standard test systems. The results show that the proposed HSGT-Net is superior to the typical methods in terms of detection accuracy, robustness and computational efficiency.

2. Model design

The HSGT-Net architecture consists of graph reconstruction, spatial feature extraction, temporal modeling, and output prediction, as illustrated in Figure 1. It first unifies branch and bus power data in the edge space. Hodge diffusion and Sparge-iTransformer capture local and global spatial features, while GRU models temporal dynamics. A fully connected layer then outputs the FDIA detection score.

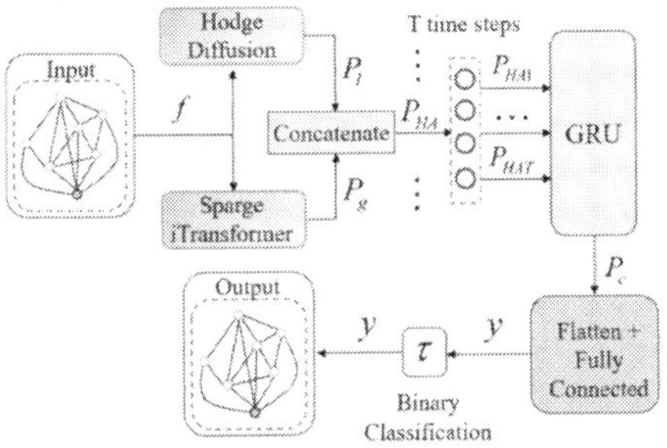

Figure 1. HSGT-Net Architecture for FDIA Detection.

2.1. Graph reconstruction

In FDIA detection, branch flow and bus injection power correspond to edge-space and node-space features. To unify them, a graph reconstruction mechanism is proposed, which introduces a super virtual node connected to all bus nodes via virtual edges. This forms a reconstructed graph $G_e = (V_e, E_e)$, where $V_e = V \cup \{N+1\}$, and $E_e = E \cup E_{\text{vir}}$, with $|E_{\text{vir}}| = N$. This design ensures spatial consistency between branch and bus power data, providing an integrated input feature f for subsequent modeling.

The input feature f combines branch flow power and bus injection power and ensures consistency with the original measurement data z used in state estimation. Ultimately, feature f serves as a unified and physically consistent foundation for the subsequent spatial feature extraction and temporal modeling.

2.2. Local spatial features

In the spatial feature extraction process, the Hodge diffusion [7] is applied to capture the local spatial dependencies in the power grid. This technique allows features to propagate through the graph based on the topology, which enables the model to account for the physical constraints of grid. The feature vector f is propagated through the graph using the diffusion operator L. The core operation of the Hodge diffusion process is as follows:

$$P_l = [f, Lf, L^2 f, \dots, L^{E_e - 1} f] \tag{1}$$

where f is the initial feature vector, L is the graph Laplacian operator that represents the graph's structure, E_e is the maximum diffusion step, and P_s is the set of spatial features obtained after applying multiple steps of diffusion.

The output feature P_l contains the multi-scale spatial information, which is then used as input for the subsequent temporal modeling stage.

2.3. Global spatial features

To address the limitations of local diffusion mechanisms in modeling long-range spatial dependencies, this paper proposes a Sparge-iTransformer module that integrates a dynamic sparsity mechanism with efficient attention operations. This module enables the global spatial modeling on power grid graphs with high accuracy and low computational overhead. The overall framework is illustrated in Figure 2.

The input edge-space features f are first linearly projected into the query, key, and value matrices:

$$Q = fW_Q, \quad K = fW_K, \quad V = fW_V \tag{2}$$

where W_Q, W_K, and W_V are learnable projection matrices.

Then, the mean pooling is applied to Q and K to obtain compressed feature representations $\overline{q_i}$ and $\overline{k_j}$, followed by the computation of initial relevance scores between edges:

$$a_{ij} = \overline{q_i} \times \overline{k_j} \tag{3}$$

To measure the information richness of each edge, the self-cosine similarity is further computed:

$$s_{self}^{q} = \text{cosine}(q_i, \overline{q}), \quad s_{self}^{k} = \text{cosine}(k_j, \overline{k}) \tag{4}$$

where $\overline{q}$ is the mean of the global query feature, $\overline{k}$ is the mean of the global key feature.

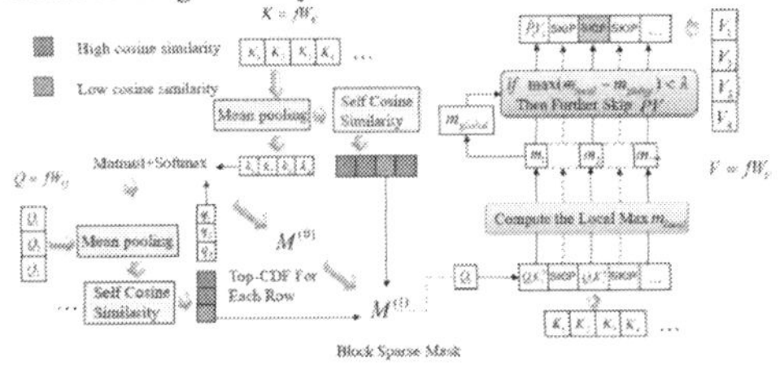

Figure 2. Sparge mechanism principle.

Based on the computed scores, a preliminary sparse mask is constructed using the Top-CDF strategy along with low self-similarity filtering [8]:

$$M_{ij}^{(1)} = \begin{cases} 1, & \text{if Top-CDF or low self-cosine similarity} \\ 0, & \text{otherwise} \end{cases} \tag{5}$$

To dynamically adjust for varying input distributions, a filtering mechanism based on the difference between local and global maxima is introduced. If the local maximum m_{local} significantly deviates from the global maximum m_{global}, the corresponding entry is retained:

$$M_{ij}^{(2)} = \begin{cases} 1, & m_{local} - m_{global} \geq \lambda \\ 0, & \text{otherwise} \end{cases} \quad (6)$$

The final sparse mask is obtained by merging both parts:

$$M_{ij} = M_{ij}^{(1)} \cup M_{ij}^{(2)} \quad (7)$$

Guided by the sparse mask, attention scores are computed only for retained positions:

$$\delta_{ij} = \frac{QK^T}{\sqrt{d}} \times M_{ij} \quad (8)$$

The attention weights are then normalized as:

$$\alpha_{ij} = \frac{\exp(\delta_{ij})}{\sum_{k \in M_t = 1} \exp(\delta_{ik})} \quad (9)$$

The final global spatial feature is aggregated as:

$$P_g = \sum_j \alpha_{ij} V_j \quad (10)$$

This mechanism enables Sparge-iTransformer to accurately model long-range spatial dependencies while significantly reducing computational complexity, thus providing high-quality global spatial features for subsequent temporal modeling and FDIA detection.

2.4. Temporal feature extraction

To capture the dynamic temporal variations in the measurement data, a GRU-based module is employed to model the temporal dependencies of spatial features across multiple time steps. The input to the GRU is a sequence of fused spatial features over T time steps, denoted as $\{P_{HA1}, P_{HA2}, \ldots, P_{HAT}\}$, where each P_{HAt} is obtained by combining the local feature P_s and the global feature P_g. The GRU updates the hidden state through two gating mechanisms. First, the update gate Z_t and the reset gate R_t are computed as:

$$Z_t = \sigma(W_z P_{HAt} + U_z H_{t-1}) \quad (11)$$

$$R_t = \sigma(W_r P_{HAt} + U_r H_{t-1}) \quad (12)$$

Then, the candidate hidden state at time step t is calculated as:

$$\widetilde{H}_t = tanh\big(W_h P_{HAt} + U_h(R_t \odot H_{t-1})\big) \quad (13)$$

where W and U are weight parameters, $\odot$ represents the Hadamard element-by-element product, and $tanh$ is the hyperbolic tangent function.

Finally, the hidden state is updated using a gated interpolation between the previous state and the candidate state:

$$H_t = Z_t \odot H_{t-1} + (1 - Z_t) \odot \widetilde{H}_t \quad (14)$$

After processing all T steps, the hidden state at the final time step H_{Tf} is selected as the output of the temporal modeling, denoted as:

$$P_c = H_{Tf} \quad (15)$$

This temporal feature P_c encodes both historical dynamic information and spatial context across time, and is subsequently used as input to the classification module for FDIA attack detection.

2.5. Others

After extracting temporal features from the GRU module, the combined temporal–spatial feature is flattened and passed through a fully connected layer to produce the final detection score y using a Sigmoid activation. The model classifies a sample as an attack if $y \geq \tau$, where τ is a predefined threshold.

A joint loss function is used during training, consisting of binary cross-entropy loss $\mathcal{L}_d$ and consistency loss $\mathcal{L}_a$, which enforces temporal feature stability. The total loss is defined as:

$$\mathcal{L} = \mathcal{L}_d + \lambda \mathcal{L}_a \quad (16)$$

where λ balances the two components.

3. Experiment results

This paper uses the Pandapower to model the IEEE-30 bus and IEEE-57 bus power systems and constructs a temporal dataset with both normal and attack states.

3.1. Dataset and experimental environment

To simulate dynamic load and generation variations, each bus's load is perturbed using a normal distribution with mean $1 + k[s_t]$ $(k = 0.1)$ and standard deviation $\sigma_s = 0.03$, where $[s_t]$ is the normalized ERCOT load data [9]. The DC optimal power flow (DCOPF) calculation is performed to obtain the bus injection power and branch flow at each time step as the baseline measurements.

For FDIA attack simulation, false data is injected during some time steps $t \in T_a$. The number of attacked buses n is randomly chosen from $[2,6]$, and the attack strength is sampled from the attack amplitude bounds $[C_{\min}, C_{\max}] = [4,5]$. Perturbations are added to the measurements with Gaussian noise $(\sigma_n = 0.012)$. The final dataset consists of 120,000 time steps, with 110,000 for training and 10,000 for testing. The attack sample ratio is 10% in training and 50% in testing. Experiments are conducted on a Windows 10 system with an NVIDIA GeForce RTX 4090 GPU, 64GB RAM, and a 2.11GHz processor, using Python 3.7 and libraries such as Pandapower, PyTorch, and Scikit-learn.

3.2. Detection performance experiment:

To evaluate the FDIA detection performance of HSGT-Net, four metrics are used: accuracy, recall, precision, and F1-

 https://doi.org/10.1051/e3sconf/202563901005

Score. We compare HSGT-Net with SVM, KNN, MLP, CNN, and GNN+LSTM, as shown in Tables 1 and 2. In both IEEE-30 and IEEE-57 systems, HSGT-Net consistently achieves the highest accuracy and F1-Score, demonstrating strong adaptability across different system scales.

Table 1. Experimental Results. (IEEE-30bus)

Model performance comparison	IEEE-30bus					
	SVM	KNN	MLP	CNN	GNN+ LSTM	HSGT-Net
Accuracy	83.50	88.83	90.29	93.20	94.56	**97.11**
Recall	82.14	88.18	90.00	92.73	94.25	**97.86**
Precision	86.79	90.65	91.67	94.44	95.63	**96.83**
F1-Score	84.40	89.40	90.83	93.58	94.94	**97.35**

Table 2. Experimental Results. (IEEE-57bus)

Model performance comparison	IEEE-57bus					
	SVM	KNN	MLP	CNN	GNN+ LSTM	HSGT-Net
Accuracy	79.8	85.22	87.68	90.34	92.68	**95.93**
Recall	77.78	84.40	86.24	89.09	91.89	**96.46**
Precision	83.17	87.62	90.38	92.80	94.44	**96.11**
F1-Score	80.38	85.98	88.26	90.91	93.15	**96.28**

Traditional machine learning models struggle to capture complex attack patterns, while deep learning methods fail to fully exploit the spatial–temporal characteristics of power systems. By integrating Hodge diffusion, Sparge attention, and GRU-based temporal modeling, HSGT-Net significantly improves FDIA detection performance.

3.3. Robustness experiment

To evaluate robustness, HSGT-Net is tested on the IEEE-30 system under varying load disturbance factor k, noise standard deviation σ_n, and attack amplitude $[C_{\min}, C_{\max}]$, using accuracy and F1-Score as metrics.

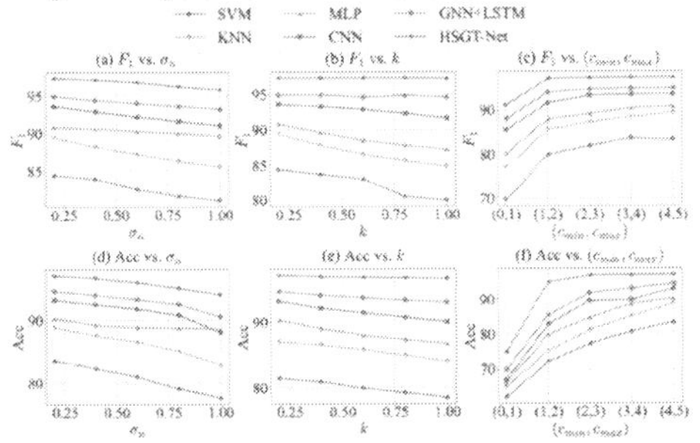

Figure 3. Detection performance comparison under various disturbance and attack conditions.

As shown in Figure 3, traditional models suffer significant performance degradation under increased disturbance and attack intensity, while HSGT-Net consistently achieves the highest accuracy and robustness, demonstrating strong adaptability and generalization in complex scenarios.

3.4. Computational cost experiment

To evaluate the training efficiency, we analyze the training time of each model on the IEEE-30 and IEEE-57 systems, as shown in Figure 4. Results show that the training time of SVM increases significantly with increasing system size, reaching 30 minutes on the IEEE-57 system, indicating poor scalability. In contrast, KNN remains relatively stable.

Among deep learning models, MLP and CNN show the fastest training speed, followed by GNN+LSTM. The HGT-Net has a higher computational cost due to the Transformer structure, with training time reaching 39 minutes per epoch on the IEEE-57 system.

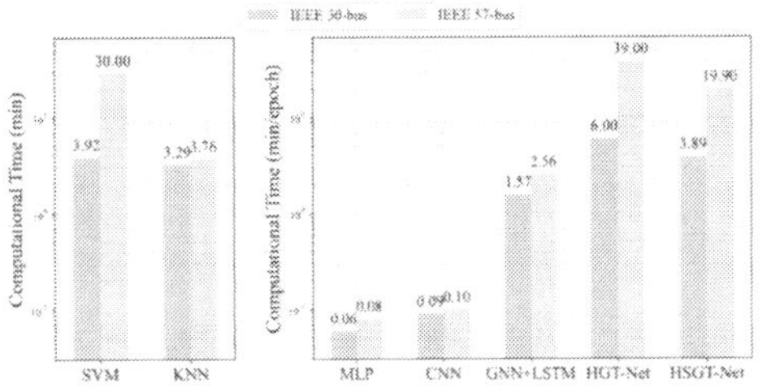

Figure 4. Comparison of computational time for different models on IEEE-30 and IEEE-57 systems.

In comparison, the proposed HSGT-Net reduces this cost to 19.90 minutes per epoch by using the Sparge module, achieving about 49% reduction. This demonstrates its training efficiency and suitability for large-scale applications.

While this section focuses on training efficiency, demonstrating HSGT-Net's reduced complexity via the Sparge module, a quantitative evaluation of inference latency and memory footprint on resource-constrained embedded devices (e.g., RTUs) was not conducted. Such analysis is crucial for assessing real-world deployment feasibility against operational requirements (like potential real-time constraints) and is planned for future work.

5. Conclusion

This paper proposes HSGT-Net, a model that jointly models spatial and temporal features to enhance FDIA detection accuracy and efficiency in smart grids. By integrating graph structures with temporal dynamics, the model addresses the limitations of traditional methods in robustness and computational cost under high-dimensional and complex scenarios. HSGT-Net demonstrates stable performance in large-scale systems and outperforms existing methods in accuracy, robustness, and efficiency. Future work will prioritize further enhancing model security by evaluating robustness against adaptive attacks targeting the Sparge mechanism's thresholds and by investigating the Hodge diffusion module's resilience to gradient masking, including the exploration of potential defense strategies.

 https://doi.org/10.1051/e3sconf/202563901005

Acknowledgment

This work is supported by the National Natural Science Foundation of China under Grant U2468203 and 62473038.

References

1. Chen, J., Wei, C., Wang, K., Yu, S., & Li, H. (2025) Cybersecurity of distributed energy resource systems in the smart grid: A survey. *Applied Energy*, 383: 125364.
2. Li, N., Wang, Q., Wang, Y., Liu, Y., Zhang, H., & Wang, J. (2025) Enhancing Detection of False Data Injection Attacks in Smart Grid Using Spectral Graph Neural Network. *IEEE Transactions on Industrial Informatics*, in press.
3. Luo, X., Zhang, Y., Liu, Y., & Li, H. (2020) Interval observer-based detection and localization against false data injection attack in smart grids. *IEEE Internet of Things Journal*, 8: 657–671.
4. Xiong, X., Li, X., Hu, Z., & Wang, T. (2022) Detection of false data injection attack in power information physical system based on SVM–GAB algorithm. *Energy Reports*, 8: 1156–1164.
5. Moayyed, H., Ghassemian, M., & Jiang, J. (2021) Image processing based approach for false data injection attacks detection in power systems. *IEEE Access*, 10: 12412–12420.
6. Kuo, P., Li, J., & Lin, Y. (2024) GNN-LSTM-based fusion model for structural dynamic responses prediction. *Engineering Structures*, 306: 117733.
7. Xia, W., Huang, Y., Li, G., Wu, J., Zhang, Y., & Liu, J. (2024) Locational detection of false data injection attacks in the edge space via hodge graph neural network for smart grids. *IEEE Transactions on Smart Grid*, in press
8. Zhang, J., et al. (2025) Spargeattn: Accurate sparse attention accelerating any model inference. *arXiv*, 2502.18137.
9. Electr. Reliability Council of Texas Corp. (2020) Backcasted (Actual) Load Profiles—Historical. [Online]. Available: http://www.ercot.com/mktinfo/loadprofile/alp/.

 https://doi.org/10.1051/e3sconf/202563901006

Research on Rural Heating Energy Reform in Heilongjiang Province: Based on a Survey of 209 Households

Shuaijun Lu, Tiansen Liu, Mengtong Dong, Hanxue Wu, and Qiaowei Zhang*

School of Economics and Management, Harbin Engineering University, 150001, China

Abstract: Since the 19th century, the continuous development of heavy industry in Northeast China has led to a yearly increase in emissions of carbon dioxide and other harmful substances, resulting in persistently high environmental pollution indices. This study collected 209 questionnaires by means of questionnaire method, and conducted descriptive analysis and empirical analysis based on the questionnaire to explore the core factors affecting the rural heating energy reform in Heilongjiang Province, and put forward conclusions and suggestions to promote the reform. By introducing indoor winter temperature(T) and annual heating expenditure(E) as independent variables and the heating technology rating(N) as a moderating variable Establish the regression equation, to exploring the influencing factors of heating satisfaction, the results reveals that villagers prefer heating methods that are environmentally friendly, easy to operate, and cost-effective. A Light GBM classification model further indicates that the promotion of clean-energy reform should not rely solely on economic subsidies but must also address public awareness and attitudes. Finally, Friedman and Wilcoxon signed rank tests were applied to examine key issues (Q) and government-driven optimization methods (V) for heating-energy structure, The results show that the government should focus on the economy, heating effect and environmental pollution of clean heating to avoid waste of resources.

1. Introduction

1.1. Research background

Over the past decade, China's emphasis on environmental protection and renewable resources has steadily increased. The 75th United Nations General Assembly in 2020 introduced the "dual carbon" targets, and in 2021 China's 14th Five Year Plan highlighted both the expansion of clean energy supply and the reduction of carbon emissions across the energy industry, fostering a green, low carbon energy consumption model. The 20th National Congress of the Communist Party of China in 2022 reaffirmed the philosophy of "lucid waters and lush mountains are invaluable assets." Under these directives, Heilongjiang Province has actively promoted green-energy reform, substituting 6.4746 million tons of bulk coal by the end of 2024[1]. However, rural winter heating in the province remains dominated by bulk coal and firewood, leading to low indoor temperatures, high energy consumption, and severe air pollution. Moreover, coal-based heating poses significant health risks: the World Health Organization (WHO) and United Nations Development Programme (UNDP) reported that approximately 1.6 million people in developing countries die annually from incomplete fuel combustion, including over 200,000 fatalities due to carbon monoxide poisoning[2].

1.2. Research objectives and significance

Heilongjiang's vast territory and harsh, prolonged winter heating season make system stability, economic viability, and energy structure optimization critical[3]. From a livelihood perspective, optimizing rural heating energy structures can eliminate safety hazards of traditional methods and improve comfort. Environmentally, it reduces carbon emissions and promotes ecological restoration. Economically, it enhances energy efficiency, lowers household heating expenses, and stimulates local industries, achieving both economic and social benefits. Policy wise, optimizing rural heating in Heilongjiang is key to energy transition and security, providing an empirical basis for more efficient energy policies.

2. Research findings

2.1. Current heating practices

Rural heating in Heilongjiang primarily relies on coal-fired kang (heated bed) and electric blankets, which are economical but suffer from low combustion efficiency and high pollutant emissions. WHO data indicates ambient air pollution causes 6.7 million premature deaths annually, with over 4.2 million linked to household and outdoor air pollution, and more than 200,000 deaths from carbon monoxide poisoning due to incomplete coal combustion[4].

* Corresponding author: *Qiaowei* Zhang, zqw9@hrbeu.edu.cn

To address this, the Heilongjiang provincial government issued the Air Quality Continuous Improvement Action Plan (2021–2025), targeting a 3.6% reduction in $PM_{2.5}$ by 2025 relative to 2020 and limiting severe pollution days in cities to 0.7% of the year. The province's abundant clean-energy resources—wind (630 million kW total capacity, 500 million kW technically exploitable, 73.26 million kW potential installed capacity), Hydropower (heilongjiang, Songhua River and other rivers with a basin area of 50km2 or more in the province, 2881 rivers, as well as Xingkai Lake, Jingpo Lake and other five lakes),biomass (18.93 million tons of corn straw and 27.73 million tons of rice straw in 2022), and livestock manure (51.55 million tons in Suihua and Qiqihar)—support large-scale clean heating alternatives[5].

2.2. Analysis of common heating technologies

A comparison of prevalent heating technologies in Heilongjiang Province is shown in Table 1[6-7].

Table 1. Compares prevalent heating technologies in Heilongjiang Province.

Heating Method	Advantages	Disadvantages	Applicability
Coal-fired heating	Low cost; stable supply	Severe pollution; policy restrictions; low efficiency	Regions rich in coal resources
Natural gas heating	High efficiency; eco-friendly; stable operation	Pipeline dependence; high construction and operational costs	Areas with pipeline coverage
Electric heating	Convenient; zero local emissions; intelligent control	High operating costs; demands on grid	Regions with low electricity prices and robust grid
Biomass heating	Renewable; low carbon; abundant resources	Fuel collection challenges; low efficiency; complex maintenance	Agricultural areas with rich biomass
Geothermal heating	High efficiency; low running costs	High upfront investment; geological constraints	Areas with geothermal potential

3. Data preprocessing and analysis

3.1. Data cleaning and reliability

The data for this study comes from the National Undergraduate Innovation Training Program (GK0000020008140901). This project focuses on heating issues in Gansu, Heilongjiang, and Anhui provinces. Therefore, when conducting a sample survey with an error margin of no more than 5%, it was calculated that at a 95% confidence level, the minimum required sample size n=385. Considering factors such as invalid questionnaires, we distributed a total of 560 questionnaires. A total of 209 questionnaires were recovered in Heilongjiang Province, 166 were valid (79.43% validity). Cronbach's alpha reliability analysis yielded α = 0.806 (standardized α), indicating good internal consistency (Table 2).

Table 2. Cronbach's Alpha Reliability Statistics.

Cronbach's alpha	Standardized Cronbach's alpha
.787	.806

3.2. Descriptive analysis

Descriptive statistics for 166 valid responses indicate that 58.44% of residents begin heating in October, with the remainder starting in September or November. Most end heating in April, yielding an average heating duration of five months, reflecting a prolonged heating season. Figure 1 illustrates the heating timeline.

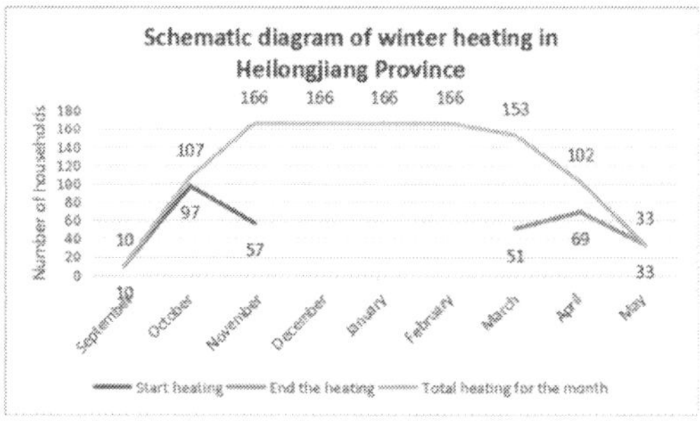

Figure 1. Schematic of Winter Heating in Heilongjiang Province.

The majority (66%) of respondents use kangs or stoves (Figure 2), followed by air conditioners and electric methods. Satisfaction ratings (Figure 3) show central heating highest, kangs second, and air conditioning lowest. Annual heating expenditure clusters in ¥1,001–2,000 and ¥2,001–3,000 (Figure 4). Over 50% report dissatisfaction with cost, pollution, and heating performance (Figure 5). Primary reform expectations are cost reduction and pollution control (Figure 6), while the chief concern regarding energy reform is high upfront investment (Figure 7). Specific suggestions for heating reform are shown in Figure 8.

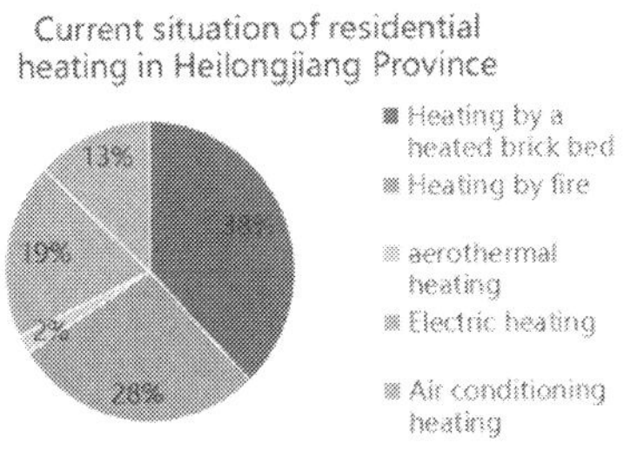

Figure 2. Current heating methods.

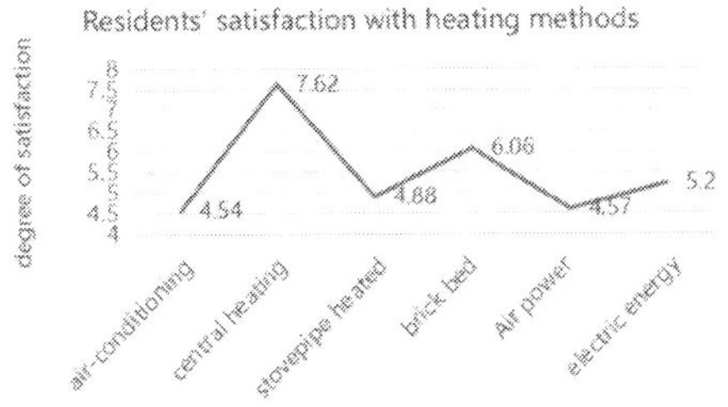

Figure 3. Resident Satisfaction with Heating Methods.

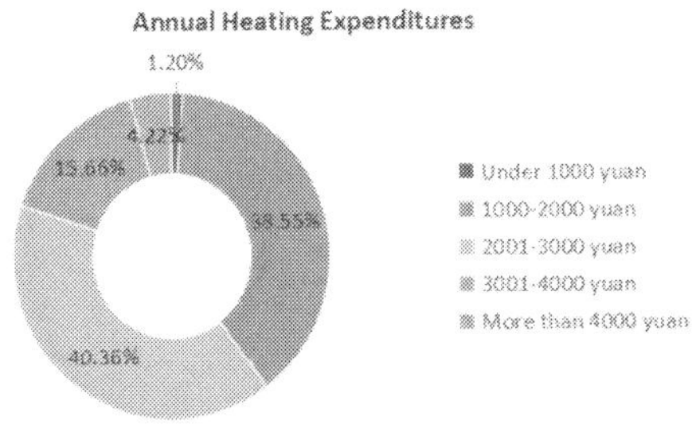

Figure 4. Annual Heating Expenditures.

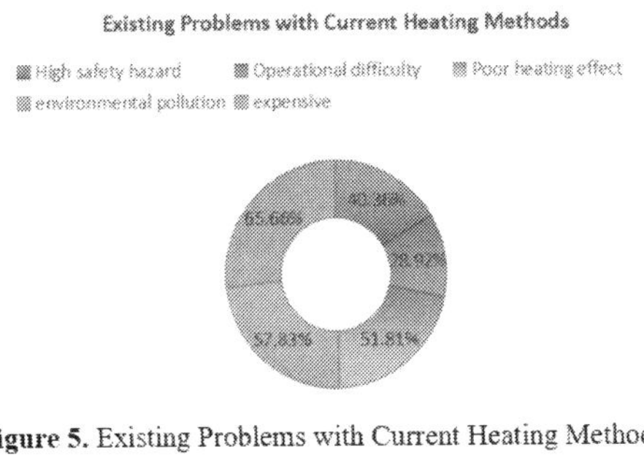

Figure 5. Existing Problems with Current Heating Methods.

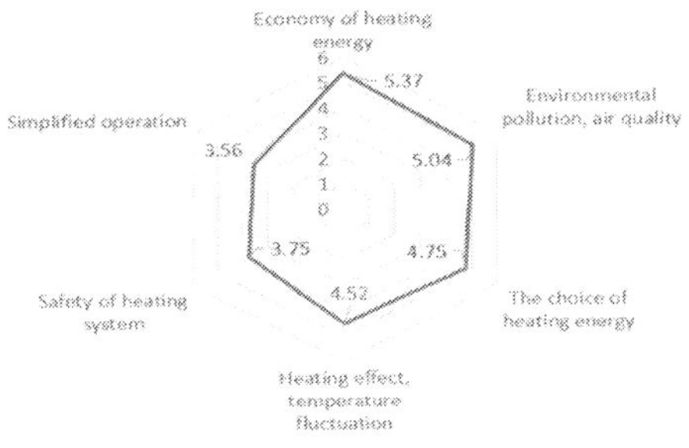

Figure 6. Issues of Greatest Concern.

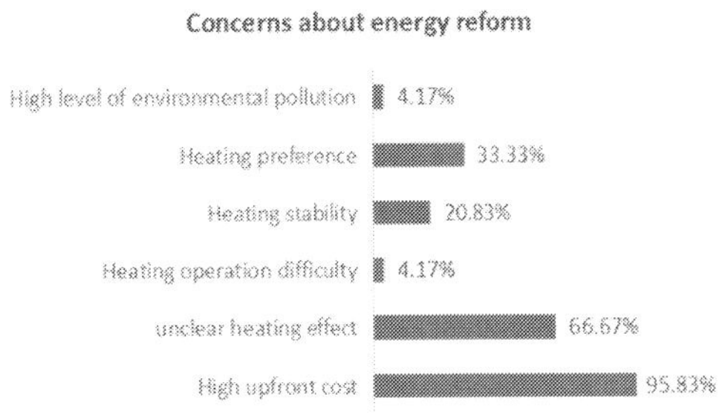

Figure 7. Concerns About Rural Heating Energy Reform.

3.3. Linear regression analysis

To investigate the primary factors influencing residents' evaluations of existing heating technologies, we designate the heating-effectiveness score as the dependent variable Y, indoor winter temperature as independent variable T, annual heating expenditure as independent variable E, and heating technology rating as independent variable N, where N comprises environmental pollution degree (W), price level (S), heating performance (C), ease of operation (K), and safety level (I). Based on this framework, the linear regression equation is formulated as:

$$Y_i = \beta_1 T_i + \beta_2 E_i + \beta_3 N_i + \mu_i \quad (1)$$

$$N_i = 5 - (W_i + S_i + C_i + K_i + I_i + \vartheta_i) \quad (2)$$

Thus:

$$Y_i = \beta_1 T_i + \beta_2 E_i + \beta_3 [5 - (W_i + S_i + C_i + K_i + I_i)\ ] + \mu_i \quad (3)$$

$$= \beta_1 T_i + \beta_2 E_i + \beta_3 [5 - (W_i + S_i + C_i + K_i + I_i)] + \mu_i \quad (4)$$

we performed separate regression analyses for each heating technology, The regression analysis results of the stove heating are shown in Table 3 below:

Table 3. Presents the results for stove heating.

	Non-standardized Coefficients B	Standard Error	Standardized Coefficients Beta	t	p
Constant	3.444	0.857	-	4.020	0.000*
N	0.537	0.195	0.318	2.759	0.008*
T	-0.326	0.242	-0.163	-1.347	0.183
E	0.700	0.250	0.334	2.799	0.007*
R 2			0.201		
Adjusted R^2			0.162		
F		F (3,62) =5.195, p=0.003			
D-W Value			1.637		

Note: Collinearity = Multicollinearity Diagnostics

Table 3, the standardized regression equation is Y=0.7E-0.326T+0.537N+3.444. This model passes the F-test and t-tests at the 5% significance level; however, variable T is not statistically significant. Consequently, we excluded T and re-estimated the model, with results shown in Table 4.

Table 4. Revised Regression Results for Stove Heating.

	Non-standardized Coefficients B	Standard Error	Standardized Coefficients Beta	t	p
Constant	2.938	0.775	-	3.791	0.000**
N	0.496	0.193	0.293	2.563	0.013*
E	0.598	0.240	0.285	2.493	0.015*
R^2			0.177		
adjusted R^2			0.151		
F		F (2,63) =6.797, p=0.002			
D-W			1.578		

Note: Collinearity = Multicollinearity Diagnostics

Thus, the optimized equation is Y=0.598E+0.496N+2.938, thus, Y=0.598E-0.496 (W+S+C+K+I) +5.418. The results indicate that heating effectiveness correlates positively with technology rating and annual expenditure and negatively with pollution, cost, safety risks, operational difficulty, and performance issues. Although we expected a positive correlation between Y and T, the insignificance of T may result from subjective survey responses or insufficient measurement precision. Overall, these findings demonstrate villagers' preference for green, user-friendly, and cost-effective heating methods.

3.4. Light GBM classification model

The complex structure of Light GBM requires enough data to train each node, and insufficient sample size may lead to inaccurate estimation of data distribution by the model, which in turn affects the reliability of the prediction results. Therefore, we used power analysis to identify the sample size before proceeding with the Light GBM classification model. Referring to the common parameters, we set the parameter Significance level to 0.05, Power (1 - β err prob) to 0.8, Number of predictors to 2, Effect size f^2 to 0.15 to carry out t-test two-tailed efficacy analysis, the calculation results are shown in Figure 8, from the results can be seen to select "importance of heating-energy optimization" and "support for clean-energy reform" as features, with poverty status as the dependent variable to construct the Light GBM model required at least 55 samples, a total of 209 questionnaire data were collected, of which 209 were valid, all of which were from rural areas in Heilongjiang Province, which met the conditions for Light GBM model analysis and were representative.

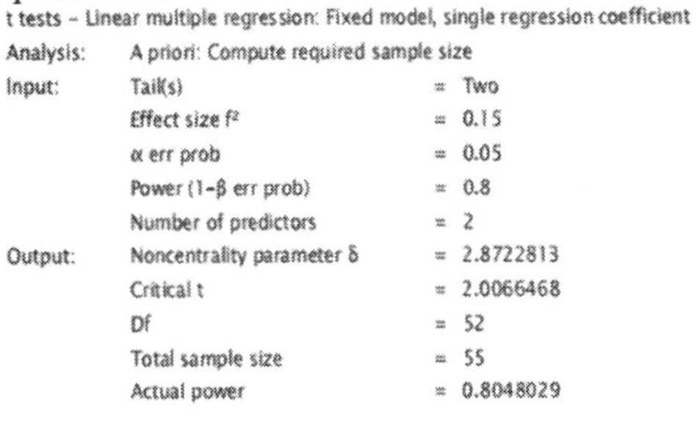

Figure 8. Effect analysis results.

Prior to training, each feature was sorted and discretized into histogram bins. After discretization, the variables were fed into the decision trees (see Figure 9) for computation, and predictions were iteratively optimized using Light GBM's leaf-wise strategy. Seventy percent of the data served as the training set, and the remaining thirty percent was used to test the trained model.

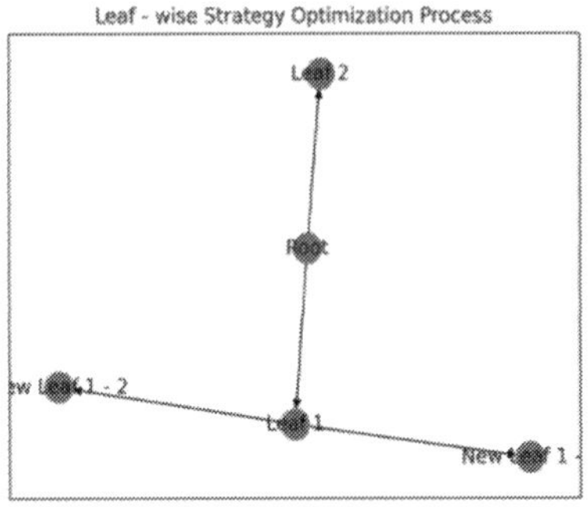

Figure 9. Leaf Node Diagram.

The resulting confusion matrix is shown in Figure 10, where diagonal entries represent correctly classified instances and off-diagonals represent misclassifications. Accuracy, precision, and recall all exceed 0.77, indicating robust classification performance. Through the Light GBM model, we determined correlations between various household characteristics and model features, thereby identifying the relationship between economic status and the demand for energy-structure optimization.

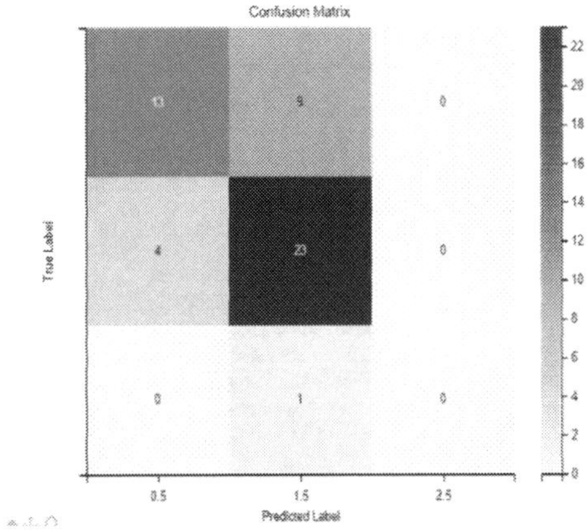

Figure 10. Light GBM Confusion Matrix.

Figure 11 reveals a moderate positive correlation between support for clean-energy reform and policy recognition, indicating that greater recognition aligns with higher support levels. In contrast, both "importance of heating-energy optimization" and "support for clean-energy reform" exhibit very weak correlations with household. economic status, suggesting limited influence of economic conditions on reform attitudes and policy endorsement. Consequently, promotion of clean-energy reform should not rely solely on economic subsidies but must emphasize public perception and attitude.

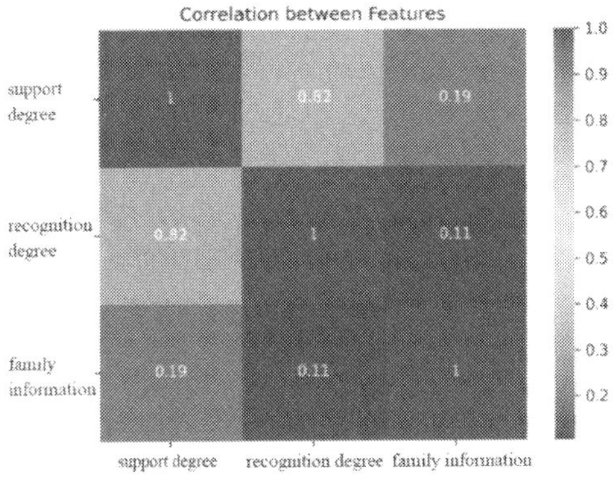

Figure11. Correlations between Household and Features.

3.5. Friedman test and wilcoxon signed-rank test

We named the measurement items for the key issues to be addressed in heating-energy structure optimization (Q) and for the central government's recommended methods to promote such optimization (V), and then conducted Friedman tests on each set. First, the Friedman test for Q is presented in Table 5.

Ordering by mean rank gives Q3 (2.63) > Q2 (2.96) > Q6 (3.25) > Q5 (3.48) > Q4 (4.25) > Q1 (4.43), indicating that Q3 and Q2 control are rated highest, while Q4 and Q1 rank lowest. The significant χ^2 ($p < 0.001$) justifies follow up pairwise comparisons via Wilcoxon signed rank tests (Table 6).

Table 5. Q-Friedman Test Results.

Measure Description	Code	N	Mean	Std. Dev.	Min	Max	Mean Rank
Heating Energy Economic Efficiency	Q3	209	2.63	1.668	1	6	2.63
Environmental and Air Pollution	Q2	209	2.96	1.636	1	6	2.96
Choice of Heating Energy	Q6	209	3.25	1.861	1	6	3.25
Heating Performance and Temperature Fluctuation	Q5	209	3.48	1.409	1	6	3.48
Safety of the Heating System	Q4	209	4.25	1.531	1	6	4.25
Simplified Operation and Ease of Use	Q1	209	4.44	1.350	1	7	4.43
Test Statistic: Chi-square χ^2 = 120.558, degrees of freedom (df) = 5, $p < 0.001$*							

Table 6. Q-Wilcoxon Signed-Rank Test Results.

	Q1	Q2	Q3	Q4	Q5
Q2	-2.874/.004				
Q3	-1.438/.150	-1.690/.091			
Q4	-.790/.430	-4.049/.000	-2.547/.011		
Q5	-5.544/.000	-7.883/.000	-6.632/.000	-5.657/.000	
Q6	-4.547/.000	-6.935/.000	-6.166/.000	-4.556/.000	-1.047/.295

At $\alpha = 0.05$, only Q1–Q3, Q1–Q4, Q2–Q3, and Q5–Q6 have $p > 0.05$ (non-significant). The most pronounced differences occur among Q3, Q2, and Q5, suggesting policymakers should prioritize these areas to avoid misallocation of resources. Next, the Friedman test for V is shown in Table 7.

Ordering by mean rank in Table 7 shows V1 (2.14) > V7 (3.92) > V2 (4.05) > V5 (4.26) > V6 (4.49) > V4(4.51)> V3 (4.63), indicating that respondents prioritize V1 and V7 most highly, while V3 and V4 are of lowest priority. The Friedman test statistic χ^2 = 158.409 ($p < 0.001$) confirms significant differences among the seven mean ranks at the 0.05 level, warranting pairwise comparisons via the Wilcoxon signed-rank test as shown in Table 8.

Table 7. V — Friedman Test Results.

Measure Description	Code	N	Mean	Std. Dev.	Min	Max	Mean Rank
Development of clean energy	V1	209	2.17	1.788	1	8	2.14
Subsidy or incentive policies	V7	209	4.56	2.093	1	8	3.92
Continued promotion of clean heating	V2	209	3.99	1.896	1	8	4.05
Replacement of industrial furnaces	V5	209	4.60	1.929	1	8	4.26
Restructuring and phasing out of coal boilers	V6	209	4.79	2.065	1	8	4.49
Construction of rural infrastructure	V4	209	4.31	1.632	1	8	4.51
Total control of coal consumption	V3	209	4.20	2.252	1	8	4.63

Table 8. V—Wilcoxon Signed-Rank Test Results.

	V1	V2	V3	V4	V5	V6
V2	-6.850/.000					
V3	-8.417/.000	-2.780/.005				
V4	-8.520/.000	-1.617/.106	-.990/.0322			
V5	-7.145/.000	-1.024/.306	-2.388/.001	-2.717/.007		
V6	-8.609/.000	-.368/.713	-2.353/.019	-1.232/.218	-1.698/.089	
V7	-8.551/.000	-1.334/.182	-.785/.433	-.021/.984	-2.651/.008	-1.582/.114

The Wilcoxon signed-rank results demonstrate that, at $\alpha = 0.05$, most pairwise comparisons yield $p < 0.05$, indicating significant differences between items. Clean-energy development (V1) emerges as the unequivocal top-priority policy, whereas V7, V3, and V6 require enhanced coordination. Policy efforts should therefore concentrate on promoting clean-energy technologies, supplemented by economic incentives to bolster public acceptance of complementary measures.

4. Conclusions and recommendations

Following the empirical analysis of Heilongjiang Province, four primary conclusions emerge. First, residents report low satisfaction with current heating methods, especially concerning cost, environmental pollution, and heating performance. Second, significant deficiencies in existing systems highlight an urgent demand for clean and efficient energy. Third, owing to Heilongjiang's abundant wind and biomass resources, clean energy reform must be tailored to local conditions. Fourth, economic viability and environmental impact are the most important factors for residents when evaluating heating energy, while safety and convenience are of lower priority.

Based on these conclusions, we propose the following recommendations to advance clean energy reform. First, promote high-efficiency clean heating equipment and improve technology ratings through collaboration between government authorities and technology providers. Second, optimize the heating cost structure to alleviate farmers' financial burdens: directly subsidize household heating expenses and explore community-level centralized heating systems to achieve economies of scale. Third, establish a government-led, society-wide cooperative mechanism, in which the government functions as both policymaker and resource allocator. Fourth, focus policy formulation on the promotion of clean energy technologies, supplemented by economic incentives to enhance public acceptance of related measures.

Acknowledgment

The all authors are sincerely appreciating the funding support from the National Social Science Fund of China (22CGL030); Subtopic of the Major Consulting Project for Sanya Yazhou Bay Science and Technology City (KY000202109260001)); National-level College Students' Innovation Training Program of China (GK0000020008140901).

References

1. Government of Heilongjiang Province. (2025). Report on the Implementation of the 2024 National Economic and Social Development Plan and Draft Plan for 2025. Heilongjiang Provincial Government Website.
2. World Health Organization & United Nations Development Programme. (2004, October 14). Statement on the Health Risks of Cooking Smoke. Press release.
3. Wei, Y. M., Fan, Y., Han, Z. Y., & Wu, G. (2006). China Energy Report (2006). Strategic and Policy Research. Beijing Science Press.
4. WHO. (2024). Ambient (Outdoor)Air Quality and Health. Retrieved from https://www.who.int/zh/news-room/fact-sheets/detail/ambient-(outdoor)-air-quality-and-health.
5. https://www.toutiao.com/article/6195354470371230210/?upstream_biz=doubao&source=m_redirect&wid=1732850948049
6. Han, T. A., & Yuan, X. (2021). Development Status of Biomass Boiler and Air-Source Heat Pump Heating Systems. Shanxi Architecture, 47(23), 96–98, 103.
7. Wang, C. B., Chen, Y. S., Zhang, L. X., Chang, Y., & Pang, M. Y. (2017). Comparative Study of Lifecycle Environmental Emissions of Straw Compressed Blocks and Coal Heating Systems. Acta Scientiae Circumstantiae,37(11), 4418–4426.

E3S Web of Conferences 639, 01007 (2025) https://doi.org/10.1051/e3sconf/202563901007
ICEMEE 2025

Indentation Relaxation Behavior of the Austenitic Heat-Resistant Steel Sanicro25

Tong Lin, Tieshan Cao*

Dalian University of Technology, 116086 Dalian, China

Abstract. Taking sanicro25, an austenitic heat-resistant steel for ultra-supercritical power stations, as the research object, a new electronic creep tester is used to conduct indentation experiments at high temperatures to evaluate the feasibility of its indentation relaxation performance; the indentation experiments are conducted using a Vickers indenter, and the high-temperature relaxation creep behavior is investigated in the range of temperatures and depths of indentation from 973-1073K and 100-180 μm, and the relaxed The results show that: the indentation at different depths is not easy. The results show that: the hardness/stress and time patterns of indentation relaxation at different depths are in accordance with the relationship similar to that proposed by Choi; the stress index is about 7 according to the power rate relationship, and the relaxation coefficient μ is about 0.3, whose inverse matches well with the stress index obtained from the indentation creep test; there is a pile-up phenomenon at the edges of the residual indentation on the surface of the specimen, which is caused by plastic flow of the material along the direction of the indenter after the formation of a completely plastic region at the bottom of the indenter. Plastic flow occurs along the indenter direction after the material forms a fully plastic region at the bottom of the indenter.

1. Introduction

Due to breakthroughs in high-temperature resistance and thermal system integration technology, China's thermal power industry has completed the leap from subcritical to ultra-supercritical units, significantly improving the efficiency of power generation per unit of coal consumption, and sanicro25 is a popular material for superheaters in ultra-supercritical boilers because of its excellent corrosion resistance, oxidation resistance, and high-temperature creep performance[1].

High temperature and stress environment, creep is the main form of material failure, so the evaluation of creep performance is critical, but the traditional creep time-consuming, large sample requirements; some components are limited by the shape, or to analyze and evaluate the localization can only be carried out with small specimens. The indentation test provides a good choice, that is, the load will be indented into the specimen, the specimen preparation is simple and size requirements are not high.

Indentation relaxation of small specimens originated from traditional hardness testing. As early as the 1980s S. N. G. CHU et al. have combined the indentation method with relaxation experiments for localized stress relaxation, and were able to obtain the same type of information as conventional relaxation. It has been shown in the literature that there is a consistency in the trend of results between relaxation and creep experiments under certain conditions in the indentation method[2]. Whereas P. Nogning Kamta [3]and others, through stress sensitivity coefficients and likelihood coefficients in Hart's theory, although linking indentation creep with indentation relaxation; only classified the material using creep and relaxation sensitivity, and did not evaluate the creep behavior, R. Mahmudi et al. [4]compared the indentation (impression, indentation) and indentation relaxation creep to match the creep behavior, and refined the creep analysis by combining microscopic images and EDS.

However, there is no perspective to evaluate the creep properties in combination with indentation relaxation, and most of them do not involve the study of the relaxation creep properties of high temperature sanicro25 steel. Evaluation of relaxation creep relationship single, relaxation experiment, strain, temperature, initial load and other conditions will affect the relaxation curve. In this paper, the indentation relaxation creep method IRCT (impression-relaxation creep test) is proposed to evaluate the creep behavior by the relaxation behavior of the specimen under different environments and mechanical conditions; observe the size of the residual indentation, morphology information, and discuss the mechanism of relaxation creep, in order to seek for a more insightful comparison of the creep behavior.

2. Experimental materials and methods

2.1. Experimental materials

The nominal chemical composition (mass fraction, %) of the experimental Sanicro25 steel (austenitic heat-resistant steel) was Ni(23.5-26.6)-Cr(21.5-23.5)W(2.0-4.0)-

* Corresponding author: tieshan@dlut.edu.cn

Cu(2.0-3.5)-Co(1.0-2.0)-Nb(0.30-0.60)-N(0.15-0.30)-C(0.04-0.1)-Mn(0.6)-Si(0.4)-P(0.03)-S (0.015) (maximum) and Fe (residual), Fig. 1 shows the initial organization of the Sanicro25 metallographic morphology, which shows that the microstructure of the Sanicro25 steel consists of the following three parts: 1) isotropic austenite grains, 2) multiple annealed twin crystals formed during the heat treatment process, 3) annealed twin crystals formed during the heat treatment, and 4) annealing twins. annealing twins, and 3) residual Nb-Cr-N composite precipitated phases[5]. The material used in the experiments was a solution (1417~1523 K) treated seamless steel pipe (OD 54 mm, wall thickness 9 mm). Preparation of samples: the steel pipe was cut out with a wire cutter to a number of specimens of 8 × 8 × 5 mm size, and then sandpapered with sandpaper of 400#, 800#, 1000#, 1200#, 1500#, and 2000# in sequence, followed by polishing treatment with polishing paste of 1.5 μm and 1 μm particle size.

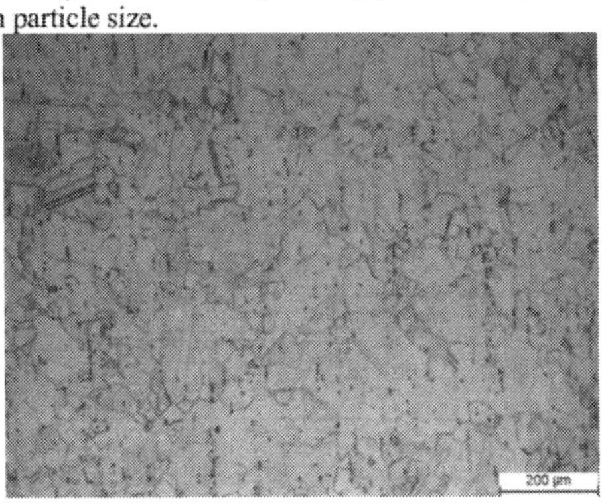

Fig. 1. Sanicro25 metallographic sketch.

2.2. Experimental methods

In the article, an electronic creep tester equipped with upper and lower zoned high-temperature ovens is used to conduct indentation relaxation tests, and the specific test procedures are as follows. The experimental conditions are shown in Table 1:

The standard Vickers indenter was fixed on the center-controlled displacement bracket, and a specimen with a thickness of 5 mm was placed directly on the carrier table, ensuring that the specimen was located in the center of the platform under the indenter. The specimen and indenter were placed in a high temperature furnace with nitrogen vented into the furnace to prevent oxidation of the specimen. Nitrogen was continuously vented into the furnace before heating and stopped 5-10 minutes before the start of the holding phase to avoid temperature fluctuations caused by gas flow that could affect the results of the experiment. The furnace was heated to the test temperature (973-1023 K), and after the surface temperature of the specimen reached the preset value, it was held for 2 hours to make the temperature uniformly distributed in the furnace. At the test temperature, an initial load of 20 N was applied to the specimen to ensure stable contact between the indenter and the specimen surface. The indenter was loaded at a constant rate of 0.5 μm/s to a preset depth (100~180 μm), and then the maximum indentation depth was kept constant. It was kept at the maximum depth for 6 h. The load change was monitored in real time by a high-precision displacement transducer (accuracy 0.1 μm), and the load, time, displacement and temperature data were recorded automatically. The load range of the tester was 0-5000 N, and the data were collected synchronously by a computer. After the holding phase is completed, the indenter is unloaded at the same rate (0.5 μm/s) and the single indentation relaxation test is completed.

Table 1. Experimental parameters of Sanicro25 steel.

Material	Temperature, T/K	depth, h/μm	Time, h
Sanicro25 Steel	973	100	6
		140	
		180	
	1023	100	
		140	
		180	

3. Results and analysis

3.1. Analysis of indentation parameters

In this paper, the most common Oliver-Pharr analytical method is used to calculate the indentation relaxation experimental data of small specimens using a Vickers (positive tetragonal cone with an apex angle of 136°) indenter in order to find the hardness and modulus of the specimens in the region of different crystal orientations, so as to investigate the effect of crystal orientation on the physical properties of the materials. The method is now briefly explained based on the load--depth curve in Fig. 2 (a) and the simplified indentation profile in Fig. 2(b). Fig. 2 (a) indicate that the load decreases with relaxation time during the maximum depth hold of the indenter, indicating that significant indentation (load) relaxation of the material occurs. The difference in the degree of relaxation for different holding times (6 h) reflects the high-temperature creep properties of the material.

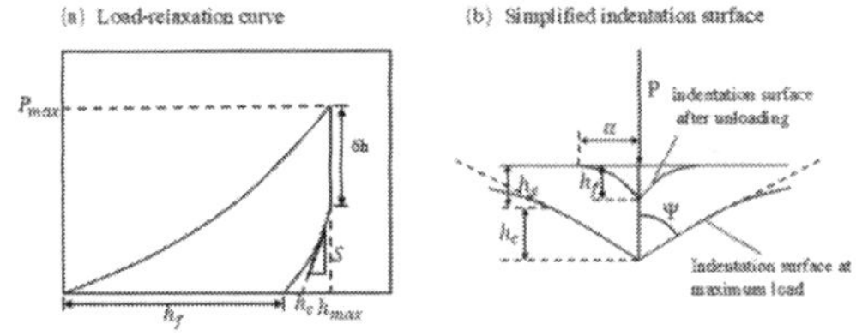

Fig. 2. (a) Schematic of load-relaxation curve (b) Schematic of indentation profile.

Oliver proposes a method to calculate the indentation area based on a simple equation for the shape of the indenter. As shown in Fig. 2(b), Pmax is the maximum indentation load; hmax is the maximum indentation depth; hc is the contact depth; hf is the residual depth; S is the contact stiffness, i.e., the slope of the unloading curve. In Fig. 2(b), P is the indentation load; α is the equivalent contact radius; Ψ is the equivalent half cone angle of the indenter; and hs is the indentation deformation.

The contact stiffness parameter S can be obtained by fitting the first 20% of the sample unloading curve according to the power exponent model given by Oliver and Pharr:

$$P = \alpha(h - h_f)^n \quad (1)$$

$$h_c = h_{max} - h_s = h_{max} - \lambda \frac{P_{max}}{S} \quad (2)$$

$$S = \frac{dP}{dh}\bigg|_{h=h_{max}} = \alpha n(h_{max} - h_s)^{n-1} \quad (3)$$

Where α and n are the fitting parameters for the unloading stage, λ is the correction coefficient about the shape of the indenter, in this paper, we use the Vickers indenter, and λs taken as 0.75 The contact area A_C can be obtained from the contact depth h_c:

$$A_C = f(h_c) = 24.56{h_c}^2 \quad (4)$$

h_c is the value of the intersection of the 20% unloading curve and the x−axis before fitting. Combined with the contact area A_C, and the indentation load Pwe can obtain the material hardness H (which represents the stress in the indentation experiment):

$$H = \frac{P}{A_C} \quad (5)$$

In addition, according to the contact stiffness S, and the contact area $A_C$we can get the Young's modulus of the material E_r:

$$E_r = \frac{\sqrt{\pi}}{2\beta}\frac{S}{A_C} \quad (6)$$

Introduction β is a coefficient related to the shape of the indenter, for the indenter selected in this paper, β is taken as 1.

According to equations (3) and (6) shown, the contact depth h_c and contact stiffness S are obtained, and the results obtained for different experimental conditions are shown in Table 2.

Table 2. Different conditions of contact depth h_c, contact stiffness S.

number	temperature /°C	high/μm	hc/mm	S	A	Er
1	700	100	0.09595	40771	0.245	72996
2	700	140	0.13065	49432	0.4802	63216
3	700	180	0.16985	59537	0.7938	59219
4	750	100	0.09595	40771	0.245	72996
5	750	140	0.13225	46241	0.4802	59135
6	750	180	0.17389	55846	0.7938	55547

Table 2 shows that the unloading depth of the instrument and the preset depth error is within 10 μm, the stiffness of the material increases with the depth of the indentation (residual indentation), and the temperature does not have much effect E_r decreases with the increase of the depth, but the effect gradually decreases after 140 μm.

3.2. Indentation relaxation behaviors

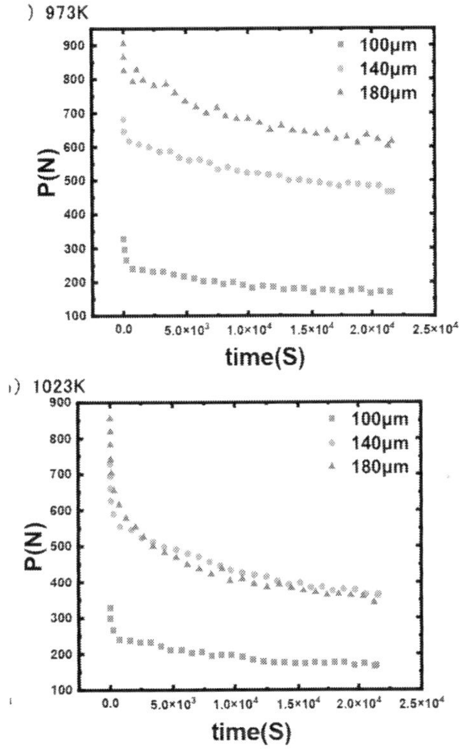

Fig. 3. Temperature 973 (a), 1023 K (b), indentation depth 100, 140, 180 μm Load-time diagrams.

Figure 3 shows the load P-time s plots at temperatures 973 (a) and 1023 K (b) with indentation depths of 100, 140, and 180 μm, respectively, and it can be noticed that relaxation of the specimens occurs during the experimental process of holding the specimens at different load maxima, and, as expected, the relaxation initiation phase imposes greater indentation depths and a correspondingly higher level of loading.

Initial stage (t<5000s): the load increases with increasing indentation depth and the initial load increases with relaxation.

Stabilization phase (t>5000 s): the 140 μm indentation depth continues to increase with no significant change in load; the load gradually stabilizes, indicating that the dynamic equilibrium is controlled by a combination of dislocation migration and grain boundary diffusion.

During relaxation testing, a constant deformation is applied and held constant. The stresses due to deformation in the initial phase of the test decay with time, a phenomenon known as stress relaxation. The total deformation can be decomposed into a plastic component ε_p and an elastic quantity ε_e the sum of which remains constant with time during the test, i.e.

$$\varepsilon_e + \varepsilon_p = \varepsilon = \frac{\sigma}{E} + \varepsilon_{SRT} + \varepsilon_p = C_1 \quad (7)$$

ε_e is the elastic strain; ε_p is the plastic strain, which remains constant during relaxation creep; ε_{SRT} is the stress relaxation test strain; σ is the applied stress; E is Young's modulus; C1 is a constant related to material properties.

$$\dot{\varepsilon}_{SRT} = -\dot{\varepsilon_e} = -\frac{\dot{\sigma}}{E} \quad (8)$$

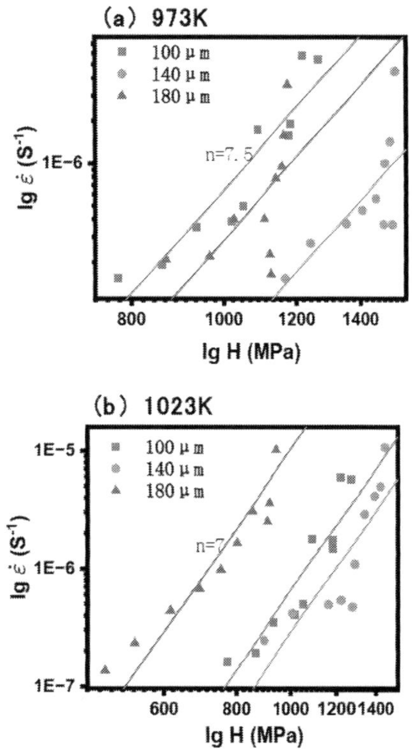

Fig. 4. Temperature 973 (a), 1023 K (b) relaxation rate-stress double logarithm plot.

According to equation (8) the relaxation curve is transformed into a double logarithmic curve of relaxation rate-hardness (stress), because of the large change of stress in the pre-relaxation period of the indentation, the unstable relaxation rate of the first 60s is excluded to obtain the curve shown in Fig. 4, the relaxation rate increases with the increase of stress, and the same stress of the same indentation depth decreases with the increase of temperature. The stress index is n=7.5 at 973 K and n=7 at 1023 K. The depth of the indentation has almost no effect on the rate of relaxation of the material, and the stress indexes are all around 7, representing that dislocation migration is the main form of relaxation.

Combining Eq.(5), which expresses stress as a function of stiffness, and considering Hooke's law ($\sigma = E\varepsilon$), which relates stress to elastic deformation, we can relate the rate of plastic deformation to the relaxation stress σ and the modulus of elasticity, E, can be obtained as follows:

$$\dot{\varepsilon}_{p_{ind}} = -\dot{\varepsilon}_{e_{ind}} = -\frac{1}{E}\frac{d\sigma}{dt} = -\frac{1}{E \cdot C}\frac{dH}{dt} \quad (9)$$

In this paper, it is suggested that the correlation between hardness/stress and time be handled using a relationship similar to that proposed by Choi et al:

$$H(t) = H_0 - A_2 t^{n^2} - B_2 t \quad (10)$$

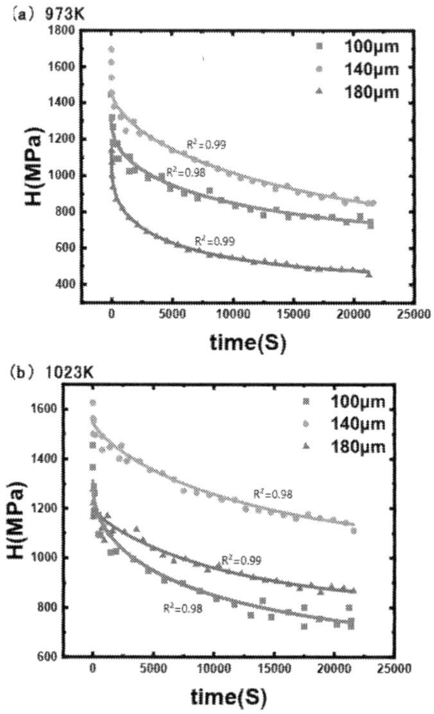

Fig. 5. (Hardness/stress)-time fit plot (a) 973K (b) 1023K.

According to Eq. (10) way to fit the results of the six conditions of 973 K and 1023 K with the above equation, as shown in Fig. 5 It is worth noting that the stress relaxation curves measured by the indentation method show similar characteristics in overall morphology with those obtained from uniaxial tensile experiments. It should be noted that the kinetic equation (Eq. 10) proposed by Nogning Kamta et al. can accurately describe the stress decay behavior with time during the indentation stress relaxation process, with an R^2 of 0.98 or 0.99 and a correlation coefficient close to 1, which suggests the suitability of the model for the experimental mechanism of high-temperature indentation relaxation in this paper.

In order to deeply investigate the relaxation behavior of materials, Hart proposed a theoretical model for describing the mechanism of intracrystalline plastic deformation in polycrystalline solids under monotonic stretching conditions. The core of the model is to establish an intrinsic relationship between the stress (σ), the inelastic deformation rate ($\dot{\varepsilon}$), and the hardness parameter (H) that characterizes the inelastic behavior of the material, with the mathematical expression: $\sigma = f(\dot{\varepsilon},\ H)$

This relationship has direct applicability in indentation experiments because indentation tests are capable of simultaneously obtaining data on the stress field distribution and the inelastic deformation rate. Hart further pointed out that the stress-strain curves exhibit a significant proportionality relationship under different strain rate conditions, and based on this, he proposed a method for constructing the master curve. The construction of the master curve is realized by translating each experimental curve along the logarithmic time axis, and the slope (μ) of the translation line is determined by the following relation:

$$\mu = \frac{\Delta \ln(\sigma)}{\Delta \ln(\dot{\varepsilon})} \tag{11}$$

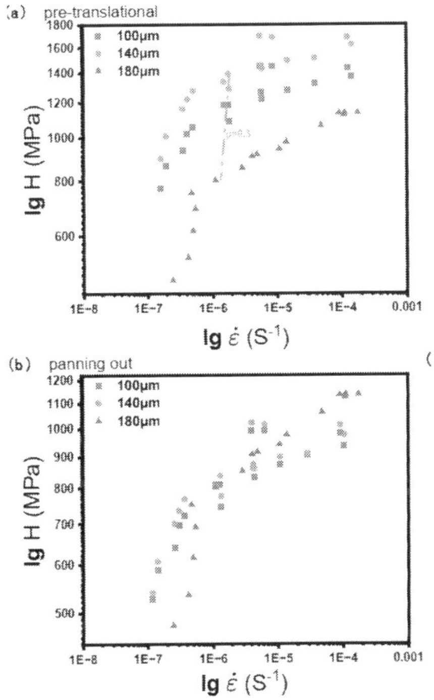

Fig. 6. Hardness change as a function of relaxation rate under three stress relaxation conditions at 1023K.

The horizontal coordinate in the translation plane is $\Delta \ln(\sigma)$and the vertical coordinate is $\Delta \ln(\dot{\varepsilon})$.

In constructing the master curve, the inverse of the translation gradient μ (1/μ) should be equivalent to the creep stress index n. The inverse of the translation gradient μ (1/μ) should be the same as the creep stress index n. By translating each relaxation curve on the logarithmic time axis so that it is iterated in segments at different indentation depths, master curves covering larger deformation regions are obtained (i.e., before and after translation, as shown in Fig. 6. The key is to convert the data from the middle and low layers (100-140 μm) to the high depth (180 μm) above 180 μm by extracting the characteristic points of the same stress decay rate (i.e., equal slopes) to realize the continuous characterization of the mechanical properties of the material over a wide range. On this basis, the creep susceptibility factor n = 1/μ = 3.3 obtained using the inversion of the translational slope μ, while the indentation relaxation creep model yields about 7, which is similar to that of T91 steel by P. Nogning Kamta et al. but which is comparable to the other experimental results of Sanicro25 and to the results of other tests by Sanicro25.

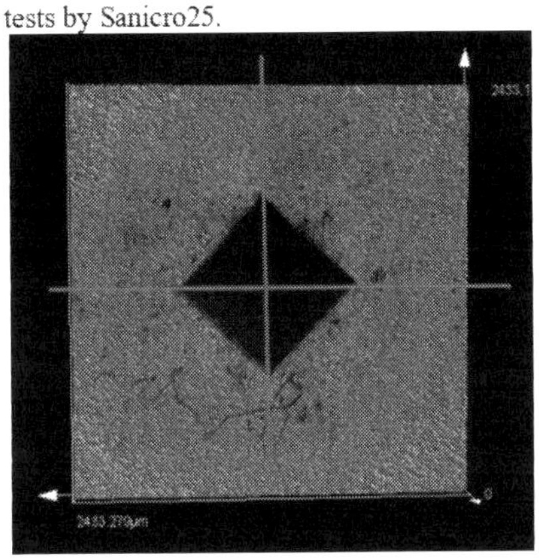

Fig. 7. Confocal surface.

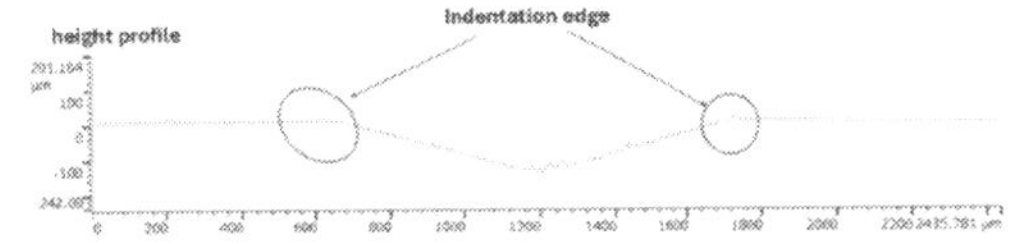

Fig. 8. 973K,140 μm two diagonal profiles confocal size.

After the indentation experiment, the surface is shown in Figure 7, leaving a diamond-shaped indentation on the surface, along the horizontal line in the figure to see the cross-section, as shown in Figure 8, to 973K, 140μm two diagonal section 3D confocal size for example, as shown in Figure 8, the red circle in the figure is the contact of the indenter on the surface edge of the specimen, this is the edge of the indentation part of the occurrence of the slight bulge, which is known as the material buildup in the surface of the specimen. Because the material in the vicinity of the indenter is compressed in the loading direction as the test progresses and is contained in the elastic-plastic region below the indenter, there is no buildup of material on the specimen surface at this time. When the area around the indentation is completely plasticized, the material begins to occur plastic flow, due to the modulus of elasticity of the indenter is much larger than the specimen, so the indenter can be regarded as a rigid body, the existence of the rigid body of the specimen has a greater constraint on the plastic flow of the specimen, resulting in the direction of flow upward, in the surface of the specimen will be manifested as a pile-up phenomenon. This pile-up phenomenon at the edge of the indenter will affect the results of E_r in the previous section, so that there is a large difference between it and the standard Young's modulus.

3.3. Relaxed tissue

Optical micrographs of the indentation creep specimen under the indenter after 100 times magnification at 973 K, 140 μm depth condition are shown in Fig. 9. The black area under the indenter in the picture is the cavity formed after the creep of the material, and it is observed that the three areas under the indentation are separated by lines, labeled as 1, 2, and 3, where area 1 is the area directly under the indenter, which is called the dead zone because there is no obvious change in the grains in this area; area 2 is called the deformation zone, which is the area in which the grains undergo more obvious plastic deformation, and area 2 is the area in which the plastic deformation occurs due to the rigid body indenter and the dead zone extrusion2. The flow in the plastic deformation zone of the region results in a buildup at the edge of the indenter; region 3 is called the region away from the indentation, which is farther away from the indentation, and no obvious signs of plastic deformation are found.

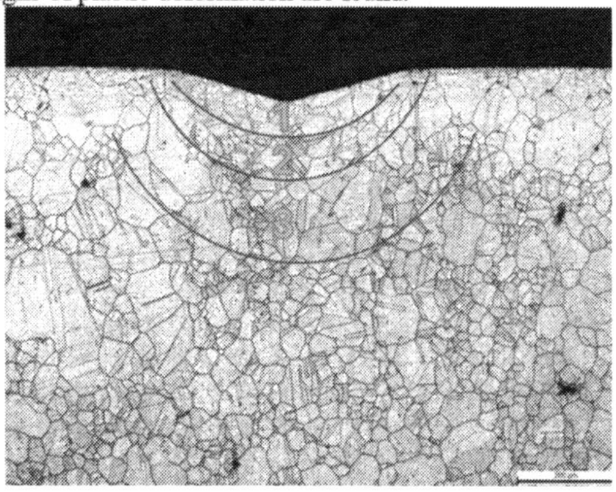

Fig. 9. 973K, 140μm indentation depth magnified 100x micrographs

4. Conclusions

Indentation depth 140 μm residual depth is basically consistent, recommended for experiments at 140 μm indentation depth.

(1) Indentation depth is the main influencing factor, as the indentation depth increases, the load increases, the stiffness increases, the Er decreases, and there is no effect on the stress exponent

(2) Temperature has little effect on the indentation parameters of this material, high temperature indentation relaxation behavior.

(3) The stress exponent in the range 973K-1023K is about 7, while the relaxation translation exponent $\mu = 0.3$, the inverse of which coincides with the stress exponent of the indentation creep experiments.

References

1. CEMPURA G, KRUK A J M C. Microstructural analysis of Sanicro 25 (42Fe22Cr25NiWCuNbN) after oxidation in steam for 25,000 h at 700° C [J]. 2025: 114751.
2. BARAL P, KERMOUCHE G, GUILLONNEAU G, et al. Indentation creep vs. indentation relaxation: A matter of strain rate definition? [J]. 2020, 781(Apr.20): 139246.1-.9.
3. KAMTA P N, CHICOT D, ROUDET F, et al. Fluage et relaxation par indentation d'aciers au chrome [J]. 2017, 105(1): 103.
4. MAHMUDI R, GERANMAYEH A R, NOORI H, et al. A comparison of impression, indentation and impression-relaxation creep of lead-free Sn–9Zn and Sn–8Zn–3Bi solders at room temperature [J]. 2009, 20(5): 312-8.
5. LI Y, WANG X J M S, A E. Strengthening mechanisms and creep rupture behavior of advanced austenitic heat resistant steel SA-213 S31035 for A-USC power plants [J]. 2020, 775: 138991.

https://doi.org/10.1051/e3sconf/202563901008

Effect of Bamboo Flour Content on the Properties of PLA-based and PBAT-Based Wood-Plastic Composites

Ling Liu[1], Chen Chen[2], Huihui He[1], Jun Li[2], Yu Su[2], Haojun Jiang[2], Zhendong Lei[2]

[1]Zhejiang Institute of Quality Science, Hangzhou 310018, China
[2]School of Materials Science and Engineering, Shanghai University of Engineering Science, Shanghai 201620, PR China

Abstract. In this study, different mass fraction (0%, 10%, 20%, 30%) of bamboo flour (BF) was blended with PLA pellets and PBAT masterbatch via extrusion compounding process. The effect of bamboo flour content on the thermal and mechanical properties of the prepared wood-plastic composites (WPCs) arecharacterized by SEM, XRD, and DSC. The results show that the addition of BF did not alter the crystal structure of PLA and PBAT. However, as the bamboo flour content increase to 20% and 30%, agglomeration and uneven distribution are observed for the composites. Additionally, the initial decomposition temperature decreases, and the tensile strength declines.

1. Introduction

In recent years, the growing demand for social and industrial development has led to a rapid increase in the production of petroleum-based plastics and their composites. However, waste plastic products are resistant to natural degradation, resulting in severe white pollution. Polylactic acid (PLA), which is polymerized from plant starch, has emerged as one of promising alternative plastics due to its excellent biodegradability and environmentally friendly. PLA is widely used as packaging materials, disposable tableware, automotive interiors, and so on. Similarly, poly (butylene adipate-co-terephthalate) (PBAT) is extensively employed as shopping bags, garbage bags, and food packaging. With the increase of environmental regulations, the demand for biodegradable materials has been surging [1]. It can provide critical insights into the structure-property relationship from the research on the degradation process of PLA/PBAT composites, which is conductive to guide the optimization of synthesis process and material performance.

Herein, wood-plastic composites (WPCs) are developed by blending different contents of bamboo flour with PLA pellets and PBAT masterbatch. The influence of BF loading on the composite's performance characteristics is systematically investigated. ith The research This research can serve as a valuable reference for studying other biodegradable materials.

2. Experimental section

2.1. Materials

PBAT: TH801T, Shanxi Jinhuizhaolong High-tech Co., Ltd.;
PLA: 4032D, America nature works.
Bamboo Flour: 80 mesh, Muxiang Weihua Flavor Co., Ltd.

2.2. Preparation of composites

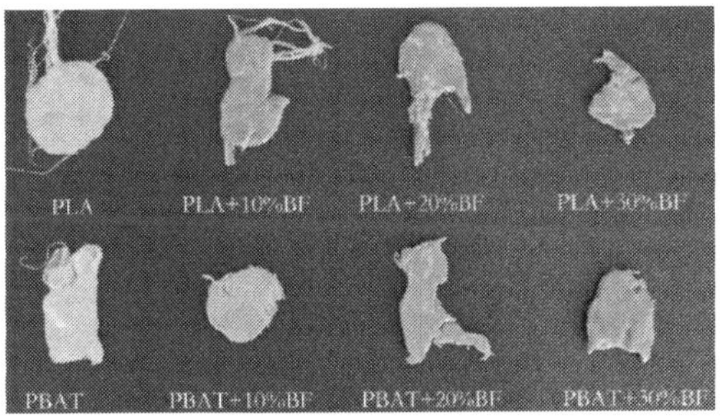

Figure 1. Photographs of the PLA/BF and PBAT/BF blends.

Firstly, PLA, PBAT, and BF were dried in a vacuum oven at 80°C for 12 h. Secondly, pre-dried PLA and PBAT were mixed with different content of BF (0%, 10%, 20%, 30%), respectively. Thirdly, the mixtures were melt-compounded using a torque rheometer at a temperature of 190°C for 360 s with a rotor speed of 50 rpm. The resulted PLA/BF and PBAT/BF blends after cooling down to room temperature are shown in Figure 1.

2.3. Characterizations

The samples were cut into 1 cm × 1 cm pieces, sputter-coated with gold for 1 minute, and then examined under a scanning electron microscope at an accelerating voltage of 5 kV to observe the material surface

*Corresponding author / E-mail: qpchenchen@126.com

morphology using Scanning electron microscope (SEM, LYRA3, America TESCAN)

The samples were cut into 1 cm × 1 cm pieces and scanned from 5° to 90° with a scan step size of 5°/min using X-ray diffractometer (XRD, Japan Rigaku SmartLab SE).

2.4. Thermogravimetric analysis (TGA)

Thermogravimetric analysis was performed under nitrogen atmosphere with the following experimental parameters: Sample weight: 10 mg, Temperature range: 30-800°C,
Heating and Cooling rate: 20°C/min applying Thermogravimetric analyzer (TGA: America TA Q600).

2.5. Tensile testing

The tensile properties of the composites were evaluated using a universal testing machine (Z005, Germany Zwick/Roell) in accordance with GB/T 1040.1-2018. Two replicates were tested for each sample group, and the average values were reported as the final results.

3. Results and discussion

3.1. Torque rheometry analysis

Figure 2 shows the torque rheology curves of PLA/ BF and PBAT/BF blends. The eventual torque stabilization represents a steady flow state, where the equilibrium torque value reflects viscosity or processability of materials [2]. The results demonstrate that the introduction of BF increases the equilibrium torque of both PLA and PBAT matrices. For PLA/BF blends, the equilibrium torque firstly rises and then declines as increasing BF content exhibiting a maximum of 7.7 N·m at 20% BF loading. In contrast, PBAT/BF blends exhibit an initial decrease followed by an increase for equilibrium torque, which shows the lowest torque value at 20% BF loading.

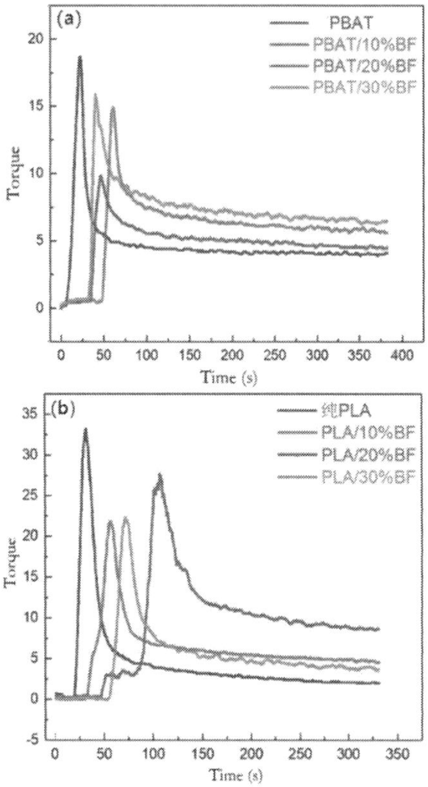

Figure 2. Torque-time curves of (a) PLA/BF and (b) PBAT/BF blends during melt compounding.

3.2. Microstructure characterization

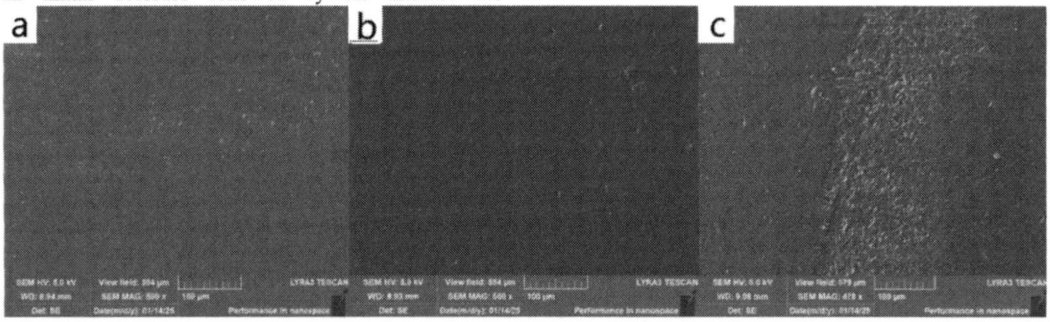

Figure 3. SEM images of PLA/BF composites with (a) 10%, (b) 20%, and (c) 30% BF.

Figure 3 shows the SEM images of PLA/BF composites with (a) 10%, (b) 20%, and (c) 30% BF loading. It can be seen that significant irregular agglomeration of BF particles and surface defects in the PLA matrix for 30% BF loading (c). Moreover, minor particle clustering is observed for 10% (a) and 20% (b) BF loadings and the overall dispersion remains relatively uniform.

 https://doi.org/10.1051/e3sconf/202563901008

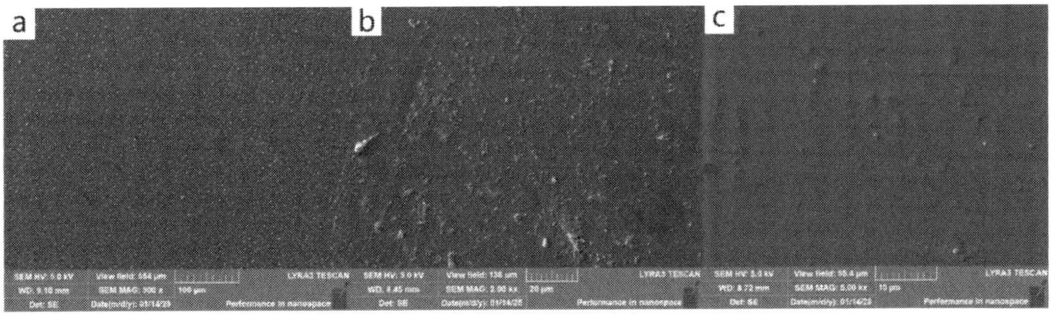

Figure 4. SEM images of PBAT/BF composites with (a) 10%, (b) 20%, and (c) 30% BF.

Figure 4 displays the SEM images of PBAT/BF composites with (a) 10%, (b) 20%, and (c) 30% BF loading. It can be observed that BF is uniformly dispersed in the PBAT matrix with only minor localized agglomeration for the composites with 10% BF loading (a). In contrast, the composites with 20% (b) and 30% (c) BF loading exhibit significant particles aggregation and non-uniform distribution, accompanied by visible surface defects.

3.3. XRD analysis

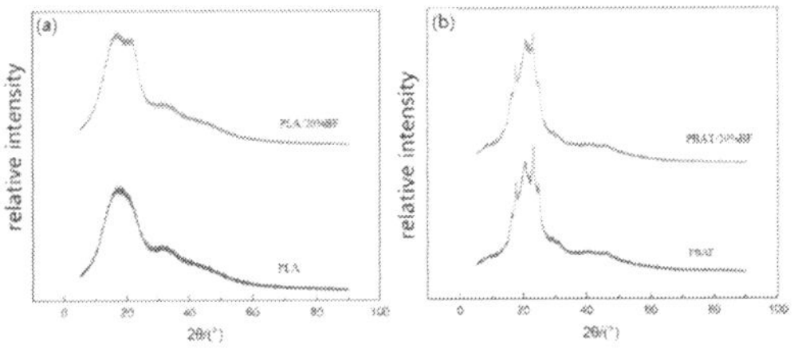

Figure 5. XRD patterns of (a) PLA and (b) PBAT before and after the introduction of BF.

Figure 5 shows the XRD patterns of PLA and PBAT before and after the introduction of BF. By comparison, there are no obvious changes of the diffraction peaks, which imply that the introduction of bamboo powder does not alter the crystal planes of PLA and PBAT. As shown in Figure 5(a), the diffraction peak for PLA at 19.2°corresponds to its (203) crystal plane [3]. After the introduction of bamboo powder, a new diffraction peak appears at 22°, which may be due to the substances in the bamboo powder [4]. As displayed in Figure 5(b), the diffraction peaks for PBAT at 16.23°, 17.79°, 21.35°, 23.54° and 25.10° are attributed to the (011), (010), (102), (100) and (111) crystal planes of PBAT, respectively [5]. After the introduction of bamboo powder, the XRD patterns becomes blurred between 21.35° and 23.54°, which may be due to the appearance of new diffraction peaks for the substances in the bamboo powder [6]. The change in the PLA pattern at this position confirms this, but in the case of PBAT, it overlaps with the characteristic peaks of PBAT [7].

3.4. Thermogravimetric test

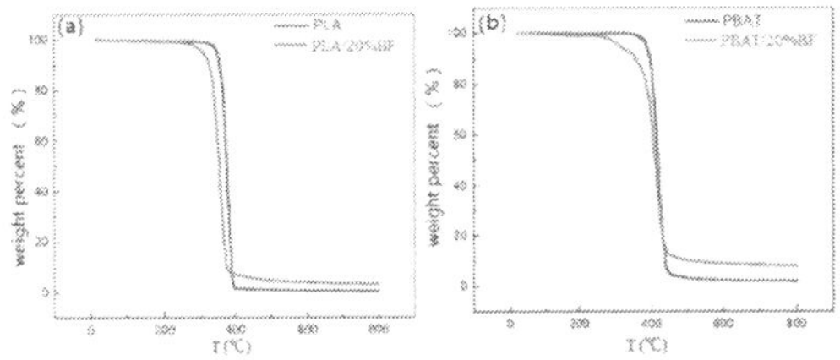

Figure 6. The thermogravimetric curves of PLA/BF (a) and PBAT/BF (b).

Figure 6 shows the thermogravimetric curves of the blends of PLA/BF composites and PBAT/BF composites. As can be seen from Figure 6, after the introduction of bamboo powder, the initial weight loss temperature for both PLA and PBAT is reduced to 332.72°C from 360.86°C, with a decrease of approximately 28°C. This phenomenon may be due to the lower decomposition temperature of bamboo powder than that of PLA and PBAT [8]. Moreover, the residual mass after the thermogravimetric process is improved approximately by 0.0286 mg after the introduction of bamboo powder. . This is because bamboo powders undergo carbonization to carbides at high-temperature And, these carbides remain together with trace amounts of inorganic impurities [9].

3.5. Mechanical properties

Table 1 and Table 2 show the mechanical properties of PLA and PBAT doped with different proportions of bamboo powder, respectively. The elongation at break of pure PBAT is too high, exceeding the maximum measurement range of this type of instrument. A specimen for the composites doped with 30% BF, which can meet the standard is difficult to prepare due to excessive defects. Thus, the above datas are not shown here.

Table 1. The mechanical properties of PLA doped with different proportions of BF.

Sample / Indicator	PLA	PLA/10%BF	PLA/20% BF	PLA/30% BF
Tensile Strength (MPa)	30.75	28.24	27.77	27.13
Elongation at Break (%)	5.06	4.33	3.90	2.88

Table 2. The mechanical properties of PBAT with different proportions of BF.

Sample / Indicator	PBAT	PBAT/ 10%BF	PBAT2 0%BF	PBAT30%B F
Tensile Strength (MPa)	\	7.98	7.41	\
Elongation at Break (%)	>1100	1089.3 2	1015.2 3	\

It can be seen that the mechanical properties for both PLA and PBAT are reduced after the introduction of bamboo powder. This may be because PLA is a hydrophobic polymer while bamboo powder contains a large number of hydrophilic hydroxyl groups (—OH) in physical blending composites [10]. As the second phase, bamboo powder possesses poor interfacial bonding force with the matrix, and the interfacial defects lead to a decrease in the efficiency of stress transfer, making it easy to form micro-cracks under the action of load [11]. Moreover, as a rigid filler, bamboo powders form a discontinuous phase in the matrix. When the content is high, the aggregation or uneven dispersion of particles can lead to local stress concentration. Thus, under the action of load, they are separated firstly from the matrix to form holes, triggering crack propagation in advance, which leads to a decrease in mechanical properties in turn [12].

4. Conclusions

In this work, PLA/BF composites and PBAT/BF composites with different mass fraction of BF are prepared through extrusion compounding process. In the XRD patterns of PLA/BF and PBAT/BF, the crystal forms of PLA and PBAT are maintained. The peak appearing at 22° should be attributed to the substances in bamboo powder. When the content of BF is 10%, it is well dispersed in both PLA and PBAT. When the content of BF increases to 20% and 30%, agglomeration and uneven distribution phenomena occur in PLA/PBAT. After the introduction of BF, the initial weight loss temperature of PLA/20%BF decreases to 332 °C. The tensile strength of PLA/BF decreases from 28.24 MPa to 27.13 MPa, and the tensile strength of PBAT/BF decreases from 7.98 MPa to 7.41 MPa.

Acknowledgements

This work was supported by Eyas Program Incubation Project of Zhejiang Provincial Administration for Market Regulation (No. CY2023214), Class III Peak Discipline of Shanghai—Materials Science and Engineering (High-Energy Beam Intelligent Processing and Green Manufacturing).

References

1. Liu Y, Yang M, Wang J, et al. Ag-CNC nanoparticles decorated thermoplastic starch/PLA/PBAT composite film with superior antibacterial property, enhanced barrier capacity and high UV-shielding effect [J]. Industrial Crops & Products, 2025, 225120510-120510.
2. AgostiniB N, KiefferZ V, SantanaC M R. Rice Husk Filled PLA/PBAT Composites for Food Containers: Processability, Rheological and Thermal Properties [J]. Macromolecular Symposia, 2024, 413(6): 2400093-2 400093.
3. Li M, Lei W, Yu W. FDM 3D Printing and Properties of WF/PBAT/PLA Composites[J]. Molecules, 2024, 29(21): 5087-5087.
4. Jonas H, Axel G, Yanling S, et al. Deformation Behavior of 3D Printed Auxetic Structures of Thermoplastic Polymers: PLA, PBAT, and Blends[J]. Polymers, 2023, 15(2):389-389.
5. Negin N, Ali K, Mohammadreza R, et al. Development of Antibacterial Biocomposites Based on Poly (lactic acid) with Spice Essential Oil (Pimpinella anisum) for Food Applications[J]. Polymers, 2021, 13(21):3791-3791.
6. M. S C, P. D F, Pilar T, et al. Active natural-based films for food packaging applications: The combined effect of chitosan and nanocellulose[J]. International Journal of Biological Macromolecules, 2021, 177241-251.
7. N Y S, Jason A, O U H, et al. Metal nanoparticles: understanding the mechanisms behind antibacterial activity. [J]. Journal of nanobiotechnology, 2017, 15(1): 65.
8. Zhou Y, Huang Z, Diao X, et al. Characterization of the effect of REC on the compatibility of PHBH and PLA [J]. Polymer Testing, 2015, 4217-25.
9. Zakikhani P, Zahari R, Sultan M, et al. Extraction and preparation of bamboo fibre-reinforced composites [J]. Materials & Design, 2014, 63820-828.
10. Dorez G, Taguet A, Ferry L, et al. Thermal and fire behavior of natural fibers/PBS biocomposites [J]. Polymer Degradation and Stability, 2013, 98(1): 87-95.
11. Luan L, Wu W, Wagner H M. Rheological behavior of lubricating systems in polypropylene/seaweed composites[J]. Journal of Applied Polymer Science, 2011, 121(4):2143-2148.
12. Song X, Wang M, Weng Y, et al. Effect of Bamboo Flour Grafted Lactide on the Interfacial Compatibility of Polylactic Acid/Bamboo Flour Composites [J]. Polymers, 2017, 9(8):323-323.

Material and Energy Exchange Dynamics in Paddy Fields Across Growth Stages in Liaohe Delta, Northeast China

Jing Miao [1], Xiyan Yi [1], Linlin Sun [1], Chao Yuan[1*], Tongyu Yang [2], Qingyu Jia [1,2], and Qiong Wu [1]

[1]Panjin National Climate Observatory, Liaoning, Panjin 124010, China
[2]Shenyang Environmental Research Institute of China Meteorological Administration, Liaoning, Shenyang 110166, China

Abstract. This study examines the material and energy exchange of rice paddies during different growth stages in the Liaohe Delta, Northeast China, during an unusually warm year (2019). Utilizing eddy covariance data, we focused on four key growth stages: transplanting, jointing, heading, and maturation. The net ecosystem exchange (NEE) peaked during transplanting, shifting from carbon emission to absorption by the heading stage. Sensible heat flux (H) was highest during jointing, while latent heat flux (LE) showed a distinct peak during transplanting, highlighting the dominance of transpiration and evaporation. Photo synthetically active radiation (PAR) and net radiation (NR) were identified as pivotal environmental drivers. Temperature and soil moisture significantly influenced these fluxes, with optimal temperatures enhancing CO_2 uptake and higher soil moisture promoting transpiration. The findings contribute to an understanding of the rice paddy ecosystem's role in carbon cycling and energy balance amidst environmental variations and contribute to informed agricultural practices in the Liaohe River Delta.

1. Introduction

Land-air interaction plays a crucial role in the climate system as it involves the exchange of matter and energy between the land surface and the atmosphere. This interaction intensity influences atmospheric motion and has a substantial impact on regional weather and climate[1,2]. Moreover, the temporal and spatial variations in land-air interaction contribute to changes in hydrology and the overall climate environment[3,4,5]. The land surface processes are influenced by various factors such as topography, altitude, vegetation cover, and soil type. Simultaneously, there is a continuous exchange of water and energy between the land and the air[6,7]. In the land-air transport process, latent and sensible heat fluxes play a crucial role and serve as significant influencing factors in the surface energy balance equation. These fluxes are the focal point of research in meteorology and atmospheric physics[8,9,10].

The farmland ecosystem plays a vital role in the regional carbon cycle[11,12]. Human activities have a significant influence on the carbon cycle within this ecosystem[13], making field management an effective approach to carbon sequestration and emission reduction. Rice fields are the predominant agricultural land use in China, accounting for a significant portion. The Asian monsoon region contributes approximately 90% of the global rice field planting area, making it a crucial factor in global climate change due to its influence on the water cycle, carbon cycle, and greenhouse gas emissions within terrestrial ecosystems[14,15]. Furthermore, photosynthetic effective radiation, leaf area index, and vapor pressure deficit (VPD) also significantly impact CO2 flux[16].

Wetland ecosystems play a crucial role in maintaining high biodiversity, influencing hydrological functions, replenishing groundwater, and facilitating nutrient cycling[17]. The study area located in the alluvial wetlands of the Liaohe River Delta, benefits from the sediment deposits brought by the Liaohe and Daliao Rivers. This river and sea wetland is the largest and most well-preserved in the Asia-Pacific region, covering a total area of 2496 km^2. The region's abundant natural resources have fostered favorable conditions for modern agriculture, establishing Panjin as a renowned rice-producing area and a hub for high-quality rice in China. In 2019, Panjin City experienced its highest recorded temperature (average temperature of 10.6°C) since meteorological records began. This temperature increase had a significant impact on the material and energy exchange processes during rice growth.

Therefore, to investigate the dynamics of water, energy, and carbon exchange within the Panjin paddy field ecosystem across different growth stages during 2019, this study utilized observational data collected from the Liaohe River Delta Wetland Ecosystem Field Station throughout the year. By examining the interactions between these factors, we aim to gain insights into how solar radiation and its related parameters influence the CO2 flux in paddy fields. Previous studies have highlighted the importance of solar radiation in agricultural systems[18,19].

*Corresponding author's email: yuanchao_mare@163.com

2. Study area overview and methodology

2.1. Study area overview

The study area is located in the Liaohe River Delta, at the junction of the Liaohe River and Daliao River estuaries. The annual mean temperature is 8.3°C, and the annual precipitation is 611.6 mm, concentrated from July to September. The study area receives 2,697.61 hours of sunshine per year[20]. Rice in Panjin (40°40′-41°27′N, 121°30′-122°30′E) is the major crop in the Liaohe River Delta wetland, located within the Shuangtai River Nature Reser[21].Long-term observations and experiments at this station have been conducted to study the functional structure of the Panjin paddy ecosystem and reveal the laws and mechanisms governing the exchange of matter and energy between the wetland ecosystem and the atmosphere[22].

2.2. Methodology and data processing

This study used observational data from the Liaohe River Delta Wetland Ecosystem Field Station in 2019, covering the full rice growth cycle. The eddy covariance method calculated CO_2 flux (Fc) and turbulent fluxes (sensible heat H, latent heat LE). Net Ecosystem Exchange (NEE) included storage flux; negative values indicate CO2 uptake. Surface energy balance was analyzed based on Rn=H+LE+G (Rn net radiation, G soil heat flux). Rn was calculated from shortwave and longwave radiation. Energy closure was assessed using the Surface Energy Balance Ratio (SEBR).

The eddy correlation data were processed by Eddypro software into 30-minute averages, excluding unreasonable values. Then, linear detrending, coordinate rotation (secondary rotation), time lag correction, frequency response correction, ultrasonic virtual temperature correction, and density effect correction were performed[23].

3. Result analysis

3.1. Climatic characteristics of rice at different growth stages

Figure 1 shows the changes in temperature, precipitation, soil temperature, and soil volumetric water content over the different growth stages of paddy. On average, the air temperature during the transplanting and ripening stages was higher than the soil temperature. In contrast, the average air temperature during the jointing and heading stages was lower than the soil temperature. The ripening stage had the lowest average air temperature, while the jointing stage had the highest.

The soil volumetric water content was highest during the transplanting stage, followed in descending order by the jointing stage, heading stage, and maturation stage. Precipitation levels showed a different pattern: the heading stage received the most precipitation, followed by the transplanting stage, jointing stage, and maturation stage, respectively (Table 1).

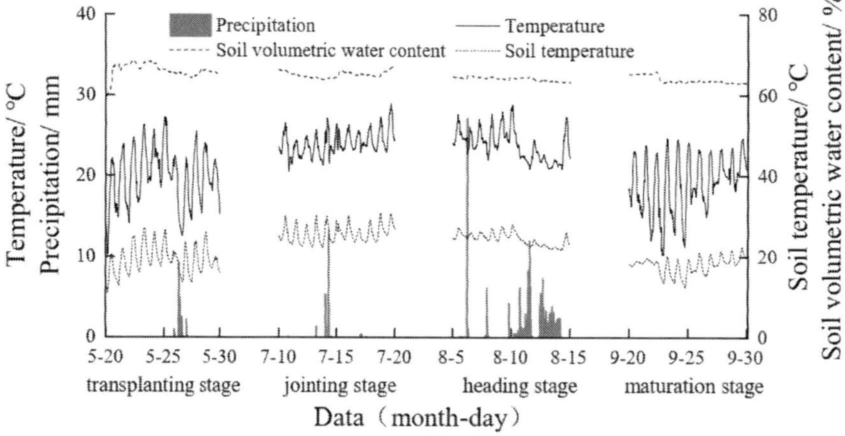

Figure 1. Climatic characteristics of different growth stages of Panjin rice in 2019.

Table 1. Temperature, precipitation, soil temperature, and relative humidity of Panjin rice at different growth stages in 2019.

	Temperature /°C	Precipitation /mm	Soil temperature /°C	Soil relative humidity /%
transplanting stage	19.89	39.7	65.97	19.15
jointing stage	24.09	31.7	65.03	25.72
heading stage	23.88	236.8	64.05	24.43
maturation stage	19.12	0	63.87	18.05

3.2. Variation characteristics of NEE

Net ecosystem exchange (NEE) differed significantly between paddy growth stages (Fig. 2a). NEE was highest during transplanting, followed by maturation, jointing and heading stages, ranging from 0.73 $g \cdot m^{-2} \cdot d^{-1}$ (May 22) to -12.51 $g \cdot m^{-2} \cdot d^{-1}$ (August 8). NEE was largely positive during transplanting but negative otherwise, indicating paddy released carbon during transplanting but absorbed carbon during other stages, especially heading.

Diurnal NEE variation exhibited a single peak during jointing, heading and ripening. Figure 2b shows the average daily variation during four distinct growth periods within the overall study period. Diurnal amplitude was small during transplanting but large during jointing and ripening (Fig. 2b). Minimum NEE occurred around 9:30 to 13:30 for all stages, with the lowest value (-0.28 $mg \cdot m^{-2} \cdot s^{-1}$) during heading.

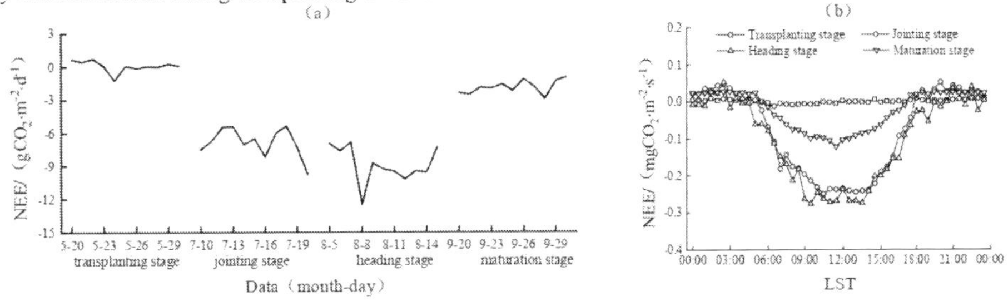

Figure 2. NEE (a) daily variation and (b) daily average variation at different growth stages of Panjin paddy in 2019.

3.3. Variation characteristics of sensible Heat flux and latent heat flux

The difference in sensible heat flux between growth stages was minor, while latent heat flux varied greatly (Fig. 3). Sensible heat flux was highest during jointing, followed by maturation, heading and transplanting, with a maximum of 3.64$MW \cdot m^{-2}$ (August 14) and minimum of -1.45$MW \cdot m^{-2}$ (May 24). In contrast, latent heat flux peaked during transplanting, then declined for heading, jointing and maturation stages, ranging from 16.4$MW \cdot m^{-2}$ (August 14) to 3.57$MW \cdot m^{-2}$ (September 29). The latent heat flux differs significantly from precipitation and air temperature. This disparity may be attributed to the fact that latent heat flux represents plant transpiration and water surface evaporation, which are influenced by the high soil moisture content during seedling transplanting. The latent heat flux is a distinct parameter that reflects these specific processes, setting it apart from precipitation and air temperature.

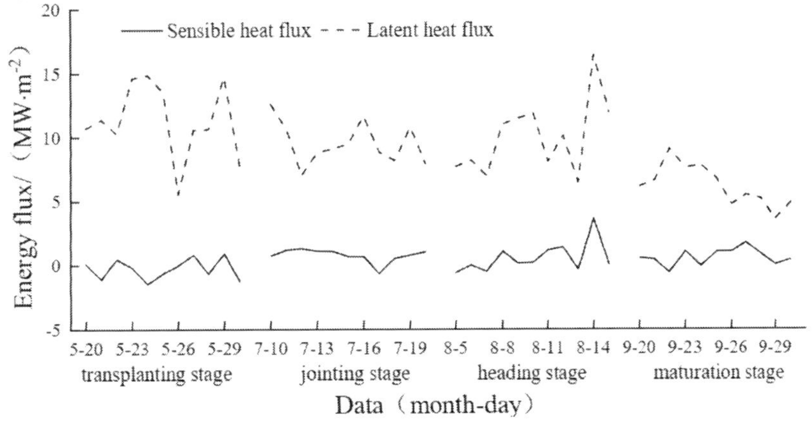

Figure 3. Daily variation of sensible heat flux and latent heat flux at different growth stages of Panjin rice in 2019.

The diurnal variations of sensible heat flux and latent heat flux during different growth stages of paddy exhibit a distinctive "inverted V" structure with a single peak (Fig.4). At night, the surface temperature was lower than air temperature near the ground due to longwave radiation from the water surface. This resulted in a downward energy transfer and negative sensible heat flux (Fig. 4a). During the day, sensible heat flux increased significantly from solar shortwave radiation, peaking around 11:30-13:00. The maximum value occurred during maturation at 115.76 $W \cdot m^{-2}$. In contrast, diurnal variation in latent heat flux between growth stages was minor (Fig. 5b). Latent heat flux remained positive as water vapor transported upward throughout the day, indicating the wetland had sufficient water, an active water cycle and a thriving ecosystem. Latent heat flux peaked around 11:30-14:00, with a maximum of 279.64 $W \cdot m^{-2}$ during transplanting.

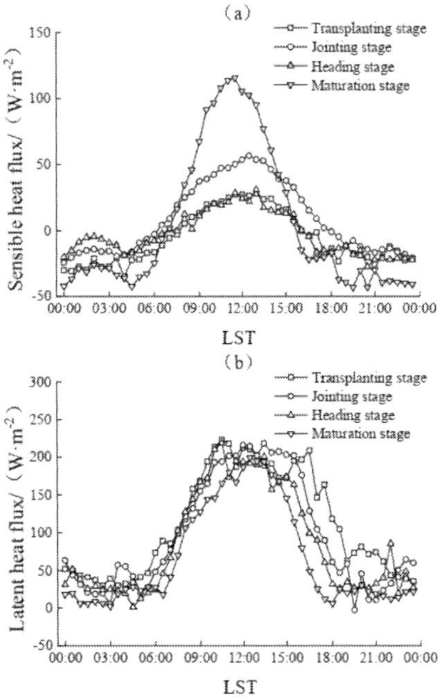

Figure 4. Daily average changes of (a) sensible heat flux and (b) latent heat flux at different growth stages of Panjin rice in 2019.

3.4. Radiation balance and changing characteristics

The daily variation in the radiation budget was similar across growth stages of paddy in Panjin, with solar shortwave radiation comprising the majority of radiation while other components exhibited little change, especially longwave radiation which approached 0W·m^{-2} (Fig. 5). The performance of net radiation at different growth stages can be ranked as follows: jointing stage > transplanting stage > heading stage > mature stage. The manifestation of shortwave radiation follows a similar pattern: jointing stage > transplanting stage > heading stage >maturation stage. Reflected shortwave radiation, on the other hand, exhibits the following order: heading stage >maturation stage > jointing stage > transplanting stage. Reflected longwave radiation shows the following pattern: jointing stage > heading stage > transplanting stage >maturation stage. Lastly, the manifestation of longwave radiation is as follows: transplanting stage > jointing stage >maturation stage > heading stage. The minimum net radiation during the mature period can be attributed to a decrease in solar shortwave radiation. Lower net radiation during the heading period is likely a result of frequent rainy days. Rain clouds block incoming solar radiation, reducing the amount of energy reaching the surface. Additionally, increased humidity associated with rain can enhance the outgoing longwave radiation emitted by the Earth's surface. These combined effects can lead to a significant decrease in net radiation during this critical growth stage, potentially impacting crop development. This cloud cover causes a decrease in solar shortwave radiation and subsequently reduces net radiation[24].

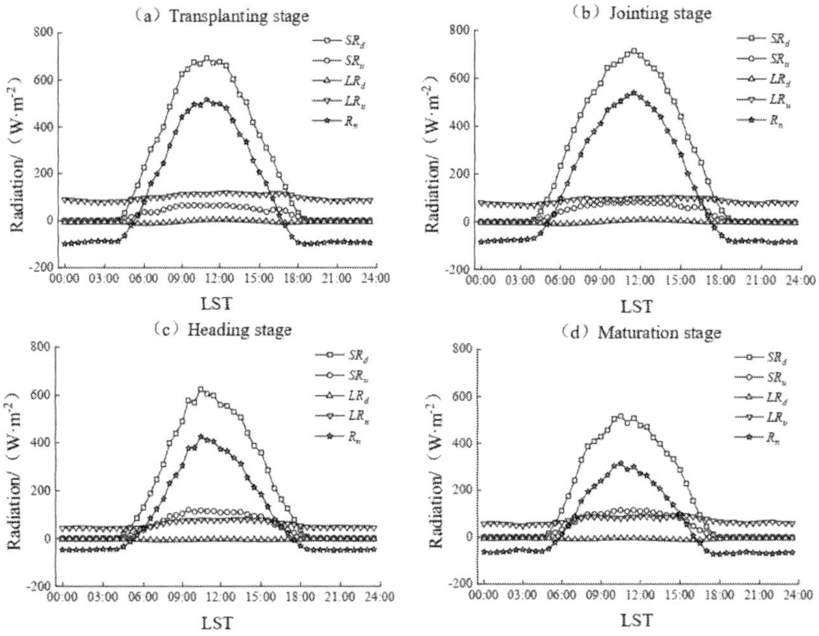

Figure 5. Daily variation of radiation energy at different growth stages of Panjin paddy in 2019 (a: transplanting stage, b: jointing stage, c: heading stage, d: maturation stage).

3.5. Energy ratio of sensible Heat flux and latent heat flux

Energy closure was good during maturation but poor in transplanting. Land-air energy exchange primarily comprises latent and sensible heat exchange. Analyzing the relationship between sensible heat flux, latent heat flux, and net radiation is key to understanding surface energy balance[25]. Energy exchange during different paddy growth stages was dominated by latent heat. The energy closure rate accounting for only sensible and latent heat flux was highest during maturation (69.63%), followed by jointing, transplanting, and heading stages (38.88%) (Fig. 6). Energy closure was good during maturation but poor in transplanting.

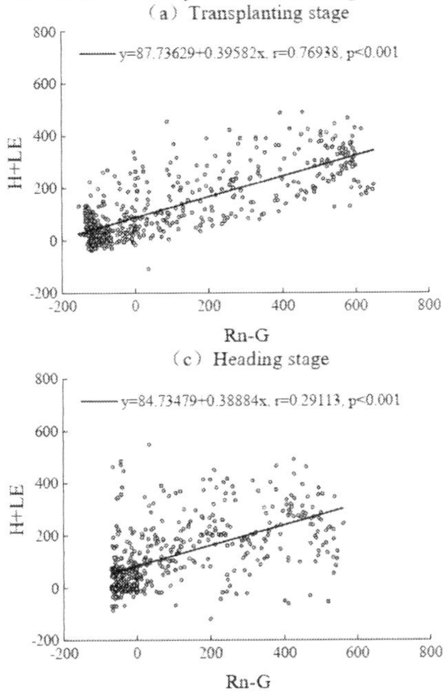

Figure 6. Energy ratio in Panjin paddy growth stage in 2019.

3.6. Factors affecting matter and energy

Correlation analysis was conducted between environmental factors and sensible heat flux (H), latent heat flux (LE), and NEE at different paddy growth stages to explore the relationship between land surface processes and the environment (Fig. 7). The analysis revealed various correlations across different growth stages. Temperature variation significantly affected energy exchange and CO2 flux. Within a suitable range, warmer temperatures enhanced CO2 uptake, but excessively high temperatures reduced photosynthesis and increased respiration, leading to higher nighttime NEE. Warmer temperatures also influenced stomatal regulation, facilitating transpiration (increased LE) in suitable conditions but causing closure and reducing LE under extreme heat stress, while generally increasing H exchange. Higher soil moisture promoted transpiration. Photosynthetically Active Radiation (PAR) and net radiation (NR) were identified as crucial environmental drivers for these fluxes. Understanding these interactions provides valuable insights into how environmental changes may affect rice growth, water use efficiency, and overall ecosystem function.

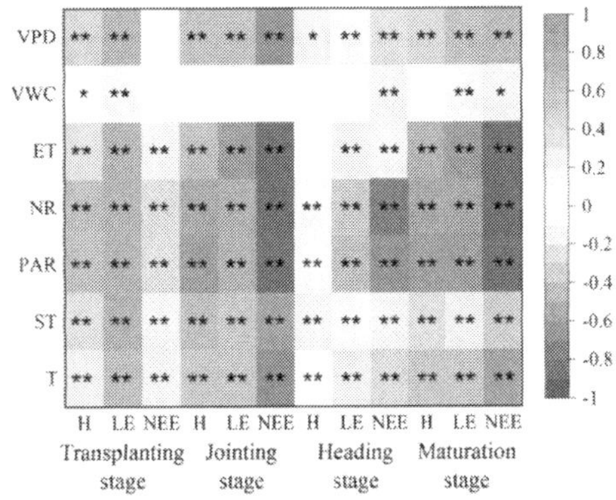

Figure 7. The relationship between material and energy and environmental factors at different growth stages of Panjin paddy in 2019 (*P<0.05; **P<0.01).

4. Summary

This study investigated the dynamics of water, energy, and carbon fluxes, as well as the radiation budget, in

Panjin paddy fields across the growth stages of transplanting, jointing, heading, and maturation in 2019. Key findings revealed distinct patterns across stages: air temperature peaked during jointing, while precipitation was highest during heading. Sensible heat flux was highest during jointing, whereas latent heat flux peaked significantly during transplanting, emphasizing the role of transpiration and evaporation. Net Ecosystem Exchange (NEE) shifted from carbon emission (positive) during transplanting to carbon absorption (negative) in subsequent stages, with sequestration most pronounced during heading. Radiation analysis showed paddy fields were mainly influenced by solar shortwave radiation; net radiation was highest during jointing but lowest during heading, largely due to rain and cloud cover impacting surface energy balance. Latent heat flux consistently played a dominant role in energy exchange, and energy closure rates were highest during maturation. Correlation analysis indicated sensible/latent heat fluxes correlated positively with environmental factors, while NEE showed negative correlations, with Photosynthetically Active Radiation (PAR) and net radiation identified as crucial drivers. These results underscore the importance of environmental factors in regulating these fluxes and confirm the paddy ecosystem's function as a carbon sink during the growing season. Ongoing research is essential to further understand long-term dynamics and support sustainable management practices.

References

1. IPCC. Chapter 2: Land–Climate interactions. In IPCC Special Report on Climate Change and Land. 2023.Retrieved from https://www.ipcc.ch/srccl/chapter/chapter-2/
2. Seneviratne S, Lüthi D, Litsch M, et al. Land–atmosphere coupling and climate change in Europe[J]. Nature, 2006, 443(7108): 205-209. doi:10.1038/nature05095
3. Zhao F, Yang P, Gao Q, et al. Effects of the long-term rice expansion on ecosystem carbon budget in the typical agricultural area of Northeast China [J]. Sustainable Production and Consumption, 2024, 52: 613-623. DOI: 10.1016/j.spc.2024.11.027.
4. Qian, H., Zhu, X., Huang, S. et al. Greenhouse gas emissions and mitigation in rice agriculture. Nat Rev Earth Environ, 2023, 4, 716–732. https://doi.org/10.1038/s43017-023-00482-1
5. Zhao X L, Cai F, Ding K K, et al. Phenology variations of spring maize and their correlations with hydrothermal conditions in Jinzhou[J]. Journal of Meteorology and Environment, 2022, 38(5):64-71. doi:10.3969/j.issn.1673-503X.2022.05.008(in Chinese)
6. Huang, J. et al. Exploration of feasible rice-based crop rotation systems to coordinate productivity, resource use efficiency and carbon footprint in central China. Eur. J. Agron[J]. 2022, 141, 126633.
7. Liu S, Li S G, Yu G R, et al. Surface energy exchanges in grassland ecosystems along a precipitation gradient[J]. Acta Ecologica Sinica, 2010,30(3):557-567. doi:10.1080/00949651003724790 (in Chinese)
8. Song, K. et al. Evaluation of methane and nitrous oxide emissions in a three-year case study on single rice and ratoon rice paddy fields[J]. J. Clean. Prod. 2021,297, 126650.
9. Wang, Y., Li, X., Lee, T., Peng, S. & Dou, F. Effects of nitrogen management on the ratoon crop yield and head rice yield in South USA[J]. J. Integr. Agric. 2021, 20, 1457–1464.
10. Alberto M C R, Wassmann R, Hirano T, et al. Comparisons of energy balance and evapotranspiration between flooded and aerobic rice fields in the Philippines[J]. Agricultural Water Management, 2011, 98(9):1417-1430. doi:10.1016/j.agwat.2011.04.011
11. Zhang Y Q, Liu C M, Yu Q, et al. Energy fluxes and the Priestley-Taylor parameter over winter wheat and maize in the North China Plain[J]. Hydrological Processes, 2004, 18(12): 2235-2246. doi:10.1002/hyp.5528
12. Anthoni P M, Knohl A, Rebmann C, et al. Forest and agricultural land-use-dependent CO_2 exchange in Thuringia, Germany[J]. Global Change Biology, 2004, 10(12): 2005-2019. doi:10.1111/j.1365-2486.2004.00863.x
13. Buyanovsky G A, Wagner G H. Carbon cycling in cultivated land and its global significance[J]. Global Change Biology, 1998, 4(2): 131-141. doi:10.1046/j.1365-2486.1998.00130.x
14. Padbhushan R., Kumar U., Sharma S., Rana D. S., Kumar R., Kohli A., et al. Impact of land-use changes on soil properties and carbon pools in India: A meta-analysis[J]. Front. Environ. Sci. 2022, 722. doi: 10.3389/fenvs.2021.794866
15. Azar C, and Johansson D. IPCC and the effectiveness of carbon sinks[J]. Environ. Res. Lett., 2022, 17, 041004 (4pp). doi: 10.1088/1748-9326/ac5bb2
16. Gao X, Gu F, Mei X, et al. Carbon exchange of a rainfed spring maize croplandunder plastic film mulching with straw returning on the Loess Plateau,China[J]. Catena, 2017, 158:298-308. doi:10.1016/j.catena.2017.07.015
17. Sun L, Song C. Evapotranspiration from a freshwater marsh in the Sanjiang Plain, Northeast China[J]. Journal of Hydrology, 2008, 352(1): 202-210. doi:10.1016/j.jhydrol.2008.01.010
18. Jiajia Xu, Ke Zhong, Yiqi Wang, Jiajian He, Yanming Kang, Influences of daily solar radiation intensity on the sea-land breeze and pollutant dispersion in coastal areas[J]. Building and Environment, 2024, 258: 111600. https://doi.org/10.1016/j.buildenv.2024.111600.
19. A. Haj Ismail, E.A. Dawi, N. Almokdad, A. Abdelkader, O. Salem. Estimation and Comparison of the Clearness Index usingMathematical Models -

Case study in the United ArabEmirates [J]. Evergreen Joint Journal of Novel Carbon Resource Sciences & Green Asia Strategy, 2023, 10(02): 863-869. https://doi.org/10.5109/6792841.

20. Jia Q Y, Yu W Y, Zhou L, et al. Atmospheric and surface condition effects on CO_2 exchange in the Liaohe delta rice, China[J]. Water, 2017, 99(10): 806. doi:10.3390/w9100806
21. Wang B L, Li G S. Quantification of the Reasons for Reference Evapotranspiration Changes over the Liaohe Delta, Northeast China[J]. Scientia Geographica Sinica, 2014, 34(10): 1233-1238. doi: 10.13249/j.cnki.sgs.2014.010.1233(in Chinese)
22. Zhou G S, Zhou L, Guan E K, et al. Brief introduction of Panjin wetland ecosystem research station[J]. Journal of Meteorology and Environment, 2006(4):1-6. (in Chinese)
23. Sun P F, Fan G Z, Qu Z, et al. Characteristics of surface-layer turbulence spectra in Xiaoxing'an mountains[J]. Plateau Meteorology, 2021, 40(2), 374-383. doi: 10.7522/j.issn.1000-0534.2020.00099 (in Chinese)
24. Zhang M L, Wang B, Wang D K, et al. The effects of rainfall on the surface radiation of permafrost regions in Qinghai-Tibet Plateau: a case study in Beiluhe area[J]. Journal of Glaciology and Geocryology, 2021, 43(4):1092-1101. doi: 10.7522/j.issn.1000-0240.2021.0073 (in Chinese)
25. Li Y, Lu P, Ding H Y, et al. Analysis of Surface Fluxes Characteristics and Energy Budget in the Cropland Surface in Chengdu Plain[J]. Plateau and Mountain Meteorology Research, 2013, 33(1):35-40. doi:10.3969/j.issn.1674-2184•2013.01.006 (in Chinese)

Sustainability Differences and Balanced Development Countermeasures of School Buildings in Urban and Rural Areas of Chengdu

Yunfei Zhang[1,2], Jeffrey C.L. Chiang[1,*]

[1]Faculty of Engineering, Built Environment and Information Technology, SEGI University,47810 Kota Damansara, PJ, Selangor, Kuala Lumpur, Malaysia
[2]School of Civil Engineering, Sichuan University of Science and Engineering, 643000 Zigong China

Abstract. As green and environmental protection concepts become increasingly popular, sustainable construction is an important development trend and direction in the construction industry. In recent years, sustainable buildings in various fields in China have developed rapidly. However, due to the differences in economic and cultural development between urban and rural areas, there is still a significant gap in the sustainability level of urban and rural buildings. The sustainability disparity between urban and rural school buildings in China was studied in this paper. In the study, taking Chengdu, Sichuan Province as an example, a series of investigations and studies were carried out. The differences in sustainability levels and characteristics of urban and rural school buildings were sorted out and analysed in detail, and countermeasures and suggestions for the balanced development of sustainable urban and rural school buildings were proposed. This study has a certain promoting effect and reference value for improving the architectural sustainability in China's education sector and the balanced development of sustainable buildings in urban and rural schools.

1. Introduction

Sustainable building, also known as green building in China, is a form of architecture with the core concepts of environmental protection, resource conservation and social responsibility. The goal of sustainable building is to minimize the negative impact of buildings on the environment while improving the economic and social benefits of buildings, which emerged under the background of dual pressure of the global environmental crisis and energy crisis. Therefore, sustainable building refers to optimizing the energy efficiency of buildings by applying green building technologies and materials in the design, construction and operation of buildings, and improving the environmental friendliness and long-term economic benefits of buildings while ensuring their functionality and economic benefits[1]. Since the 1980s, with the acceleration of industrialization and urbanization in China, the excessive consumption of resources and the negative impact of the construction industry on the environment have become increasingly prominent. At present, buildings account for 40% of energy consumption and 30% of carbon emissions in China. How to reduce energy consumption and environmental pollution in the construction industry has become a important challenge in China, and sustainable buildings have become a feasible path to solve this problem in recent years. In this context, the Chinese government has promoted sustainable development in the construction sector and issued a series of certification standards related to building sustainability, including the Green Building Evaluation Standard (GB/T 50378-2019), Green Building Star Rating, Green Building Design and Evaluation Code (JGJ/T 345), Building Energy Efficiency Evaluation Standard (GB/T 50376), Sustainable Building Evaluation Standard (GB/T 51393), etc. These standards and systems provide a development standard and framework for China's construction industry, and promote the application and promotion of sustainable buildings. However, due to a series of historical reasons, there is still a significant gap between China's cities and rural areas in terms of the number and proportion of sustainable buildings, sustainability levels, etc., which affects the sustainable development level of China's construction industry[2]. The gap between urban and rural areas in China is also very obvious when it comes to the sustainability level of school buildings.

Chengdu is an important economic development center in southwest China. With the continuous advancement of local urbanization, the city's resource agglomeration effect is becoming increasingly obvious. In the process of sustainable development of school buildings, although the Chengdu Municipal Government has adopted a series of policy tools and measures to promote the balanced development of urban and rural school buildings, there are still obvious differences in infrastructure sustainability, green building technology resources, environmental education and awareness cultivation, supporting resource support, environmental impact level, and teacher and student participation. The gap in the sustainability level of school buildings in

* Corresponding author: jeffreychiang@segi.edu.my

Chengdu was sorted out and analyzed in this paper, and suggestions for the balanced development of sustainable buildings in urban and rural schools were subsequently put forward, which has a positive role in narrowing the level of sustainable development of urban and rural buildings and improving the environmental friendliness of local school buildings.

2. Research design

Chengdu has jurisdiction over 20 districts (including counties and cities). There are many schools at all levels in Chengdu. As of 2024, there are more than 1,300 educational institutions including primary schools, secondary schools and universities in Chengdu. As the concept of green environmental protection gradually becomes popular, the issue of sustainable development of buildings in the education system has also received great attention from the government and all sectors of society. According to statistics and evaluations by the Chengdu Municipal Housing and Urban-Rural Development Bureau (CMHURD), the sustainability level of internal buildings in all types of schools in Chengdu has been significantly improved in recent years. However, CMHURD also pointed out that there is still a certain gap in the sustainability level of buildings in urban schools and rural schools. Therefore, this paper mainly analyzed the specific differences in the sustainability level of urban and rural school buildings in Chengdu. Through investigation and analysis of the current status of school buildings in urban and rural areas of the city, the development status of the sustainability level of urban and rural school buildings is sorted out and compared in detail, and targeted suggestions and measures are put forward to promote the balanced development of the sustainability level of urban and rural schools in Chengdu.

In the study, through field investigations in more than 1,300 schools in urban and rural areas of Chengdu, and combined with the statistical and evaluation data of CMHURD, the sustainability levels of buildings in urban and rural areas of Chengdu were counted and analyzed, including the energy utilization and energy-saving configuration of buildings, the level of water resource management, the environmental protection level of building materials and construction, the indoor environmental quality control level of buildings, the environmental level of buildings, the environmental awareness of students and teachers, and the differences in sustainable resources of buildings.

Based on the field survey and data collation, the survey data was classified and statistically analyzed, and statistical analysis methods were used to analyze and compare the differences in the building sustainability levels of urban schools and rural schools in Chengdu from multiple perspectives, and to comprehensively analyze and evaluate the differences in the level of sustainable building development in urban and rural schools.

3. Sustainability differences between urban and rural school buildings in Chengdu

In terms of the sustainability level of school buildings, urban schools in Chengdu are significantly better than those in rural areas. According to statistics from CMHURD and Chengdu Municipal Education Bureau (CMEB), by 2024, there are 18,842 and 4,418 school buildings and other infrastructure in urban and rural areas of Chengdu. Of these, 15,842 and 1,533 have passed the green building certification in the Green Building Evaluation Standard (GB/T 50378-2019), accounting for 84.1% and 34.7% respectively. Rural schools are significantly worse than urban schools.

Further, according to the five dimensions of green building certification in the "Green Building Evaluation Standard" (GB/T 50378-2019), namely energy utilization and energy-saving facility configuration, water resource management level, environmental friendliness of building materials and construction, indoor environmental quality control, and environmental friendliness of buildings, a comparison was made between urban and rural school buildings in Chengdu to analyze the differences in the sustainability levels of urban and rural school buildings. At the same time, the environmental awareness of teachers and students in urban and rural schools in Chengdu and the resource investment in sustainable development of buildings were also investigated and counted, further exploring the gap in the sustainability level of urban and rural school buildings from the perspective of the external environment.

3.1. Comparison of energy utilization and energy-saving facilities configuration in urban and rural school buildings in Chengdu

The energy consumption of buildings, such as renewable energy utilization rate, energy saving rate, intelligent lighting and power system configuration rate, energy-saving air conditioning and ventilation system configuration rate, are important factors affecting the sustainability level of buildings. By conducting a statistical analysis on the energy utilization of urban and rural school buildings in Chengdu, the results shown in Figure 1 were obtained.

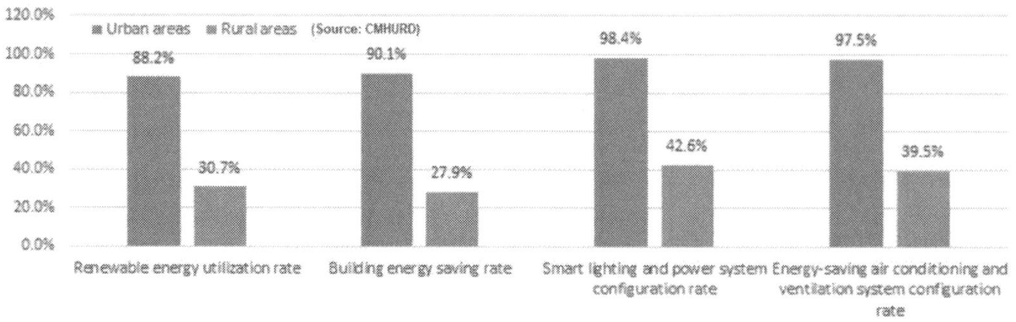

Figure 1. Comparison of energy utilization and energy-saving facilities configuration in urban and rural school buildings in Chengdu.

According to Figure 1, rural school buildings in Chengdu lag behind urban schools in terms of energy utilization and energy-saving facilities configuration:

- In terms of renewable energy utilization, the average utilization rate of renewable energy such as solar energy and wind energy in urban school buildings reached 88.2%, while that in rural school buildings was only 30.7%, there is a difference of 57.5 percentage points.
- In terms of building energy saving rate, the energy saving rate of urban school buildings reached 90.1%, while that of rural schools was only 27.9%, there is a difference of 62.2 percentage points.
- 98.4% of urban school buildings are equipped with intelligent lighting and intelligent power supply systems, while the configuration rate of rural school buildings is only 42.6%, there is a difference of 55.8 percentage points.
- 97.5% of urban school buildings are currently equipped with energy-saving facilities such as energy-saving air conditioners and ventilation systems, while only 39.5% of rural school buildings are equipped with the above energy-saving facilities, there is a difference of 58 percentage points.

3.2. Comparison of water resource management levels in urban and rural school buildings in Chengdu

Water resource management capability has an important impact on the ecological and environmental load and operating costs of buildings, and is also an important criterion for evaluating the sustainability level of buildings[3]. By investigating and analyzing the water resource management capabilities of urban and rural school buildings in Chengdu, the results shown in Figure 2 were obtained.

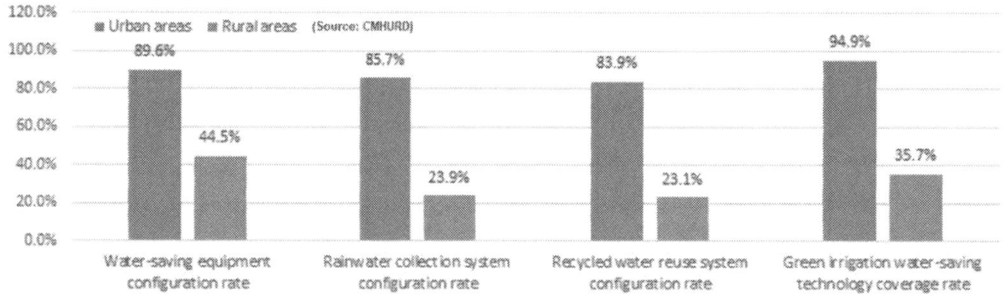

Figure 2. Comparison of water resource management levels of urban and rural school buildings in Chengdu.

As shown in Figure 2, there is a clear gap in water resource management capabilities between urban and rural school buildings in Chengdu, which mainly including the following aspects:

- In terms of water-saving equipment configuration of buildings, 89.6% of urban school buildings have been equipped with water-saving equipment, while only 44.5% of rural school buildings have been equipped, there is a difference of 45.1 percentage points.
- In terms of rainwater collection systems, 85.7% of urban school buildings have installed rainwater collection systems, which effectively reduces the pressure on the water supply system, while only 23.9% of rural school buildings have installed such facilities, there is a difference of 61.8 percentage points.
- In terms of recycled water reuse systems, 83.9% of urban school buildings are currently equipped with water reuse systems that can reuse wastewater, while this proportion of rural school buildings is only 23.1%, there is a difference of 60.8 percentage points.

- In terms of the application of water-saving technology in greening irrigation, 94.9% of urban schools have adopted more environmentally friendly water-saving technologies in the process of campus greening irrigation, while only 35.7% of rural schools have adopted such technologies, there is a difference of 59.2 percentage points.

3.3. Comparison of building materials and environmental performance of urban and rural schools in Chengdu

During the construction phase of a building, whether the building materials and construction methods are environmentally friendly is also an important criterion for measuring the sustainability level of a building[4]. By choosing low-energy, renewable building materials from the local market, the transportation costs and pollution of building materials during the construction process can be effectively reduced. Through the investigation and statistics on the use of environmentally friendly materials and the application of environmentally friendly technologies in the construction process of urban and rural school buildings in Chengdu, the results are shown in Figure 3.

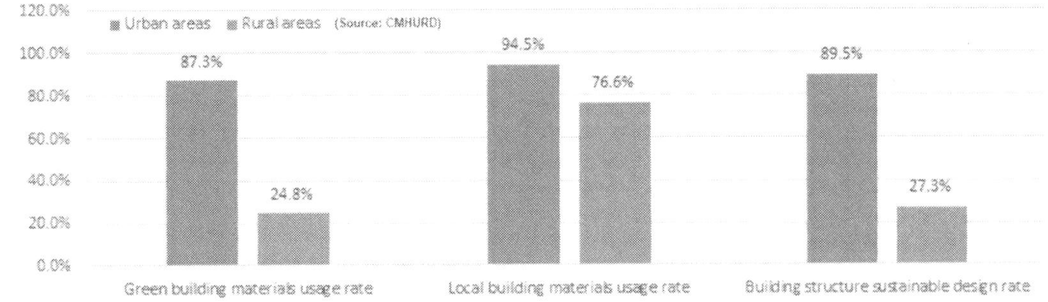

Figure 3. Comparison of building materials and construction environmental performance of urban and rural schools in Chengdu.

As shown in Figure 3, the sustainability of building materials and construction techniques in urban school buildings in Chengdu is significantly better than that in rural schools:

- In terms of the use of green building materials, 87.3% of urban school buildings used more environmentally friendly green building materials during construction, while only 24.8% of rural schools did so, there is a difference of 62.5 percentage points.
- In terms of the use of local building materials, both urban and rural schools used a large amount of building materials from the local market, but the use rate of urban schools was still 17.9 percentage points higher, and rural schools still lagged significantly behind urban schools.
- In terms of sustainable design and construction technology during construction, 89.5% of urban schools adopted more energy-saving, green and environmentally friendly sustainable design solutions and construction technology during building design and construction, while only 27.3% of rural schools did so, there is a difference of 62.2 percentage points.

3.4. Comparison of Indoor environmental quality control levels of urban and rural school buildings in Chengdu

The quality of the indoor environment, such as lighting and ventilation, directly determines the user's experience of the building and is also an important evaluation criterion for the sustainability level of the building[5]. By investigating and analyzing the indoor environmental quality management status of urban and rural school buildings in Chengdu, the results shown in Figure 4 were obtained.

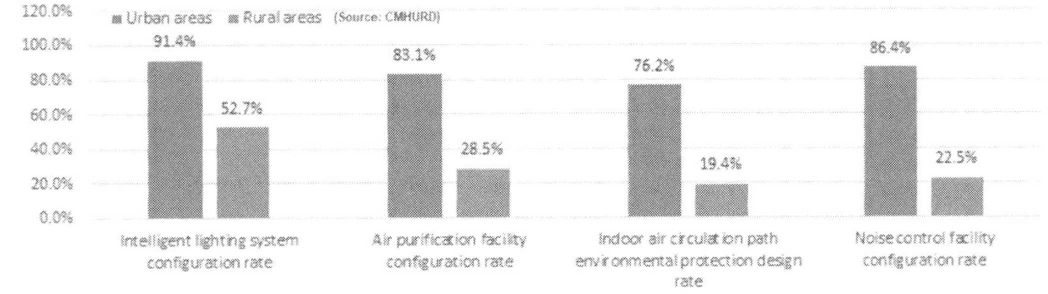

Figure 4. Comparison of indoor environmental quality control levels of urban and rural school buildings in Chengdu.

As shown in Figure 4, there is also a significant gap in indoor environmental control between urban and rural school buildings in Chengdu. The indoor environmental control level of urban school buildings is significantly better than that of rural schools:

- In terms of internal lighting, 91.4% of urban school buildings have installed smarter indoor lighting systems, while only 52.7% of rural schools have installed such systems, there is a difference of 38.7 percentage points.
- In terms of air purification capacity inside buildings, 83.1% of urban schools have installed air purification facilities such as air purifiers and pollutant filters, while only 28.5% of rural schools have installed such facilities, there is a difference of 54.6 percentage points.
- In terms of the environmental protection of the indoor air circulation path design of buildings, 76.2% of urban school buildings have adopted a more scientific and environmentally friendly indoor air circulation path design scheme during the design stage, while only 19.4% of rural schools have adopted a similar design scheme, there is a difference of 56.8 percentage points.
- In terms of noise control inside buildings, 86.4% of urban schools have installed more effective noise reduction facilities, while only 22.5% of rural schools have installed such facilities, there is a difference of 63.9 percentage points.

3.5. Comparison of the environmental friendliness of urban and rural school buildings in Chengdu

During the use of school buildings, supporting campus greening, rainwater infiltration treatment, shading design, etc. also have an important impact on the sustainability of the building and are important external indicators in the green building evaluation process[6]. By investigating and evaluating the environmental friendliness of supporting facilities such as the campus greening rate of urban and rural schools in Chengdu, the results shown in Figure 5 were obtained.

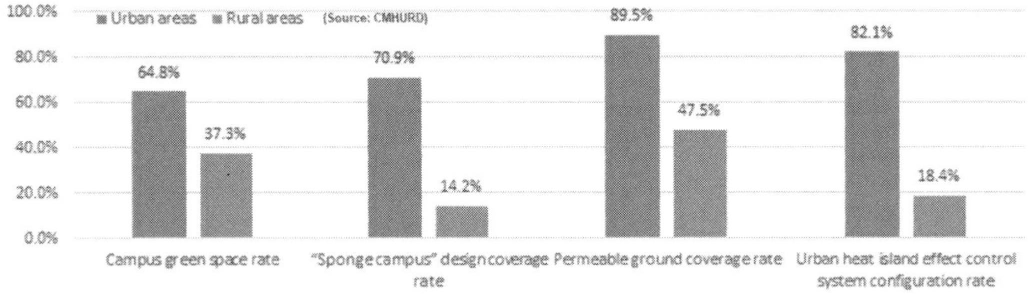

Figure 5. Comparison of environmental friendliness of urban and rural school buildings in Chengdu.

As shown in Figure 5, the environmental friendliness of supporting facilities such as campus greening rate and rainwater infiltration treatment in urban schools in Chengdu is also lower than that of rural schools:

- In terms of campus greening, the average greening rate of schools in Chengdu's urban areas reached 64.8%, while that of rural schools was only 37.3%, there is a difference of 27.5 percentage points.
- 70.9% of urban schools adopted the overall design of "sponge campus", which can effectively cope with extreme weather such as heavy rain, while only 14.2% of rural schools adopted a similar plan, there is a difference of 57.6 percentage points.
- In terms of ground permeability, the average permeable ground coverage rate of urban schools reached 89.5%, while that of rural schools was only 47.5%, there is a difference of 42 percentage points.
- 82.1% of urban schools were equipped with a "heat island effect control system" specifically for hot weather, while this proportion of rural schools was only 18.4%. Urban schools performed significantly better than rural schools in regulating campus microclimate.

3.6. Comparison of environmental awareness among teachers and students in urban and rural schools in Chengdu

Compared with urban schools, rural schools in Chengdu not only have a large gap in the sustainability level of school buildings, but also have relatively backward environmental awareness among teachers and students. Through field research on urban and rural schools in Chengdu, the following results were obtained:

- 87.2% of urban schools have incorporated sustainable buildings, energy conservation and environmental protection into their daily teaching system, and have set up special teaching content in science, geography, labor technology and other courses to enhance students' awareness of environmental protection and green energy conservation. However, only 26.7% of rural schools have opened similar teaching courses, there is a difference of 60.5 percentage points.
- In terms of green campus practice activities, 86.7% of urban schools have carried out relevant on-campus practical activities such as "photovoltaic power generation experiment observation" and

"rainwater collection system experience" in recent years, while the proportion of rural schools is only 19.5%, there is a difference of 67.2 percentage points.

- In terms of internal school management, 68.3% of urban schools have established on-campus organizations such as student clubs related to environmental protection, and organized a series of collective activities related to energy-saving inspections, water-saving publicity, garbage classification, etc. to cultivate students' environmental awareness. However, only 25.1% of rural schools have established similar on-campus student organizations, there is a difference of 43.2 percentage points.

3.7. Comparison of sustainable resource inputs in urban and rural buildings in Chengdu

In terms of supporting financial investment and policy resource support for the sustainable development of school buildings, although the Chengdu Municipal Government has continued to increase resource investment in recent years, urban schools have received more financial funds and policy support than rural schools, resulting in insufficient resource guarantees for the sustainability of rural school buildings.

From 2020 to 2024, Chengdu's annual fiscal budget for green and environmentally friendly renovation of urban and rural school buildings and new green buildings will reach more than 1 billion yuan. However, according to statistics based on the number of urban and rural schools and the amount of allocated fiscal funds, urban schools received significantly more fiscal funds than rural schools, as shown in Figure 6.

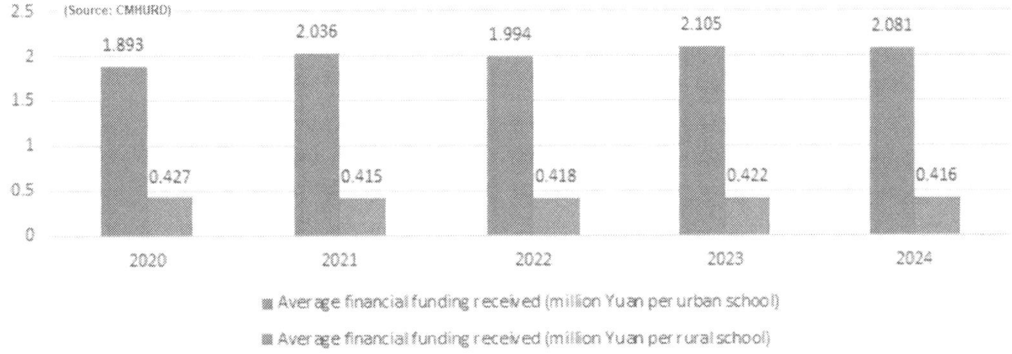

Figure 6. Average fiscal funding for sustainable building development in urban and rural schools in Chengdu (2020-2024).

As shown in Figure 6, from 2020 to 2024, the average annual fiscal funds received by each urban school in Chengdu for improving building sustainability will remain at around 2 million yuan, while that of rural schools will be around 400,000 yuan. The fiscal funds received by urban schools are about five times that of rural schools, the gap is very obvious.

In terms of supporting policy support, there is also a large gap in the level of policy support in urban and rural schools. At present, relevant departments in Chengdu have gradually incorporated the improvement of the sustainability level of urban school buildings into the green development system of urban communities, and have issued targeted policy documents in terms of land approval, technical support, tax and fee concessions, etc. However, due to the relatively scattered geographical distribution of rural schools and the certain economic development differences between urban and rural areas, Chengdu has not yet provided supporting policy support for the sustainable development of rural school buildings. This is one of the important reasons why the sustainability level of rural school buildings lags significantly behind that of urban schools.

4. Strategies for sustainable and balanced development of urban and rural schools in Chengdu

4.1. Strengthen policy guidance and financial support

To promote the sustainable development of green buildings in rural schools and to narrow the gap with urban areas, it is recommended that the government consider increasing financial investment and establishing dedicated subsidy mechanisms to provide stable funding support for green building renovation projects. For example, a "Rural School Green Building Special Fund" could be created, with an annual investment scale of no less than 1 billion yuan, specifically used to support initiatives such as energy-saving facility upgrades and building energy efficiency improvements in rural schools. In addition, it would be beneficial for the government to introduce corresponding tax incentive policies to encourage enterprises and social capital to actively participate in rural green campus construction. Measures such as offering tax exemptions or providing policy-based loans to enterprises that donate to green building projects could be explored to stimulate broader social investment. It is also suggested to formulate unified development standards for green buildings across both urban and rural schools, clearly specifying technical requirements for energy saving, the proportion of environmentally friendly materials, and models for energy consumption management. This would help ensure fairness and consistency in sustainable building development between different regions. Furthermore, strengthening the supervision and evaluation mechanisms for the use of green building funds in rural schools may also be considered. By ensuring transparency, improving fund utilization efficiency, and preventing resource waste or inefficient investment, the government could effectively promote the widespread adoption and long-term success of sustainable building practices in rural education settings.

4.2. Enhance the promotion and application of green technologies

Considering that rural schools are often constrained by limited construction funds and technical capabilities, it is advisable to prioritize the adoption of low-cost, high-efficiency green technologies in the process of building transformation. For instance, installing solar water heating systems on school rooftops could significantly reduce coal or electricity consumption, with savings exceeding 40%, while the promotion of LED energy-saving lighting may help lower lighting energy consumption by more than 50%. In addition, passive energy-saving design strategies could also be encouraged as an important approach to improving building sustainability. By optimizing building orientation, incorporating insulation layers, and making rational use of natural ventilation, schools may effectively reduce their reliance on air conditioning and heating equipment, thereby enhancing overall energy efficiency. Beyond hardware upgrades, attention could also be given to strengthening the training of building technicians in rural areas to improve the long-term maintenance and management of sustainable building systems. It would be beneficial for government agencies or social organizations to provide free training programs, offering rural building managers and technical workers opportunities to participate in specialized courses on green construction techniques and energy-efficient equipment maintenance. To further support knowledge dissemination, distance education platforms could be leveraged, enabling rural school managers to access advanced green building management practices and continuously enhance their professional skills in sustainable building operation and maintenance.

4.3. Improve school building management mechanism

At present, many rural schools face challenges due to the lack of professional management systems and technical support, leading to excessive energy consumption and low operational efficiency of building facilities. In order to promote the long-term sustainable development of campus infrastructure, it is suggested that government and education departments consider encouraging both urban and rural schools to establish a unified green building management framework, covering key aspects such as energy consumption monitoring, equipment maintenance, and building operation optimization. For instance, the introduction of building energy consumption monitoring systems could be promoted to enable the automatic collection and analysis of data related to water, electricity, and gas usage, allowing for timely detection of anomalies and rapid adjustment to reduce unnecessary energy waste. Furthermore, schools might benefit from setting up dedicated green campus management teams responsible for the regular inspection and maintenance of energy-saving facilities, ensuring that systems such as photovoltaic power generation, energy-efficient air conditioners, and rainwater collection devices can operate at optimal efficiency. In addition, leveraging information technology to develop intelligent management platforms, such as campus energy management applications, could allow managers to monitor campus building energy consumption in real time and perform remote control operations when necessary. Through optimizing management models and enhancing operational efficiency, the sustainable building practices of urban and rural schools could gradually evolve towards greater intelligence and digitalization.

4.4. Strengthen green environmental awareness and capacity building

In order to promote the sustainable development of green buildings in urban and rural areas, it is suggested that Chengdu actively incorporate the concept of sustainable construction into the school curriculum, enabling teachers and students to systematically understand green building technologies, energy-saving measures, and the broader principles of sustainable development. At the same time, schools may strengthen the environmental practice abilities of teachers and students by offering experimental courses and launching green campus initiatives. For instance, activities such as a "Green Campus Energy Saving Challenge" could be organized to encourage student participation in energy audits, waste sorting, water-saving campaigns, and other hands-on activities, helping to embed environmental protection concepts more deeply into students' daily learning and lifestyle. In addition to cultivating environmental awareness among teachers and students, it is equally important to strengthen the green consciousness of government agencies, school management teams, and students' parents. By promoting coordinated environmental education across these key groups, a more supportive atmosphere for sustainable practices can be created. Joint efforts from all stakeholders would further reinforce the development of green campuses and ensure the long-term success of sustainable building initiatives.

4.5. Promote coordinated development of urban and rural integration

To accelerate the balanced development of green buildings between urban and rural areas, it is recommended that urban schools fully leverage their resource advantages to establish long-term cooperation mechanisms with rural schools, promoting experience sharing and technical support. In addition, it would be beneficial for urban schools to share high-quality online educational resources with rural counterparts, such as offering distance learning courses focused on green building technologies and energy-saving management practices, enabling rural teachers and students to access advanced knowledge and improve their practical skills. Furthermore, encouraging enterprises and social organizations to participate in the urban-rural cooperation mechanism could be considered, with actions such as donating energy-efficient equipment, providing financial support, and jointly promoting green building initiatives in rural schools. For example, urban schools and enterprises might collaborate to establish a "Rural Green Campus Fund," dedicated to supporting energy-saving renovations and building maintenance projects in rural schools, thereby contributing to the overall improvement of sustainable building standards in rural education environments.

5. Implementation plan AND feasibility analysis for sustainable development of urban and rural school buildings in Chengdu

5.1. Implementation plan

Step 1: Policy Formulation and Fund Allocation (Months 1–6)

- Establish a Leading Task Force: Led by the Chengdu Municipal Education Bureau in collaboration with departments such as finance, housing and urban-rural development, and ecology and environment.
- Issue Policy Documents: Develop guidelines such as the Chengdu Urban-Rural School Green Building Development Directive and Special Fund Management Measures.
- Set Up a Special Fund: Establish a “Rural Green Building Special Fund” with clearly defined usage (e.g., energy-saving renovation, equipment procurement, training).
- Launch Incentive Policies: Introduce tax relief or low-interest policy loans to encourage enterprises and social organizations to support rural green campus construction.

Step 2: Pilot Renovation and Technology Promotion (Months 6–18)

- Select Pilot Schools: Choose 10–20 rural schools with strong foundations and high motivation to participate as pilot units.
- Implement Low-Cost Green Renovations: Prioritize practical technologies such as solar water heating systems, LED lighting, thermal insulation, and natural ventilation.
- Provide Technical Support: Organize expert teams to conduct assessments and guide pilot schools in developing and executing green renovation plans.

Step 3: Management System Building and Smart Platform Development (Months 12–24)

- Develop Unified Management Standards: Standardize energy consumption monitoring and equipment maintenance procedures across urban and rural schools.
- Install Energy Monitoring Systems: Deploy systems in pilot schools to automatically track water, electricity, and gas usage.
- Build Smart Management Platforms: Develop or adopt digital platforms for remote supervision, anomaly detection, and data analytics in green campus operations.

Step 4: Educational Integration and Capacity Building (Ongoing and Concurrent)

- Revise Curriculum: Integrate green building and sustainability topics into subjects such as science, geography, and technical education.
- Organize Environmental Activities: Launch campus programs like energy-saving challenges and eco-volunteer activities to enhance student awareness and engagement.
- Conduct Targeted Training: Offer annual training sessions for rural school administrators and technicians, focusing on green building operations and maintenance.

Step 5: Urban-Rural Collaboration Mechanism (From Month 12 Onward)

- Establish Pairing Programs: Form partnerships between urban and rural schools, with urban schools providing technical guidance and management support.
- Develop Resource-Sharing Platforms: Facilitate access to online courses, management tools, and operational manuals for rural educators and students.
- Engage Social Stakeholders: Encourage businesses and non-profits to donate energy-saving equipment and participate in green campus initiatives.

5.2. Feasibility analysis

The Chengdu municipal government has long placed high importance on the balanced development of urban and rural education. Policies related to green buildings and energy conservation are already being promoted in various sectors. With the establishment of dedicated special funds and implementation guidelines, there is a solid policy foundation for sustainable development in school construction.

By first implementing green renovations in a select number of pilot rural schools, the plan avoids the risks associated with large-scale simultaneous rollout. This phased approach allows for continuous optimization based on feedback, which increases the overall success rate and replicability of the initiative.

At the same time, many green building technologies—such as LED lighting, solar water heaters, and passive insulation techniques—are now technically mature, cost-effective, and easy to maintain. These characteristics make them suitable for implementation in rural schools with limited funding and technical capacity. And by establishing support mechanisms between urban and rural schools, the project leverages existing technical and managerial expertise in cities. This effectively compensates for the shortcomings in rural school personnel and technical resources, greatly improving the feasibility and sustainability of the project.

Finaly, the initiative encourages involvement from enterprises, social organizations, and communities. Through diversified funding and collaborative mechanisms, it brings together policy support, capital investment, technical services, and public participation, providing comprehensive assurance for implementation.

6. Conclusion

Based on the data of CMHURD and a field survey on the sustainability level of urban and rural school buildings in Chengdu, it is found that there is a large gap in the capacity and level of sustainable building development between urban and rural schools, mainly including the configuration level of building energy utilization and energy-saving settings, water resource management capabilities, environmental protection level of building materials and construction technology, indoor environmental quality control level, and environmental friendliness level of buildings. At the same time, there are also obvious urban-rural differences in the cultivation of environmental awareness among teachers and students, as well as the supporting funds and policy support obtained, between urban and rural schools in Chengdu. To address these issues, the government needs to increase financial support, promote applicable green building technologies, improve management mechanisms, strengthen sustainable development education, and promote coordinated urban and rural development in order to promote the balanced development of sustainable building levels in urban and rural schools and ultimately achieve a fair and just educational environment.

References

1. Q. Hu, J. Xue, R. Liu, GQ. Shen, F. Xiong, Energy Build., **278**, 112641(2023). DOI: 10.1016/j.enbuild.2022.112641.
2. Y. Liu, RD. Chang, J. Zuo, F. Xiong, N. Dong, Constr. Build. Mater., **30**, 805(2023). DOI: 10.1108/ECAM-06-2021-0538.
3. C. Hepburn, Y. Qi, N. Stern, B. Ward, C. Xie, D. Zenghelis, Environ. Sci. Ecotechnol., **8**, 100130 (2024). DOI: 10.1016/j.esee.2021.100130.
4. N. Murtagh, L. Scott, J. Fan, J. Clean. Prod., **268**, 122264 (2024). DOI: 10.1016/j.jclepro.2020.122264.
5. AF. Kineber, AE. Oke, M. Elseknidy, MM. Hamed, FS. Kayode. Sustain. Cities Soc., **15**, 825 (2025). DOI: 10.3390/su15010825.
6. M. L. Tseng, S. X. Li, C. W. R. Lin, A. S. Chiu, J. Ind. Prod. Eng., **40**, 35(2024). DOI: 10.1080/21681015.2023.2460059.

Researching the Mechanism Underlying the Low Cellulose Pretreatment Efficiency at High DMSO/Ionic Liquid Ratios

Xinyu Guo, Jiming Yang*, Wei He*, and Zheng Fang

College of Biotechnology and Pharmaceutical Engineering, Nanjing Tech University, Nanjing 211816, PR China

Abstract. High DMSO/[Amim]Cl ratios impair cellulose pretreatment efficiency. In DMSO/[Amim]Cl systems (0–60 wt% DMSO), cellulose solubility and saccharification rates declined gradually (0–50 wt%) then sharply (50–60 wt%). Structural changes intensified at 50–60 wt% DMSO, with greater decrystallization (20.93% vs 2.85%) and hydrophobicity (23.08° vs 1.86°) reductions. SEM revealed denser, smoother surfaces at 60 wt% DMSO, reducing accessibility. Kamlet-Taft analysis showed nonlinear decreases in hydrogen bond acidity/basicity (α/β: 0.59/1.25 → 0.24/0.73), suggesting weakened hydrogen bond disruption at 60 wt% DMSO, lowering pretreatment efficiency.

1. Introduction

With the development of human society, the drawbacks of petrochemical - based resources have become more and more prominent[1]. Cellulose is an abundant renewable resource on earth. After being enzymatically hydrolyzed into glucose, it can be further converted into various materials and energy sources[1]. However, natural cellulose has strong resistance to enzymatic hydrolysis[2]. However, natural cellulose has strong resistance to enzymatic hydrolysis[3]. To boost the efficiency of cellulose enzymatic hydrolysis, it is essential to pretreat cellulose to change its structure.

Ionic liquids are a category of ionic compounds predominantly composed of organic cations paired with either inorganic or organic anions[4]. These substances exhibit minimal vapor pressure and strong thermal stability. Ionic liquids are regarded as ideal green solvents[5]. Since Rogers reported that ionic liquids could dissolve cellulose[6], a range of ionic liquids have been reported to be used for pretreating cellulose[7]. The pretreatment of cellulose with ionic liquids is a feasible method. However, high viscosity, slow mass transfer, and cost hinder large-scale application[8]. Researchers have attempted to add DMSO to ionic liquids to construct a cellulose pretreatment system with low viscosity and relatively low cost[9]. A series of research results show that when the DMSO content is low, the effect of the binary system in pretreating cellulose is comparable to that of pure ionic liquids[10]. However, when the amount of DMSO is high, there is a big gap in the performance of the binary system in pretreating cellulose compared with that of pure ionic liquids[11].

Studying (1) the variations in crystallinity, surface morphology, and hydrophobicity of regenerated cellulose with increasing DMSO concentration, will help us gain a profound insight into the mechanism responsible for the low cellulose pretreatment efficiency at high DMSO/Ionic liquid ratios; and (2) the changes in hydrogen bond acidity/basicity of the DMSO/ionic liquid system with increasing the content of DMSO.

2. Experimental

2.1. Saturation solubility of microcrystalline cellulose

A 10.0 g mixture of DMSO/[Amim]Cl was loaded into a 25 mL Shrek tube, followed by the addition of dried microcrystalline cellulose (MCC, 1 wt% of the solvent mass). The tube was purged with N_2 and sealed. Dissolution was conducted at 95 °C (350 rpm), with MCC added incrementally. Samples were monitored microscopically every 0.5 h; saturation was reached when newly added MCC remained undissolved after 1 h. DMSO was then introduced to reduce viscosity. The mixture was centrifuged (9000 rpm, 6 min), yielding supernatant A (containing dissolved cellulose) and precipitate A (undissolved MCC). Supernatant A was mixed with 10 mL H_2O and centrifuged to regenerate cellulose. Both precipitate A and regenerated cellulose were washed five times with H_2O ($AgNO_3$-confirmed Cl^- removal) and freeze-dried for 24 h. The saturated solubility of cellulose is computed based on Eq. 1:

$$Saturation\ solubility = \frac{A-B}{10} \tag{1}$$

Among them, 10 is the total mass of the solvent system (g), A was the total mass of cellulose injected into the shrek tube, and B was the mass of undissolved cellulose.

* Corresponding author: jmyang@njtech.edu.cn

2.2. Pretreatment of MCC

A 10.0 g mixture of DMSO/[Amim]Cl and 0.4 g MCC was added to a 25 mL Shrek tube under N_2 atmosphere. The tube was sealed and heated at 95 °C (350 rpm, 5 h) for cellulose pretreatment. After pretreatment, 10 mL H_2O was added, and the mixture was centrifuged to obtain regenerated cellulose. The product was washed five times with 10 mL H_2O ($AgNO_3$ testing confirmed complete Cl^- removal). Finally, the samples were freeze-dried for 24 h. To study DMSO's effect, the DMSO/[Amim]Cl mass ratio was varied while maintaining a total mass of 10 g.

2.3. Component analysis

Sugar content was analyzed using an Agilent 1260 HPLC system (USA) with a Bio-Rad Aminex HPX-87H column and refractive index detector. Operating conditions included a column temperature of 65 °C, detector temperature of 50 °C, and a mobile phase of 0.5% H_2SO_4 at 0.4 mL/min. The sugar standard curve exhibited a high correlation coefficient (R^2 = 0.999). Cellulose purity was determined for raw microcrystalline cellulose, undissolved cellulose, and five regenerated cellulose samples.

2.4. Enzymatic hydrolysis

To prepare the enzyme solution, the cellulase was added into an acetate/sodium acetate buffer with a pH of 4.8. 1.5 mL cellulase solution and 30 mg cellulose sample were transferred into a 5 mL centrifugation tube. The enzyme load was 0.035 FPU/g cellulose. The enzymatic hydrolysis was processed at 50 °C with a stirring speed of 300 rpm and sampled every 24 hours. The solution sampled was diluted and purified by means of a 0.20 μm syringe filter.

The concentration of glucose in the hydrolysate was tested using HPLC. All the enzymatic hydrolysis experiments were tested three times, and the mean value was documented. The enzymatic hydrolysis rate was computed using Eq. 2:

$$Enzymatic\ digestion\ rate\ (\%) = \frac{C}{D \times 1.11} \times 100 \quad (2)$$

Among them, C is the mass of glucose in the hydrolysis solution; D is the mass of cellulose in the sample; 1.11 is the conversion factor for cellulose to glucose.

2.5. Characterization of regenerated cellulose

(1) Cellulose specimens underwent X-ray diffraction (XRD) analysis via an X-ray diffractometer (Ultima IV, Japan). The instrument utilized Cu radiation as its source, operating at an emission current of 40 mA and an acceleration voltage of 40 kV. Scanning was conducted across a range of 5 to 40 degrees, at a scanning speed of 2 degrees per minute. The crystallinity index (CI) was ascertained by the Segal method according to Eq. 3:

$$CI = (I_{002} - I_{am})/I_{002} \times 100\% \quad (3)$$

In the formula: I_{am} was the diffraction intensity of the amorphous peak (2θ=18.0°); I_{002} was the diffraction intensity of the 002 crystal plane in the crystalline part (2θ=22.52 °).

The decrystallization rate was computed using Eq. 4:

$$Decrystallization\ rate = (E - F)/E \times 100 \quad \% \ (4)$$

In the formula: E was the crystallinity of raw cellulose and F was the crystallinity of regenerated cellulose.

(2) Surface contact angles were measured using a contact angle measuring instrument (Kunshan Shengding SDC 350KS) at room temperature. A 5 μL cellulase solution droplet was deposited on the sample surface, and the contact angle was immediately recorded using the sessile drop method.

(3) The surface structure of raw MCC and regenerated cellulose was investigated using SEM (ZEISS Sigma 360, German). Before loading the samples, a gold coating was applied to the sample surfaces for 45 seconds via ion sputtering coating technology to enhance conductivity. The samples were observed under an accelerating voltage of 3 kV.

2.6. Determination of K-T values of [Amim]Cl/DMSO systems

0.2mg 4-nitroaniline, 0.2mg N, N-diethyl-4-nitroaniline and 0.75mg 2,6-dichloro-4 (2,4,6-triphenyl-1-pyridinium) phenolate (Reichardt's dye 33) were dissolved in 5 g [Amim]Cl/DMSO mixture in 3 glass bottles, respectively. The dye solutions were filled into a 2mm quartz colorimetric ware which would be placed in the sample slot of an UV visible spectrometer. At 25 °C, the solution was scaned and the maximum absorption wavelength λmax was obtained. The α (hydrogen bond donating ability) and β (hydrogen bond accepting ability) of the [Amim]Cl/DMSO system was calculated as follows:

$$v(dye) = \frac{1}{\lambda_{max}(dye) * 10^{-4}}$$

$$E_T(30) = 0.9986 * \left(\frac{28592}{\lambda_{max(Reichardt\ 's\ dye\ 33)}}\right) - 8.6878$$

$$\alpha = 0.0649E_T(30) - 2.03 - 0.72\pi^*$$

$$\beta = \frac{\left(1.035v_{(N,N)} + 2.64 - v_{(4-nitroaniline)}\right)}{2.80}$$

3. Results and discussion

3.1. Saturation solubility and enzymatic digestion

Previous studies showed that cellulose solubility in DMSO/IL systems exhibited varying trends with DMSO content depending on temperature: at 40 °C and 60 °C, solubility initially increased before declining, while at 80 °C, it decreased monotonically[12]. In this study, the trend at 95 °C (Fig. 1) aligned with the 80 °C observations—solubility consistently decreased with increasing DMSO content. Specifically, solubility declined from 17.17 wt% (0% DMSO) to 9.50 wt% (50% DMSO), with the sharpest drop occurring between 50–60% DMSO. This drastic reduction at high DMSO concentrations may correlate with shifts in solvent polarity parameters (α/β), as later analyzed via Kamlet-Taft measurements.

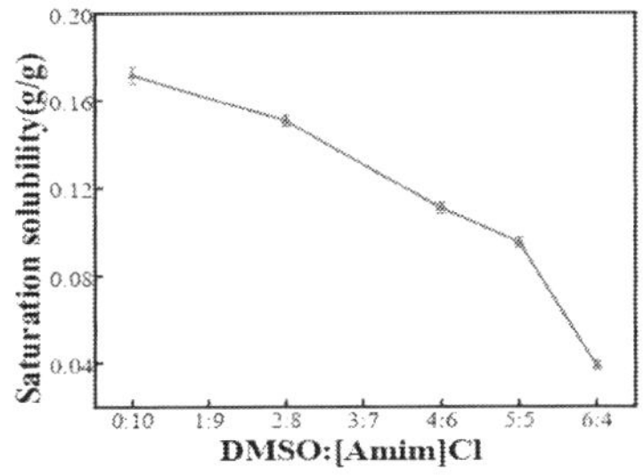

Fig. 1. Saturation solubilities of cellulose in different DMSO/[Amim]Cl systems.

Cellulose was pretreated with DMSO/[Amim]Cl systems at varying IL/DMSO ratios. The enzymatic hydrolysis rate of untreated MCC was low (16.05%), but pretreatment significantly enhanced it. As shown in Fig. 2, the hydrolysis rate declined with increasing DMSO content, dropping from 58.67% to 43.13% as DMSO increased from 0 to 50 wt%. A sharper decrease occurred at 50–60 wt% DMSO, with pure [Amim]Cl yielding a hydrolysis rate 1.87 times higher than the 60 wt% DMSO system[13].

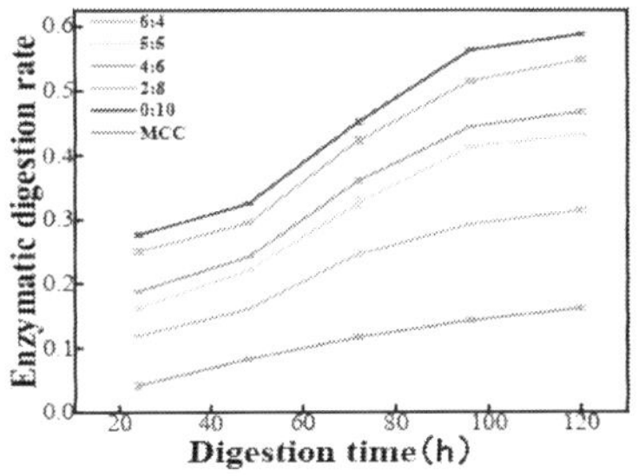

Fig. 2. Enzymatic digestion rate of regenerated cellulose.

3.2. Characterization analysis of regenerated cellulose

X-ray diffraction (XRD) technique is commonly used to evaluate the crystalline characteristics of cellulose substrates. According to previous research, the diffraction peaks of MCC at 2θ=14.50°, 17.50°, 22.0° and 34.6° belong to cellulose I crystals[14], while the characteristic peaks of cellulose II appear at around 2θ=12.05°, 20.3° and 22.0°. Cellulose II can be obtained through decrystallization and recrystallization of natural cellulose[15]. As shown in Fig. 3, regenerated cellulose pretreated with DMSO/[Amim]Cl systems displayed a peak at 20.3°, confirming the transition from cellulose I to II. Crystallinity (CI) and decrystallization rates varied nonlinearly with increasing DMSO content. Pure [Amim]Cl pretreatment yielded a CI of 21.76% and a decrystallization rate of 71.41% (Table 1). At 50 wt% DMSO, CI increased to 32.61%, while decrystallization decreased to 57.15%. A sharp rise in crystallinity (48.54%) and a drop in decrystallization (36.22%) occurred at 60 wt% DMSO.

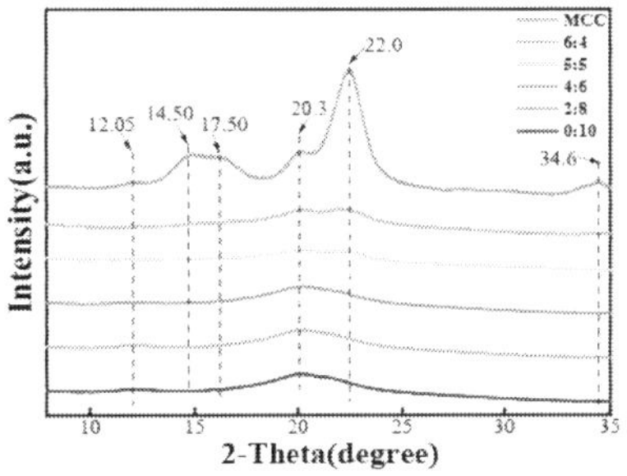

Fig. 3. XRD analysis of regenerated cellulose samples.

Table 1. CI of regenerated cellulose samples.

DMSO:[Amim]Cl	CI cellulose I(%)	Decrystallization rate (%)
MCC	76.10	0.00
6:4	48.54	36.22
5:5	32.61	57.15
4:6	28.62	62.39
2:8	23.59	69.00
0: 10	21.76	71.41

The contact angles of cellulose pretreated with DMSO/[Amim]Cl systems are shown in Fig. 4. Raw microcrystalline cellulose (MCC) exhibited strong hydrophilicity (contact angle = 0°), consistent with previous studies[16]. After pretreatment, hydrophilicity decreased, and the static contact angle increased, likely due to the disruption of cellulose's crystalline and surface-oriented structure[17]. As DMSO concentration increased, the crystallinity index (CI) of regenerated cellulose rose, while the contact angle gradually decreased. This reduced hydrophobicity could weaken enzyme binding and lowering enzymatic hydrolysis efficiency.

CA:33.42 CA: 56.50 CA:65.78

DMSO:[Amim]Cl =6:4 DMSO:[Amim]Cl =5:5 DMSO:[Amim]Cl =0: 10

Fig. 4. The contact angles of regenerated cellulose samples.

SEM analysis (Fig. 5) shows that raw MCC displays a smooth, compact, and well-ordered structure, typical of its rigid fiber morphology. After [Amim]Cl pretreatment, the cellulose surface becomes rough and porous with pronounced wrinkles, indicating structural disruption that enhances cellulose accessibility and enzymatic hydrolysis. In contrast, at 60 wt% DMSO, MCC exhibits only partial structural disruption, with a denser and smoother surface compared to pure [Amim]Cl treatment, likely reducing cellulose accessibility and enzymatic efficiency.

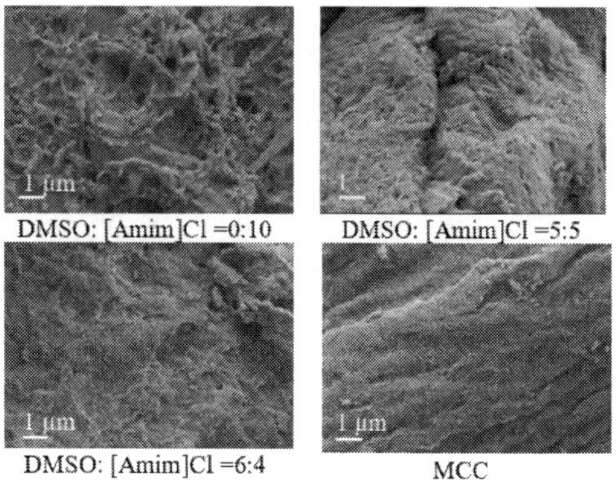

Fig. 5. The SEM images of cellulose samples.

3.3. Analysis of Kamlet–Taft parameters

It has been reported that α and β of ionic liquid based solvents have a great influence on the destruction of hydrogen bond network in cellulose[9]. As shown in Fig. 6, increasing DMSO content (0–60 wt%) led to a decline in α (0.59→0.24) and β (1.25→0.73), with the sharpest drop occurring at 50–60 wt% DMSO (α: 0.40→0.24; β: 1.02→0.73). Since higher α and β values enhance cellulose hydrogen bond disruption and dissolution[18,19], the pronounced reduction at 60 wt% DMSO weakened the system's pretreatment efficiency, consistent with solubility and XRD results. This explains the inferior performance compared to pure [Amim]Cl.

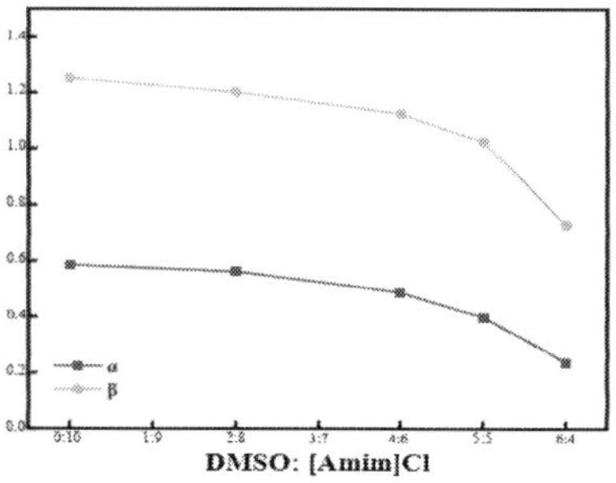

Fig. 6 Kamlet-Taft parameters of DMSO/[Amim]Cl systems

4. Conclusion

The pretreatment efficiency of cellulose is significantly impaired when the DMSO concentration exceeds 50 wt%, attributable to the dual constraints arising from structural modifications and solvent interactions.

DMSO/[Amim]Cl systems were formulated with DMSO mass fractions ranging from 0 wt% to 60 wt%. In DMSO/[Amim]Cl systems (0–60 wt% DMSO), cellulose solubility decreased from 17.17% (0 wt%) to 9.50% (50 wt%), then sharply dropped to 3.87% (60 wt%). Similarly, enzymatic hydrolysis (120 h) declined from 58.67% (0 wt%) to 43.13% (50 wt%) and further to 31.37% (60 wt%). Decrystallization rate (-20.93% vs. -2.85%/10 wt% at 0–50 wt%) and contact angle (-23.08° vs. -1.86°/10 wt%). SEM revealed smoother, denser morphologies at 60 wt% DMSO, reducing cellulose accessibility. Kamlet-Taft parameters (α/β) declined nonlinearly (0.59/1.25 to 0.40/1.02 at 50 wt%, then 0.24/0.73 at 60 wt%), indicating weakened hydrogen bond disruption at 60 wt% DMSO, lowering pretreatment efficiency.

The National Key Research and Development Program of China (2021YFC2101904); The research has been supported by The National Natural Science Foundation of China (22178168, 21776130, and 22078150); Key Research and Development Plan of Jiangsu Province (BE2022791 and BE2021083); The Top-notch Academic Programs Project of Jiangsu Higher Education Institutions; Postgraduate Research & Practice Innovation Program of Jiangsu Province (No. SJCX23_0454).

References

1. B.W. Liu, H.B. Zhao, Y. Z. Wang, Advanced flame-retardant methods for polymeric materials, Adv. Mater. **34(46)**, 2107905 (2022).
2. I. Elfaleh, F.Abbassi, M.Habibi, F. Ahmad, M. Guedri, M. Nasri, C. Garnier, A comprehensive review of natural fibers and their composites: An eco-friendly alternative to conventional materials, Results Eng. **19**, 101271 (2023).
3. A.E.K. Afedzi, P. Parakulsuksatid, Recent advances in process modifications of simultaneous saccharification and fermentation (SSF) of lignocellulosic biomass for bioethanol production, Biocatal. Agric. Biotechnol. **54**, 102961 (2023).
4. G. Kaur, H. Kumar, M. Singla, Diverse applications of ionic liquids: A comprehensive review, J. Mol. Liq. **351**, 118556 (2022).
5. B. Azimi, H. Maleki, V. Gigante, R. Bagherzadeh, A. Mezzetta, M. Milazzo, L. Guazzelli, P. Cinelli, A. Lazzeri, S. Danti, Cellulose-based fiber spinning processes using ionic liquids, Cellulose, **29(6)**, 3079-3129 (2022).
6. R.P. Swatloski, S.K. Spear, J.D. Holbrey, R.D. Rogers, Dissolution of cellose with ionic liquids, J. Am. Chem. Soc. 124, 4974-4975 (2002).
7. A.S. Norfarhana, R.A. Ilyas, N. Ngadi, M.H.D. Othman, M.S.M. Misenan, M.N.F. Norrahim, Revolutionizing lignocellulosic biomass: A review of harnessing the power of ionic liquids for sustainable utilization and extraction, Int. J. Biol. Macromol. **256**, 128256 (2024).
8. L. Zhou, F. Pan, Y. Liu, Z. Kang, S. Zeng, Y. Nie, Study on the regularity of cellulose degradation in ionic liquids, J. Mol. Liq. **308**, 113153 (2020).
9. F. Ren, J. Wang, J. Yu, C. Zhong, F. Xie, S. Wang, Dissolution of cellulose in ionic liquid–DMSO mixtures: roles of DMSO/IL ratio and the cation alkyl chain length, ACS omega, **6(41)**, 27225-27232 (2021).
10. Y. Tomimatsu, H. Suetsugu, Y. Yoshimura, A. Shimizu, The solubility of cellulose in binary mixtures of ionic liquids and dimethyl sulfoxide: Influence of the anion, J. Mol. Liq. **279**, 120-126 (2019).

https://doi.org/10.1051/e3sconf/202563901011

11. H. Wu, B. Wang, T. Li, Y. Wu, R. Yang, H. Gao, Y. Nie, Efficient recycle of waste poly-cotton and preparation of cellulose and polyester fibers using the system of ionic liquid and dimethyl sulfoxide, J. Mol. Liq. **388**, 122757 (2023).
12. D.L. Minnick, R.A. Flores, M.R. DeStefano, A.M. Scurto, Cellulose solubility in ionic liquid mixtures: temperature, cosolvent, and antisolvent effects, J. Phys. Chem. B. **120(32)**, 7906-7919 (2016).
13. S. Zhang, N. Sun, X. He, X. Lu, X. Zhang, Physical properties of ionic liquids: database and evaluation, J. Phys. Chem. Ref. Data. **35(4)**, 1475-1517 (2006).
14. S. Naduparambath, T.V. Jinitha, V. Shaniba, M.P. Sreejith, A.K. Balan, E. Purushothaman, Isolation and characterisation of cellulose nanocrystals from sago seed shells, Carbohydr. Polym. **180**, 13-20 (2018).
15. K. Aghmih, A. Bouftou, M. El Bouchti, A. Boukhriss, S. Gmouh, S. Majid, Synthesis and application of functionalized ionic liquids-based imidazolium as solvent for cotton fibre cellulose dissolution, Cellulose **30(3)**, 1467-1481 (2023).
16. K. Olonisakin, R. Li, X.X. Zhang, F. Xiao, J. Gao, W. Yang, Effect of TDI-assisted hydrophobic surface modification of microcrystalline cellulose on the tensile fracture of MCC/PLA composite, and estimation of the degree of substitution by linear regression, Langmuir. **37(2)**, 793-801 (2021).
17. J. Shuai, X. Gao, J. Zhao, W. Ge, M. Tian, Z. Lei, X. Wang, Dissolution and regeneration of cellulose using superbase-based dicarboxylic ionic liquids with tailored amphiphilicity, Chem. Eng. J. **495**, 153280 (2024).
18. B. Lu, A. Xu, J. Wang, Cation does matter: how cationic structure affects the dissolution of cellulose in ionic liquids, Green. Chem. **16(3)**, 1326-1335 (2014).
19. A. Xu, J. Wang, H. Wang, Effects of anionic structure and lithium salts addition on the dissolution of cellulose in 1-butyl-3-methylimidazolium-based ionic liquid solvent systems, Green. Chem. **12(2)**, 268-275 (2010).

Numerical Simulation Study on Wormhole Propagation Morphology in Fractured Carbonate Rocks during Acidizing

Weiren Mo*, and Yongming Li*

State Key Laboratory of Oil & Gas Reservoir Geology and Exploitation, Southwest Petroleum University, Chengdu, Sichuan, 610500, China

Abstract. Carbonate reservoirs are key targets for global oil and gas exploration, accounting for more than 50% of both reserves and production. However, due to the widespread existence of fractures, vugs, and strong heterogeneity in these reservoirs, they are prone to damage during drilling and production operations, which can lead to a decline in well productivity. Matrix acidizing has become a core technique for restoring and enhancing the productivity of carbonate reservoirs. This paper presents a non-isothermal, three-dimensional acidizing model for fractured carbonate rocks. By integrating a three-dimensional embedded discrete fracture model (EDFM) with the two-scale continuum model, the model accurately captures the three-dimensional topological structure of natural fracture networks. A fully coupled thermo-hydro-chemical model is employed to investigate the effects of acid-rock reaction heat, injection temperature, and formation thermal conduction on the reaction rate. The results indicate that injection rate, acid concentration, and temperature have significant impacts on the morphology of wormhole propagation. Furthermore, the study reveals the critical role of fracture architecture in controlling acid fluid diversion and the competitive growth of wormholes.

1. Introduction

Carbonate reservoirs represent a significant focus for global oil and gas resource development, with their reserves and production accounting for over 50%[1]. Due to the common presence of fractures, vugs, and strong heterogeneity in these reservoirs, they are susceptible to damage from mud invasion and mineral precipitation during drilling and production operations, leading to blockage of flow pathways and a decrease in productivity[2]. Matrix acidizing technology, which involves injecting acid into the formation to dissolve reservoir minerals and create high-conductivity wormhole networks, has become a crucial method for restoring and enhancing the productivity of carbonate reservoirs[3]. In recent years, however, with the exploitation of deep and ultra-deep carbonate oil and gas reservoirs, acidizing technology faces challenges such as high temperatures, complex fracture-vug systems, and increased heterogeneity. There is an urgent need to optimize operational designs through numerical simulation approaches[4].

In recent years, significant progress has been made in the numerical simulation of matrix acidizing, particularly in model construction and multi-physics coupling mechanisms. Researchers commonly employ a two-scale continuum model combined with thermal-hydraulic-mechanical-chemical (THMC) coupling methods to reveal the influence of heterogeneity, temperature, and mineral composition on wormhole morphology and acid injection parameters[5-8]. For instance, Yang et al.[5] developed an acidizing program for heterogeneous reservoirs based on the TSC model, demonstrating that low heterogeneity leads to fewer but wider wormhole branches, while high heterogeneity results in more but narrower branches. Jia et al.[9]improved the accuracy of predicting radial flow acid injection rates by optimizing the Navier-Stokes-Darcy equations. Research focus has shifted towards characterizing complex pore structures and simulating competitive acid flow in fractured carbonate rocks. Mou et al.[10] constructed a pore model incorporating matrix, vugs, and natural fractures using geostatistical methods, finding that fracture density and vug spatial distribution dominate wormhole propagation paths. Asl et al.[11] simulated acidizing in triple-porosity media using a pseudo-fracture approach, highlighting that intersecting fractures can significantly reduce the pore volume of acid required to reach the breakthrough (PV_{BT})[11-13]. Moreover, studies on optimizing acid injection parameters have been further advanced. Rodrigues et al.[14] confirmed through micro-CT experiments that oil saturation reduces acid consumption, while Yoo et al.[15] improved the wormhole tip propagation model using CT visualization techniques, significantly reducing prediction errors in acid injection volumes. Karimi et al.[16] proposed a scale-independent model based on the finite element method, providing a rapid computational framework for large-scale 3D core acidizing simulations, whereas Mahmoud et al.[17] extended their mechanistic model to radial flow via semi-empirical formulas,

*Corresponding author
Email address: 2622681634@qq.com (Weiren Mo);
swpifrac@163.com (Yongming Li)

validating the sensitivity of acidizing efficiency to temperature and mineral composition[18].

Although recent studies have made progress in two-dimensional or quasi-three-dimensional simulations of fractured carbonate acidizing, most existing models are still limited by 2D simplifications. For instance, the quasi-3D Discrete Virtual Internal Bond model proposed by Zhu et al.[19] can characterize fracture aperture variations but does not couple the real 3D in-situ stress field with acid-rock reaction kinetics. Moreover, high-fidelity representation of 3D fracture networks remains a challenge: Wang et al.[20] generated statistical fracture models using the Monte Carlo method, but these were projected onto 2D grids. Liu et al.[21] employed the finite volume method to simulate the influence of single fracture dip angles; However, they did not quantify the mechanisms by which 3D fracture topology affects acid diversion[20, 22]. Although the THMC models developed by Ma et al.[7] and Liu et al.[8] incorporate interactions between rock deformation and wormhole branching, the simulation of acid-rock reactions within 3D fracture systems is still in its early stages, particularly lacking characterization of the spatial competition between natural fractures and wormholes[7, 23]. Furui et al.[24] simulated spontaneous wormhole formation in homogeneous rocks using a phase-field approach, yet the induced effects of heterogeneous fracture networks were not considered. Therefore, three-dimensional numerical simulation of acidizing processes in carbonate reservoirs containing complex fracture networks remains underdeveloped.

In this study, we develop a non-isothermal, three-dimensional acidizing model for fractured carbonate rocks by integrating the 3D Embedded Discrete Fracture Model with the dual-scale continuum framework and incorporating heat transfer mechanisms. The model reveals how injection rate, acid concentration, and injection temperature influence wormhole morphology during acidizing of fractured carbonate reservoirs.

2. Three-dimensional mathematical model of fractured carbonate rock acidizing

2.1. Seepage field

Governing equation of pressure:

$$\frac{\partial \phi_m}{\partial t} - \nabla \cdot \left(\frac{K_m}{\mu} \cdot \nabla p_m\right) + \frac{\psi^{mf}}{V_m} = 0 \quad (1)$$

$$\frac{\partial \phi_f}{\partial t} - \nabla \cdot \left(\frac{K_f}{\mu} \cdot \nabla p_f\right) + \frac{\psi^{fm}}{V_f} + \frac{\psi^{ff}}{V_f} = 0 \quad (2)$$

ϕ_m and ϕ_f are the porosity of matrix and fracture respectively, and the subscripts m and f represent matrix and fracture respectively, hereinafter the same; K_m is the permeability of matrix, m^2;μ is the viscosity, Pa·s; p_m and p_f are the fluid pressure in the matrix and fracture respectively, Pa; ψ^{mf} and ψ^{fm} is the flux transfer function between the damaged matrix and the fractures, m^3/s; ψ^{ff} is the flux transfer function between intersecting fracture,m^3/s; V_m and V_f are the grid volumes of the matrix and the fracture respectively, m^3;

$$\psi^{mf} = \frac{T_{mf}}{\mu}(p_m - p_f) = -\psi^{fm} \quad (3)$$

$$\psi^{ff} = \frac{T_{ff}}{\mu}(p_{f1} - p_{f2})\# \quad (4)$$

T_{mf} is the transmissibility factor between the matrix and fracture plane, m^3; T_{ff} is the transmissibility factor between the intersecting fracture plane, m^3. p_{f1} and p_{f2} are the pressure between the intersecting fracture planes, Pa.

T_{mf} can be calculated as:

$$T_{mf} = \frac{2A_f K_m}{d_{mf}} \quad (5)$$

$A_f$is the fracture plane area in the grid block, m^2; d_{mf} is the average normal distance between the fracture plane and the matrix block, m;

T_{ff} can be calculated as:

$$T_{ff} = \frac{T_1 T_2}{T_1 + T_2}; T_1 = \frac{k_{f1} w_{f1} L_{int}}{d_{f1}}; T_2 = \frac{k_{f2} w_{f2} L_{int}}{d_{f2}} \#(6)$$

$$T_{mf} = \frac{2A_f K_m}{d_{mf}} \quad (7)$$

k_{f1} and k_{f2} are the permeability of fracture segments, m^2; w_{f1} and w_{f2} are the width of intersecting fracture planes, m; L_{int} is the length of the intersection line, m.

2.2. Chemical reaction field

$$\frac{\partial}{\partial t}(\phi_m C_m) + \nabla \cdot (U_m C_m) - \nabla \cdot (\phi_m D_{e\xi} \cdot \nabla C_m) + \psi^{mf}\frac{C^{up}}{V_m} + R(C_{sm})a_{vm} = 0 \quad (8)$$

$$\frac{\partial}{\partial t}(\phi_f C_f) + \nabla \cdot (U_f C_f) - \nabla \cdot (\phi_f D_m \cdot \nabla C_f) + \psi^{fm}\frac{C^{up}}{V_f} + \psi^{ff}\frac{C_f}{V_f} + R(C_{sf})a_{vf} = 0 \quad (9)$$

Here, C_m and C_f are acid concentration in the matrix and fracture plane respectively, kmol/m^3; U_m and U_f are fluid velocities of matrix and fracture plane respectively, m/s. $D_{e\xi}$ is the effective diffusion tensor in the different directions and ξ represents x, y and z, m/s. D_m is the effective diffusion, m/s; a_{vm} and a_{vf} are the interfacial area of the matrix and fracture plane respectively, m^2/m^3; $R(C_{sm})$ and $R(C_{sf})$ are reaction kinetics in the matrix and fracture plane respectively, kmol/(s·m^2).

$$C^{up} = \begin{cases} C_m, \psi^{mf} > 0 \\ 0, \psi^{mf} = 0 \\ C_f, \psi^{mf} < 0 \end{cases} \quad (10)$$

C^{up} is a function of ψ^{mf}, kmol/(s·m^2).

$$R(C_{si}) = \frac{k_{ci}k_{si}}{k_{ci}+k_{si}}C_i a_{vi} \quad (11)$$

Here, the subscript i represents matrix m or fracture plane f, the same as below; k_{ci} is the mass transfer coefficient, m/s; k_{si} is reaction rate constant, with units of velocity, m/s;

$$k_{ci} = \frac{\left(Sh_{\infty} + \frac{0.7}{m^{\frac{1}{2}}}Re_p^{\frac{1}{2}}Sc^{\frac{1}{3}}\right)D_m}{2r_{pi}} \quad (12)$$

Here, Sh_{∞} is the asymptotic Sherwood number; Re_p is the pore Reynolds number, expressed as $Re_p = \frac{2\rho_l|U_i|r_{pi}}{\mu}$. Sc is the Schmidt number, expressed as $Sc = \frac{\mu}{D_m\rho_l}$; m is the ratio of pore length to pore diameter; r_{pi} is average pore radius, m; ρ_l is acid density, kg/m^3; $|U_i|$ is the magnitude U_i, m/s.

$$D_{e\xi} = \alpha_{os}D_m + \frac{\lambda_{\xi}PeD_m}{\phi_m} \quad (13)$$

α_{os} and λ_{ξ} (λ_x= 0.5, λ_y= 0.1 and λ_z= 0.1) is constant related to the pore structure; Pe is Peclet number, expressed as $Pe = \frac{2|U_i|r_P}{D_m}$. r_p is the average pore radius, m.

$$\frac{K_i}{K_{i0}} = \frac{\phi_i}{\phi_{i0}}\left[\frac{\phi_i(1-\phi_{i0})}{\phi_{i0}(1-\phi_i)}\right]^{2\beta}$$
$$\frac{r_{pi}}{r_{pi0}} = \sqrt{\frac{K_i\phi_{i0}}{K_{i0}\phi_i}} \quad (14)$$
$$\frac{a_{vi}}{a_{vi0}} = \frac{\phi_i r_{pi0}}{\phi_{i0} r_{pi}}$$

$$\frac{\partial\phi_i}{\partial t} = \frac{R(C_{si})a_{vi}\alpha}{\rho_s} \quad (15)$$

Here, K_{i0} is the initial permeability, m^2; ϕ_{i0} is the initial porosity; β is the pore-broadening parameter; r_{pi0} is the initial values of average pore radius, m; a_{vi0} is the initial interfacial area respectively, m^2/m^3; α is the dissolving power of the acid, kg/kmol; ρ_s is rock density, kg/m^3.

2.3. Temperature field

Governing equation of acid temperature:

$$\frac{\partial}{\partial t}(\phi_m\rho_l C_{pl}T_m) + \nabla\cdot(U_m\rho_l C_{pl}T_m) + (\psi^{mf}\rho_l C_{pl})\frac{T^{up}}{V_m} = \nabla\cdot(\phi_m k_{el}\cdot\nabla T_m) + h_c a_{vm}(T_{sm}-T_m) \quad (16)$$

$$\frac{\partial}{\partial t}(\phi_f\rho_l C_{pl}T_f) + \nabla\cdot(U_f\rho_f C_{pl}T_f) + \psi^{fm}\rho_l C_{pl}\frac{T^{up}}{V_f} + \psi^{ff}\rho_l C_{pl}\frac{T_f}{V_f} = \nabla\cdot(\phi_f k_{el}\cdot\nabla T_m) + h_c a_{vf}(T_{sf}-T_f) \quad (17)$$

$$T^{up} = \begin{cases} T_m, \psi^{mf} > 0 \\ 0, \psi^{mf} = 0 \\ T_f, \psi^{mf} < 0 \end{cases} \quad (18)$$

Here, C_{pl} is the specific heat capacity of the acid, J/(kg·K); T_m and T_f are the acid temperatures in the matrix and the fracture plane, respectively, K; k_{el} is thermal conductivity of the acid, W/(m·K); h_c is the acid-rock heat transfer coefficient, W/(m^2·K). T^{up} is a function of ψ^{mf}, similar to the function C^{up}.

Governing equation of rock temperature:

$$\frac{\partial}{\partial t}\left((1-\phi_m)\rho_s C_{ps}T_{sm}\right) = \nabla\cdot\left((1-\phi_m)k_{es}\cdot\nabla T_{sm}\right) + \Delta H_r(T_{sm})a_v R(C_{sm}) - h_c a_{vm}(T_{sm}-T_m) \quad (19)$$

$$\frac{\partial}{\partial t}\left((1-\phi_f)\rho_s C_{ps}T_{sf}\right) = \nabla\cdot\left((1-\phi_f)k_{es}\cdot\nabla T_{sf}\right) + \Delta H_r(T_{sf})a_v R(C_{sf}) - h_c a_{vf}(T_{sf}-T_f) \quad (20)$$

Here, C_{ps} is the specific heat capacity of rock, J/(kg·K); T_{sm} and T_{sf} are the temperatures of matrix rock and fracture wall, respectively, K; $\Delta H_r(T_{sm})$ and $\Delta H_r(T_{sf})$ are the acid-rock molar reaction heat in matrix and fracture plane respectively, J/mol; k_{es} is thermal conductivity of the rock, W/(m^2·K).

2.4. Empirical formula:

The standard molar reaction heat of HCl with $CaCO_3$ affected by temperature can be expressed as[25]:

$$\Delta H_r(T_{si}) = -13692 + \left(-6.443\times10^{-3}T_{si}^2 + 16.075T_{si} - \frac{17.406\times10^5}{T_{si}}\right) \quad (21)$$

The empirical formula of reaction rate between HCL and $CaCO_3$ was obtained by using previous experiments[26]:

$$k_{si} = 0.015e^{\left(-\frac{18616}{RT_{si}}\right)}C_i^{1.1997} \quad (22)$$

2.5. Boundary condition

Fig. 1 is three-dimensional fractured carbonate rock acidizing model schematic. At the inlet, a constant acid injection flow rate is maintained. The right boundary is set as a zero-pressure boundary, while all other boundaries are zero-flux boundaries. The initial pressure of the porous medium is 0 Pa; L is the length of the model, m; W is the width of the model, m; H is the height of the model, m. At the inlet, acid concentration is kept at a fixed value, while all other boundaries are zero-flux boundaries. The initial acid concentration of the porous medium is 0 kmol/m^3. At the inlet, acid temperature is kept at a fixed value, while all other boundaries are zero-flux boundaries. The rock temperature field boundary is also a zero-flux boundary.

 https://doi.org/10.1051/e3sconf/202563901012

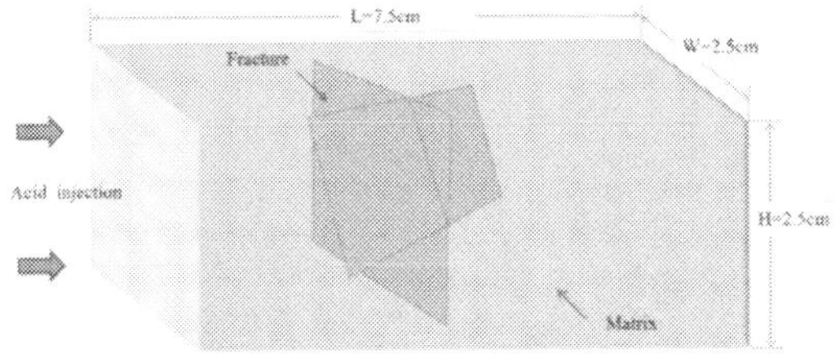

Fig. 1. Schematic diagram of the model.

3.Numerical simulation results

This section investigates the impact of acid injection rate, concentration, and temperature on wormhole morphology by simulating the three-dimensional acidizing process of fractured carbonate rocks. Table 1 provides a detailed list of the primary simulation parameters used in this study. During model initialization, the matrix grid is divided into 90×30×30 cells, with 75 fractures randomly generated and further subdivided into 5651 individual grid cells. The entire model thus comprises 86,651 grid cells, as illustrated in Fig. 4.

Table 1. Simulation Parameters.

Parameters	Symbol	Values
Core length	L	0.075m
Core width	W	0.025m
Core height	H	0.025m
Acid injection concentration	C_{inj}	4.4178kmol/m^3(15wt%HCl)
Flow rate of acid injection	Q_{inj}	1–50ml/min
initial values of matrix porosity	ϕ_{m0}	0.2, (0.05-0.35)
initial values of the fracture plane porosity	ϕ_{f0}	0.8
initial values of matrix permeability	K_{m0}	1.0×10^{-14} m^2
initial values of the fracture plane permeability	K_{f0}	2.5×10^{-13} m^2
the temperature of the acid injected	T_{inj}	293.15 K
initial values of the temperature	T_{sm0}	293.15K
initial values of matrix interfacial area	a_{vm0}	5000m^2/m^3
initial values of the fracture plane interfacial area	a_{vf0}	5000m^2/m^3
Rock density	ρ_s	2710kg/m^3
acid density	ρ_l	1080kg/m^3
Molecular diffusion coefficient	D_m	3.6×10^{-9}m^2/s
Pore-broadening parameter	β	1
Constant	α_{os}	0.5
Asymptotic Sherwood number	Sh_∞	3.66
Fluid viscosity	μ	0.001Pa·s
Solubility of acid	a	50kg/kmol
heat capacity of acid	C_{pl}	4,180J/(kg·K)
heat capacity of rock	C_{ps}	999J/(kg·K)
thermal conductivity of acid	k_{el}	0.6508J/(kg·s·K)
thermal conductivity of rock	k_{es}	5.2J/(kg·s·K)

3.1. Discussion on wormhole propagation morphology in fractured carbonate rocks during acidizing under different acid injection rates

This section primarily discusses the development of wormholes under different acid injection rates. In Fig. 3, the acid injection rate corresponding to the at the lowest point of the curve is called the optimal injection rate, which is particularly important in the field acidizing construction. The optimal injection rate is 5 ml/min. Fig. 4 illustrates the evolution of wormholes at an injection rate of 5 ml/min for various pore volumes (PV) of injected acid. It can be observed that at PV = 0.1, most of the acid remains near the injection point, dissolving the matrix around the injection boundary, with the wormhole extending only about 0.01 m. Observing the dimensionless pressure curve (Fig. 2), it is noted that when the wormhole connects with fractures, the injection pressure rapidly decreases. The acid penetrates through natural fractures and leaks along their edges, increasing the porosity of both the fracture and surrounding matrix. After encountering non-fractured matrix regions, the rate of decrease in the dimensionless injection pressure curve slows down, which differs from typical pressure drop curves seen in acidizing of non-fractured carbonate rocks; this pressure curve exhibits more fluctuation. As the pore volume of injected acid increases, the wormhole further extends. Due to the high permeability characteristics of natural fractures, the acid preferentially flows through low-resistance paths, as shown in Fig. 4. The development morphology of acid wormholes is significantly influenced by the distribution of natural fractures. Fig. 4 and Fig. 5 clearly demonstrate the dynamic coupling characteristics between wormhole propagation and acid concentration distribution during the acidizing process of fractured carbonate rocks. In the initial stage, the acid preferentially invades along the natural fracture network, forming localized dissolution channels at fracture intersections where acid concentrations remain high within the fractures, while only weak seepage occurs in the matrix areas. As the amount of injected acid increases, significant diversion of acid occurs at fracture intersections, with the main wormhole tip branching along secondary fractures. High concentration regions of acid gradually migrate deeper into the fractures, resulting in a tree-like extension of the dissolution zone. By the breakthrough stage, interconnected fracture networks dominate the acid migration pathways, forming pervasive preferential dissolution channels. Acid becomes highly concentrated in the main fractures, while dissolution in the matrix area almost ceases due to concentration depletion.

This process reveals the strong control of fracture structures on wormhole morphology: fractures not only serve as preferred pathways for acid migration but also regulate acid diversion and concentration gradients through their spatial distribution (density and connectivity), driving wormhole development from localized dissolution to directed expansion, ultimately determining the final wormhole morphology achieved through acidizing treatments.

Fig. 6 clearly illustrates the dynamic regulation mechanism of acid injection rate on fracture transport characteristics and dissolution patterns: At low injection rates, the acid flow rate is below the threshold of fracture conductivity, prolonging the residence time of acid within the fractures, which leads to conical dissolution along the fracture walls, making the fractures themselves the primary objects of dissolution. In contrast, at high

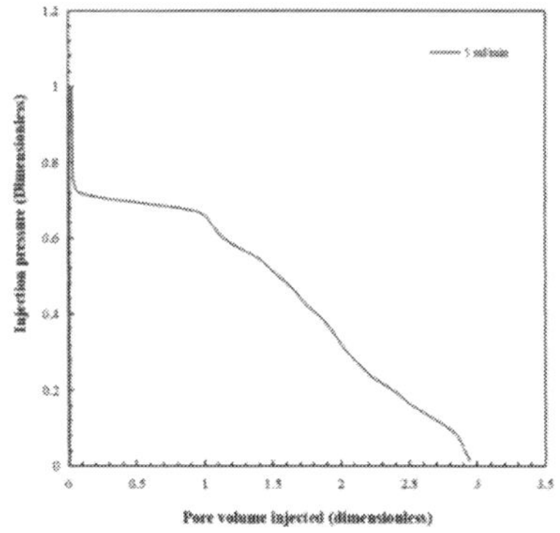

Fig. 2. Dimensionless injection pressure change curve.

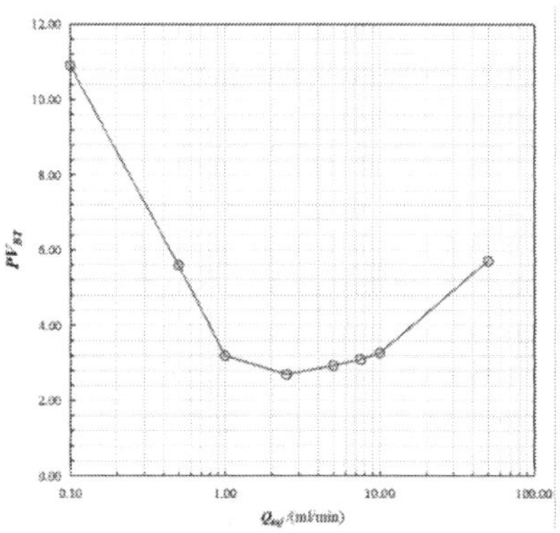

Fig. 3. wormhole breakthrough curve.

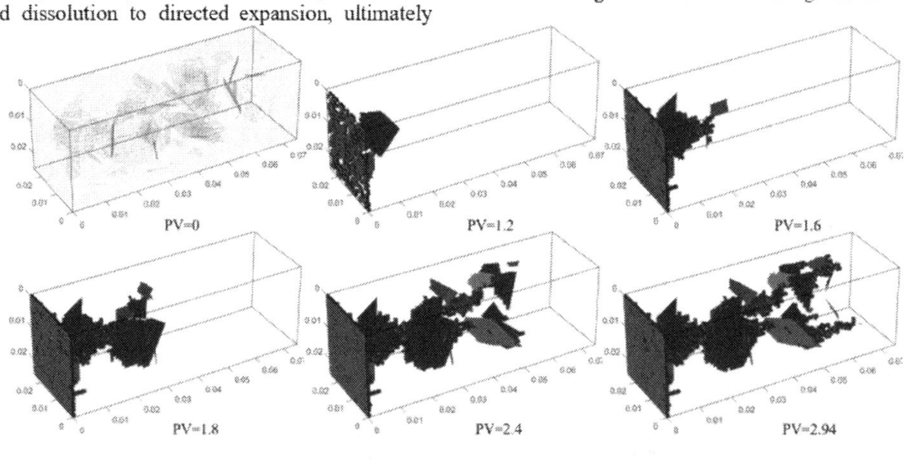

Fig. 4. Different injection pore volume wormhole propagation.

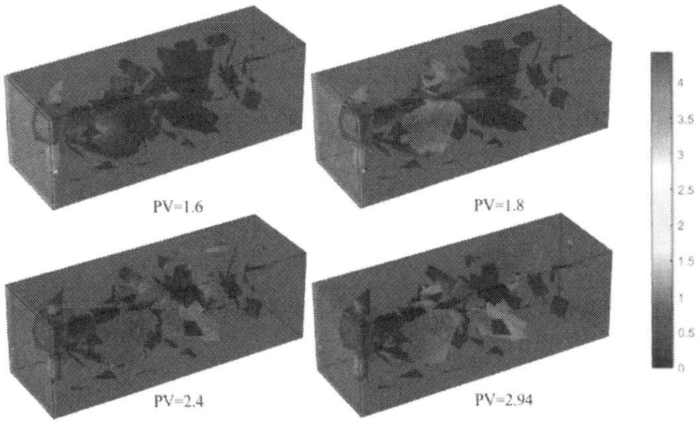

Fig. 5. Different injection pore volume acid concentration distribution.

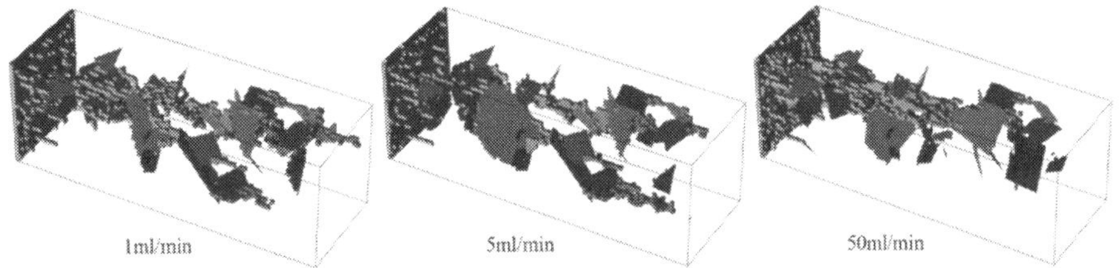

Fig. 6. Different injection rate wormhole propagation.

injection rates, inertial flow dominates, exceeding the fracture transport threshold, resulting in divergent dissolution of the matrix. Under these two extreme conditions, the conductive potential of fractures is not effectively utilized, and the dissolution effect of acid on the fractures overshadows their natural transport advantages. By comparison, near the optimal injection rate, the acid flow rate precisely matches the fracture conductivity, allowing acid to be efficiently transported along the fracture network. Acid leaks from the edges of fractures rather than directly dissolving or traversing them, keeping the fracture surfaces smooth. Furthermore, by controlling the direction of acid diversion and concentration gradients, the fractures significantly enhances control over the branching angles and extension paths of wormholes.

3.2. Discussion on wormhole propagation morphology in fractured carbonate rocks during acidizing under different acid concentrations

Fig. 7 presents the wormhole morphology under different acid concentrations at an injection rate of 5 ml/min. The wormhole patterns formed under the three acid concentrations are generally similar. The wormhole volumes (defined as regions with porosity greater than 0.9) resulting from acidizing with 10 wt%, 15 wt%, and 20 wt% HCl are 1.0349 cm³, 1.0241 cm³, and 1.0117 cm³, respectively. However, the corresponding breakthrough pore volumes (PV_{BT}) differ significantly, being 4.39, 2.94, and 2.14, respectively. As the acid concentration increases, the PV_{BT} decreases, but the rate of reduction diminishes. The overall development patterns of the wormholes are comparable, primarily governed by the distribution of high-permeability pores and fractures. Differences in acid concentration mainly influence local features of the wormholes, such as the main channel diameter and branch density. In contrast, the macroscopic framework of the overall morphology remains predominantly controlled by the fracture structure. This observation suggests that, during the acidizing of fractured carbonate reservoirs, the porosity distribution and natural fracture system serve as the "basic skeleton" for wormhole propagation, while parameters such as acid concentration act to "carve details" on this skeleton by modulating dissolution efficiency.

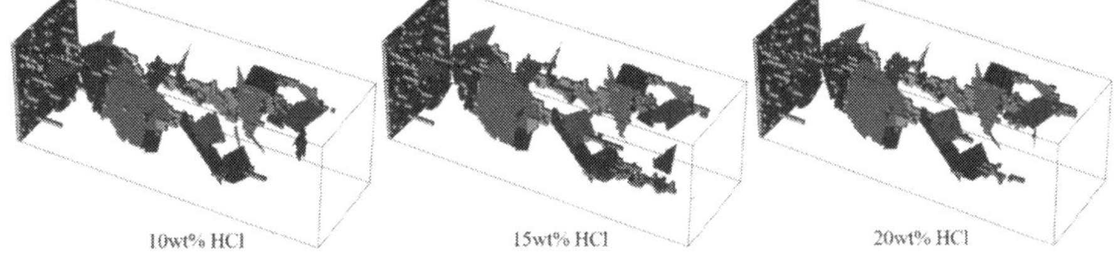

Fig. 7. Different injection concentration wormhole propagation.

3.3. Discussion on wormhole propagation morphology in fractured carbonate rocks during acidizing under different acid temperatures

Fig. 8 illustrates the wormhole propagation morphology and the post-breakthrough temperature distribution within the rock at an injection rate of 5 ml/min under different acid injection temperatures. The initial acid injection temperature was set to 293.15 K, and the study examined the characteristics of temperature changes on the rock surface and interior due to the molar reaction heat between acid and rock. Results show that the highest temperature point always occurs at the tip of the wormhole. As more acid is injected, new reaction fronts continuously form, releasing heat through reactions between the acid and the rock surface, causing the rock temperature to rise via solid-phase thermal conduction. At an injection rate of 5 ml/min, the maximum temperature increase induced by the molar reaction heat between acid and rock can reach up to 7.00 K. Previous studies that did not consider the influence of natural fractures showed relatively continuous reaction heat distributions[27]. However, in contrast, our model accounts for how variations in pore radius and local convective velocity over time and space affect the Reynolds number, Prandtl number, and heat transfer coefficients. These parameters exhibit more significant changes, particularly in fracture zones and their vicinity, leading to non-uniform heat distribution patterns. This indicates that the presence of fractures significantly increases the complexity of the reaction heat distribution. Nonetheless, heat generally diffuses outward from the wormholes. For the three different acid injection temperatures, the average rock temperatures after wormhole breakthrough were 294.76 K, 298.75 K, and 302.55 K, with corresponding PV_{BT} values of 2.94, 2.88, and 2.83, respectively. The results indicate that as temperature increases, there is a slight reduction in acid consumption, but the impact on wormhole morphology is minimal.

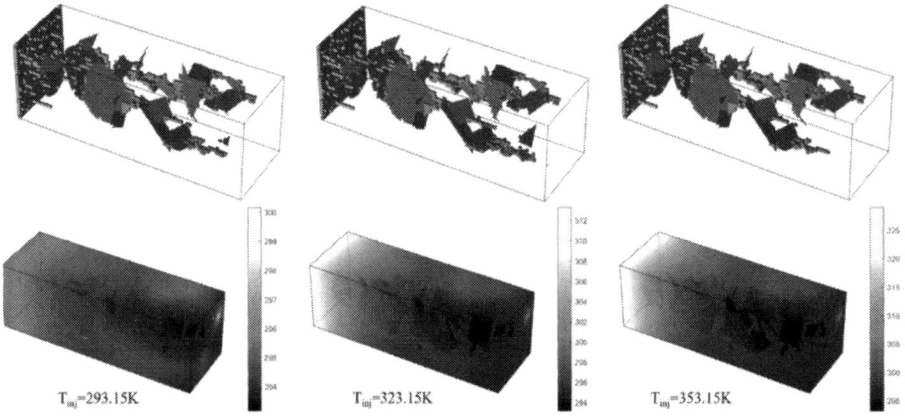

Fig. 8. Different injection temperature wormhole propagation and rock temperature distribution after wormhole breakthrough.

4. Main conclusions

Based on the developed non-isothermal three-dimensional acidizing model for fractured carbonate rocks, this study investigated the effects of acid injection rate, concentration, and temperature on wormhole morphology. The main conclusions are as follows:

There exists an optimal range of acid injection rates where the acid flow rate matches the fracture conductivity most effectively, achieving maximum acidizing efficiency with minimal acid consumption. Both excessively high and low injection rates result in suboptimal acidizing performance.

Although the wormhole volumes formed under different acid concentrations are similar, the pore volume of acid required for breakthrough decreases with increasing acid concentration. Fracture distribution determines the basic framework of wormhole propagation, while acid concentration primarily influences local features such as the diameter of the main channel and branch density.

Increasing temperature slightly reduces acid consumption but has a minor impact on wormhole morphology. The presence of fractures increases the complexity of reaction heat distribution; However, heat generally diffuses outward from the wormholes.

Fractures not only serve as preferred pathways for acid migration but also regulate acid diversion and concentration gradients through their spatial distribution. This drives wormhole propagation from localized dissolution to directed expansion, ultimately determining the final wormhole morphology achieved through acidizing treatments.

Acknowledgements

This study was supported by the National Natural Science Foundation of China (No.52174033; No. U23B20156).

References

1. Burchette TP. Carbonate rocks and petroleum reservoirs: a geological perspective from the industry. 2012.
2. Shepherd M. Oil field production geology: AAPG

Memoir 91: AAPG; 2009.

3. McLeod HOJJoPT. Matrix acidizing. 1984;36:2055-69.
4. Chang FF, Nasr-El-Din HA, Lindvig T, Qiu X. Matrix acidizing of carbonate reservoirs using organic acids and mixture of HCl and organic acids. SPE Annual Technical Conference and Exhibition?: SPE; 2008. p. SPE-116601-MS.
5. Yang G, Wu X, Hou J, Zhou F, Nie F. Study on the Influence of Heterogeneity of Low-Permeability Reservoirs on Wormhole Morphology and Acidizing Process Parameters %J Processes. 2024;12:2740-.
6. Yang J, Cheng Y, Han S, Han Z, Yang H, Zhang Z, et al. Numerical simulation of the acidification of two mineral components of carbonate rocks considering the temperature effect %J Geoenergy Science and Engineering. 2024;243:213360-.
7. Ma G, Chen Y, Jin Y, Wang H. Modelling temperature-influenced acidizing process in fractured carbonate rocks %J International Journal of Rock Mechanics and Mining Sciences. 2018;105:73-84.
8. Piyang L, Chaoping H, Lijing J, Weijing J, Zhao Z, Kai Z. Numerical Simulation of the Wormhole Propagation in Fractured Carbonate Rocks during Acidization Using a Thermal-Hydrologic-Mechanics-Chemical Coupled Model %J Water. 2022;14:4117-.
9. Cunqi J, Kamy S, Zhaoqin H, Haiyang Z, Jun Y. Numerical studies and analysis on reactive flow in carbonate matrix acidizing %J Journal of Petroleum Science and Engineering. 2021;201.
10. Mou J, Gao B, He J, Zhang R, Zhang Y, Zhang L. Wormholing behavior in matrix acidizing in vuggy, naturally fractured carbonate reservoirs %J Geoenergy Science and Engineering. 2024;235:212732-.
11. Ali MA, Behnam S, Alireza EK. Fracture and vug effects on wormhole pattern during acidizing of triple porosity carbonate rocks %J Geoenergy Science and Engineering. 2024;232.
12. Cunqi J, Kamy S, Haiyang Z, Jun Y. Numerical Studies and Analyses on the Acidizing Process in Vug Carbonate Rocks %J Frontiers in Earth Science. 2021;9.
13. Cunqi J, Tao H, Jun Y, Hongchuan X, Haiyang Z. Effect of Isolated Fracture on the Carbonate Acidizing Process %J Frontiers in Earth Science. 2021;9.
14. Rodrigues NJ, Santos LCRd, da SDNN, de CJST, Palharini SM, Pandava APT. Assessing the impact of oil saturation on wormhole morphology in carbonate acidizing %J Fuel. 2024;358.
15. Hyunsang Y, Tan N, Jeonghwan L. Improved wormhole prediction model considering propagation characteristics of wormhole head in carbonate acidizing %J Journal of Petroleum Science and Engineering. 2022;216.
16. Masoud K, Shahab A. A new insight into wormhole propagation prediction in carbonate matrix acidizing based on experimental approaches %J Gas Science and Engineering. 2023;116.
17. Ali M, Ziauddin M. Carbonate acidizing: A mechanistic model for wormhole growth in linear and radial flow %J Journal of Petroleum Science and Engineering. 2020;186:106776-.
18. Gomez CO, Maysam P. Matrix Acidizing in Carbonate Formations %J Processes. 2022;10:174-.
19. Zhu T, Zhao B, Zhang Z. Quasi-3D fracture acidizing simulation based on discrete virtual internal bond method %J Computers and Geotechnics. 2024;173:106537-.
20. Lei W, Jianye M, Siqi M, Bing Z, Zhiyuan L, Xuxin T. Modeling matrix acidizing in naturally fractured carbonate reservoirs %J Journal of Petroleum Science and Engineering. 2019;186:106685-.
21. Liu P, Yao J, Couples GD, Ma J, Huang Z, Sun H. Modeling and simulation of wormhole formation during acidization of fractured carbonate rocks %J Journal of Petroleum Science and Engineering. 2017;154:284-301.
22. Liu M, Zhang S, Mou J, Zhou F. Wormhole Propagation Behavior Under Reservoir Condition in Carbonate Acidizing %J Transport in Porous Media. 2013;96:203-20.
23. Guowei M, Yun C, Huidong W, Tuo L, Wen N. Numerical analysis of two-phase acidizing in fractured carbonate rocks %J Journal of Natural Gas Science and Engineering. 2022;103.
24. K. F, T. A, T. W, K. Y. Phase-field modeling of wormhole formation and growth in carbonate matrix acidizing %J Journal of Petroleum Science and Engineering. 2022;209.
25. Green DWJH. Perry's chemical engineers'. 2008.
26. Liu P, Xue H, Zhao L, Zhao X, Cui M. Simulation of 3D multi-scale wormhole propagation in carbonates considering correlation spatial distribution of petrophysical properties. Journal of Natural Gas Science and Engineering. 2016;32:81-94.
27. Xue H, Huang Z, Zhao L, Wang H, Kang B, Liu P, et al. Influence of acid-rock reaction heat and heat transmission on wormholing in carbonate rock. 2018;50:189-204.

Ring-Opening Copolymerization of ε-Caprolactone and L-Lactide Catalyzed by Aluminum Complexes

Mengxue Zhang[1], Sen Xiang[1], Shaofeng Liu[1,2ba*], Ze Kan[1,2,b*]

[1] Key Laboratory of Biobased Polymer Materials, College of Polymer Science and Engineering, Qingdao University of Science and Technology, Qingdao, 266042, China
[2] Shandong Key Laboratory of High Performance Polyolefin Materials and Recycling, Qingdao University of Science and Technology, Qingdao, 266042, China

Abstract. In recent years, poly(ε-caprolactone) (PCL) and poly (lactic acid) (PLA) have been widely adopted as biodegradable materials, with applications ranging from simple packaging to sophisticated biomedical devices. Although PLA possesses notable advantages such as biocompatibility and high mechanical strength, it suffers from several limitations including brittleness, poor elasticity, low thermal stability, and inadequate permeability. In this study, we synthesized a series of aluminum complexes (Al1−Al6) bearing different substituent structures. These metal complexes were thoroughly characterized by nuclear magnetic resonance (NMR) spectroscopy, elemental analysis, and single-crystal X-ray diffraction. Subsequently, these aluminum complexes (Al1−Al6) were employed as catalysts for the ring-opening polymerization (ROP) of L-lactide (L-LA), followed by the ring-opening copolymerization (ROCOP) of L-LA with ε-caprolactone (ε-CL). Through these polymerization processes, we successfully synthesized a series of high-performance copolymers with excellent properties.

1. Introduction

Poly(ε-caprolactone) (PCL), as a highly important polymer, exhibits outstanding mechanical properties, excellent miscibility with other polymers, and biodegradability[1]. These characteristics have led to its widespread applications in drug delivery systems[2] (particularly for contraceptive agents), adhesives in microelectronics[3], and packaging materials[4]. Structurally, PCL is an aliphatic polyester composed of repeating hexanoate units. It is a biodegradable, low-melting ($T_m \approx 60$ °C), semi-crystalline polymer[5] with a crystallinity of up to 69%. In recent years, poly(lactic acid) (PLA) has attracted significant attention in both academic and industrial research[6-8]. Due to the presence of methyl side groups, PLA exhibits excellent hydrophobicity, enabling its use in diverse applications. For instance, its combination of favorable physical properties and biodegradability has led to widespread adoption in disposable packaging and commercial products. Additionally, PLA serves as a biodegradable implant material for bone fracture fixation and joint stabilization[6]. Consequently, combining these monomers through copolymerization confers excellent performance characteristics to the polymer. Metal complexes are widely employed as catalysts or initiators in the ring-opening polymerization (ROP) of polyesters. Commonly used metals in these complexes include titanium, zirconium, aluminum, magnesium, zinc, tin, and rare-earth metals. Recently, aluminum complexes have achieved notable success in polyester ROP due to their high catalytic activity, low toxicity, and excellent control over polymer molecular weight (M_w), molecular weight distribution ($Đ$), and regio-/stereoselectivity. The performance of aluminum-based catalytic systems can be finely tuned by modifying the electronic and steric properties of the ligands, making ligand design a critical factor in optimizing these catalysts.

In this study, a series of aluminum complexes bearing amino-quinoline ligands have been synthesized and characterized through ^{1}H NMR spectroscopy. Subscequently, ring-opening homopolymerizations of L-LA and copolymerizations of L-LA with ε-CL were conducted using these aluminum complexes (Al1−Al6) in combination with BnOH. The polymerization results demonstrate that these complexes not only exhibit living polymerization characteristics in L-LA homopolymerization, but also produce copolymers with tunable properties in the L-LA/ε-CL copolymerization. This highlights the effectiveness and versatility of the catalytic system in regulating copolymer microstructure and associated thermal properties.

2. Experimental section

2.1. General procedure for ring-opening copolymerization of L-LA and ε-CL

The polymerization was performed in tubes connected

* Corresponding author
[a]shaofengliu@qust.edu.cn; [b]zkan@qust.edu.cn

through a double-manifold system under inert atmosphere. Schlenk tube 1 was charged with Al1 (0.020 mmol) dissolved in anhydrous toluene (2 mL), while Schlenk tube 2 was loaded with BnOH (0.020 mmol), L-LA (2.0 mmol), and ε-CL (2.0 mmol). After connecting the twin Schlenk tubes to the preheated double manifold at the polymerization temperature, the aluminum catalyst solution in tube 1 was rapidly transferred via cannula into tube 2 containing the initiator/monomer mixture to initiate the polymerization. The reaction was allowed to proceed for a predetermined time (4-12 hours) before being quenched with acetic acid. A small aliquot of the quenched reaction mixture was dissolved in $CDCl_3$ for ^1H NMR analysis to determine monomer conversion. The remaining solution was poured into cold methanol (100 mL) to obtain polymer.

2.2. Calculations for L-LA and ε-CL copolymerization

A small aliquot of the quenched polymerization solution was dissolved in $CDCl_3$ for ^1H NMR characterization to determine monomer conversion. The conversion rate was calculated by dividing the remaining monomer amount by the initial monomer quantity. The M_w and $Đ$ of polymer were determined by gel permeation chromatography (GPC) using THF as the mobile phase at 40 °C. The theoretical molecular weight (Mn, cal) of the copolymers was calculated using the equation: Mn, cal = MLA × ([LA]:[BnOH]) × conversion + MCL × ([CL]:[BnOH]) × conversion + MBnOH, where MLA and MCL represent the molecular weights of the LA and CL repeating units, respectively, and MBnOH denotes the molecular weight of the initiator BnOH.

3. Result and discussion

3.1 Synthesis and Characterization of Ligands and Complexes

The ligands L1–L6 were synthesized via Pd-catalyzed coupling reactions between 8-bromo-2-(2,6-R^1-C_6H_2-N=C-4-R^2)-quinoline and 1 equivalent of R^3-SH under standard cross-coupling conditions, according to procedures reported previously (Fig. 1)[9-11]. Their ^1H NMR spectrum were shown in Fig. 2.

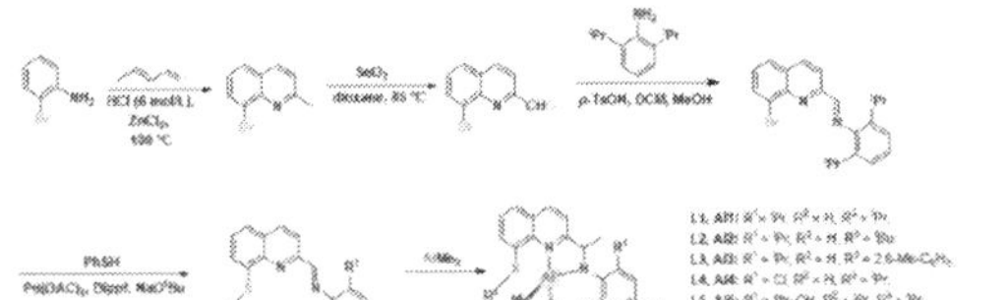

Fig. 1. Synthesis routes of the ligands and metal complexes.

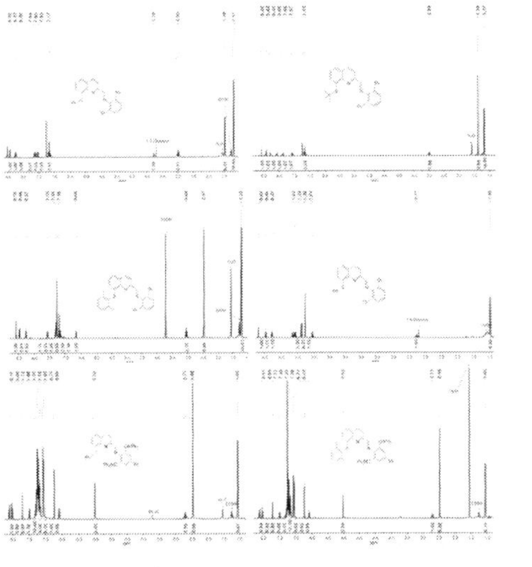

Fig. 2. ^1H NMR spectrum of **L1**-**L6**.

The metal complexes were synthesized according to previously reported procedures by reacting the ligands with 1.1 equivalents of trimethylaluminum ($AlMe_3$) (Fig. 1)[12]. Their ^1H and ^{13}C NMR spectrum were shown in Fig. 3.

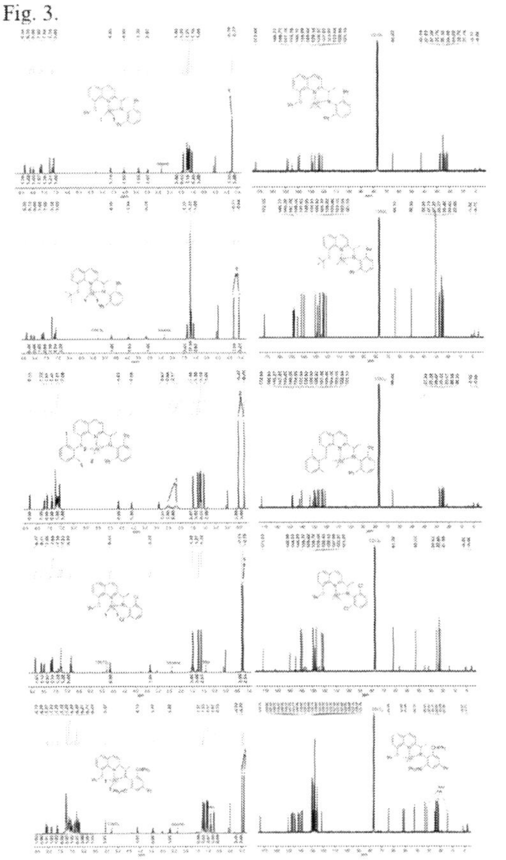

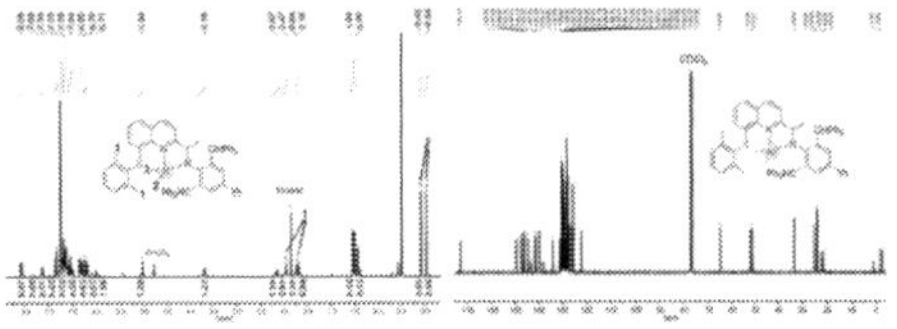

Fig. 3. ^{1}H NMR and ^{13}C spectrum of **Al1-Al6**.

3.2. Ring-opening polymerization of L-LA

A series of L-LA ring-opening polymerizations (runs 1-6; Table 1) were conducted using complexes Al1-Al6 as catalysts. Under the conditions of $[L\text{-}LA]_0$: $[BnOH]_0$:$[Al]_0$ = 100:1:1, all polymerizations were quenched after 4 hours. ^{1}H NMR analysis revealed high conversions (>88%) in all cases, with the resulting PLAs exhibiting narrow molecular weight distributions. However, GPC-measured molecular weights were consistently slightly lower than theoretical values (runs 1-6, Table 1), likely due to intramolecular transesterification occurring during prolonged L-LA conversion. As shown in Fig. 4 (a), the GPC traces of PLA displayed monomodal distributions.

Table 1. Ring-opening polymerization of L-LA using Al1−Al6/BnOH [a].

run	cat.	conv.(L-LA) (%)	M_n GPC × 10^{-4}	M_n (cal.) × 10^{-4}	$Đ$
1	Al1	89	0.81	1.29	1.22
2	Al2	89	0.92	1.29	1.13
3	Al3	96	0.81	1.39	1.13
4	Al4	96	0.79	1.39	1.31
5	Al5	93	0.89	1.35	1.19
6	Al6	93	0.85	1.33	1.17

[a] conditions: 20 μmol of Al1−Al6 as catalysts, BnOH as initiator, [L-LA] = 1 M (mol/L) in toluene at 100 °C, with $[LA]_0$:$[Al]_0$:$[I]_0$ = 100:1:1.

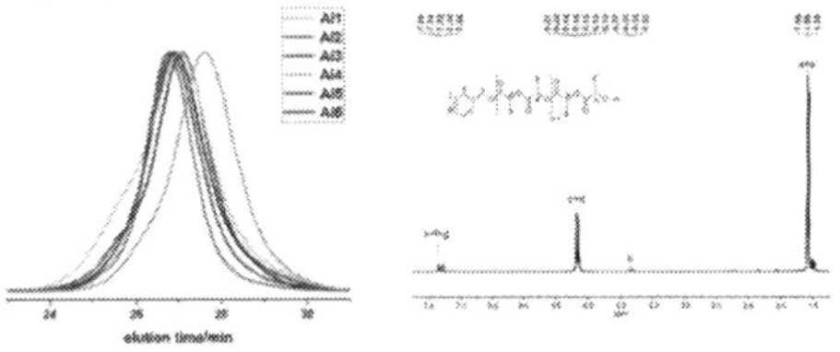

Fig. 4. (a) GPC curves of PLA obtained by Al1−Al6 with $[L\text{-}LA]_0$: $[BnOH]_0$: $[Al]_0$ = 100:1:1 (Runs 1−6, Table 1); (b) ^{1}H-NMR spectrum of PLA in $CDCl_3$ obtained by reaction of L-LA and Al3 in ratio [monomer]: [Cat.]: [BnOH] = 100: 1: 1 at 100 °C in toluene.

The ring-opening polymerization of L-LA was conducted in toluene at 100 °C using Al3 as the catalyst with a molar ratio of $[L\text{-}LA]_0$:$[BnOH]_0$:$[Al]_0$ = 100:1:1, yielding low-molecular-weight PLA oligomers. End-group analysis of the resulting poly(lactic acid) was performed by ^{1}H NMR spectroscopy. The ^{1}H NMR spectrum (Fig. 4 (b)) confirmed that the PLA chains were terminated with BnO-($PhCH_2O$) groups, demonstrating BnOH effectively initiated the ring-opening polymerization. To further verify the end-group structure and intramolecular architecture, MALDI-TOF mass spectrometry was conducted (Fig. 5, run 3 in Table 1). The mass spectrum displayed a series of ion peaks spaced at intervals of $\Delta m/z$ = 72, corresponding to the repeating unit mass. The observed pattern followed the equation: M/z = 72n + 108.14 (BnOH) + 39.0 (K+) g·mol^{-1}, which verified the linear PLA structure with BnOH end groups. Notably, the mass difference between adjacent peaks (72 g·mol^{-1}) represents exactly half the molecular weight of the lactide monomer (144 g·mol^{-1}), providing direct evidence for intramolecular transesterification during polymerization. This mass spectral data unambiguously confirmed the proposed PLA structure and substantiated the occurrence of transesterification. Despite the competing transesterification side reaction, the Al1-Al6 catalyzed ring-opening polymerization of L-LA maintained characteristic living polymerization features, as evidenced by the linear relationship between M_w and $Đ$ (< 1.15).

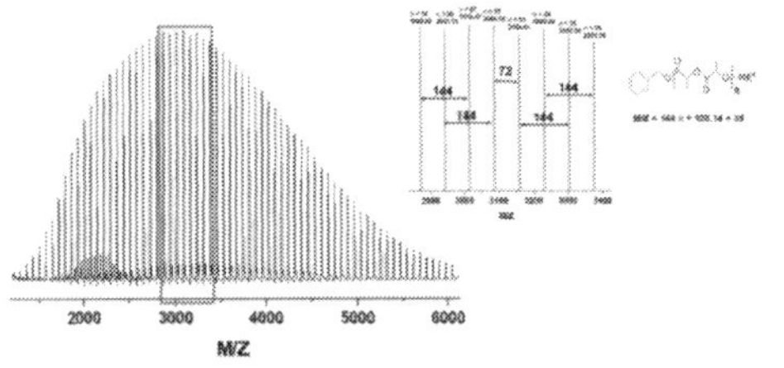

Fig. 5. MALDI-TOF mass spectra of PLA prepared with $[L\text{-}LA]_0$: $[Al]_0$: $[BnOH]_0$ = 100:1:1 (Run 3, Table 1).

3.3. Influence of reaction parameters on catalytic behavior

The ring-opening copolymerization of ε-CL and L-LA was performed in toluene at 100 °C using aluminum complexes Al1−Al6 as catalysts with 1 equivalent of BnOH as initiator. The resulting copolymers were thoroughly characterized by ^{1}H NMR spectroscopy, GPC and DSC.

Table 2. Ring-opening copolymerization of ε-CL and L-LA using Al1−Al6/BnOH [a]

run	cat.	$[\varepsilon\text{-CL}]_0$: $[L\text{-}LA]_0$: $[Al]_0$: $[BnOH]_0$	conv. (ε-CL, L-LA) (%)	M_n GPC × 10^{-4}	M_n (cal.) × 10^{-4}	$Đ$
1	Al1	100:100:1:1	90, > 99	2.90	2.47	1.67
2	Al2	100:100:1:1	74, > 99	2.55	2.29	1.52
3	Al3	100:100:1:1	36, > 99	1.38	1.86	1.14
4	Al4	100:100:1:1	90, > 99	2.13	2.47	1.38
5	Al5	100:100:1:1	84, > 99	3.04	2.40	1.45
6	Al6	100:100:1:1	20, 92	1.01	1.55	1.09
7	Al5	100:100:1:5	56, > 99	2.25	2.08	1.36
8	Al5	50:150:1:1	80, 98	2.31	2.33	1.60
9	Al5	150:50:1:1	93, > 99	2.94	2.50	1.67

[a] conditions: 20 μmol Al1−Al6 as catalysts, BnOH as initiator, [ε-CL] = [L-LA] = 1 M in toluene at 100 °C.

The copolymerization results catalyzed by Al3 and Al6 demonstrated complete consumption of L-LA but

 https://doi.org/10.1051/e3sconf/202563901013

significant residual ε-CL (conversion <40%, runs 3 and 6 in Table 2), with GPC analysis (Fig. 6 (a)) revealing monomodal molecular weight distributions that were notably narrower than those obtained with other aluminum complexes. In contrast, Al1, Al2, Al4, and Al5 catalysts achieved full L-LA consumption within 12 hours, with Al1 and Al4-catalyzed systems exhibiting particularly high ε-CL conversions exceeding 90% (runs 1 and 4, Table 2). Among these, the chloro-substituted Al4 (R^3 = Cl) showed optimal performance, achieving nearly equivalent consumption of ε-CL and L-LA while maintaining excellent control over polymer characteristics, as evidenced by the well-controlled molecular weight (Mn = 2.13 × 10^4 Da) and narrow dispersity (*Đ* = 1.38, run 4 in Table 2). Al6 exhibited superior controllability in the ring-opening copolymerization process, while the aluminum complexes Al1, Al2, Al4, and Al5 demonstrated relatively poorer control with a higher propensity for transesterification side reactions. Although transesterification typically introduces undesirable side reactions in polymerization, the occurrence of transesterification during L-LA ring-openning polymerization has simultaneously opened new pathways for the copolymerization of L-LA with ε-CL, providing an alternative approach to tailor copolymer microstructures.

To further investigate the influence of reaction conditions on the aluminum-catalyzed ring-opening copolymerization of cyclic esters, systematic studies were conducted. When employing excess benzyl alcohol, the resulting polymers exhibited better-controlled molecular weights and narrower dispersities compared to those obtained with 1 equivalent of BnOH (run 7, Table 2), suggesting that increased initiator loading suppresses transesterification side reactions. Subsequent variation of monomer feed ratios (runs 8 and 9, Table 2) demonstrated tunable polymer properties, where DSC analysis revealed composition-dependent thermal transitions. Notably, a distinct melting endotherm emerged at CL:LA = 150:50 (Fig. 6 (b)), while all compositions showed measurable differences in glass transition temperatures.

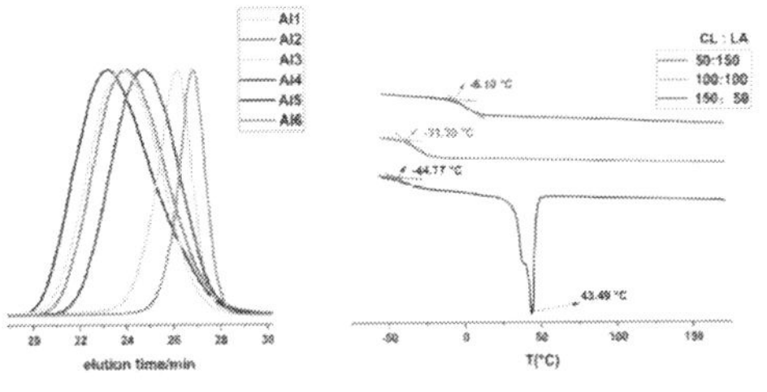

Fig. 6. (a) GPC curves of tapered copolymer P(LA-grad-CL)-b-PCL L obtained by Al1–Al6 with $[CL]_0$: $[L\text{-}LA]_0$: $[BnOH]_0$: [Al]0 = 100: 100: 1: 1 (runs 1–6, Table 2); (b) DSC curve of monomers with different ratios tapered PLA-grad-PCL copolymer was prepared by Al5 (runs 5 and 7–8, Table 2).

Comparative ^{1}H NMR analysis (Fig. 7) of homopolymers and copolymers obtained from Al4-catalyzed ring-opening polymerization of L-LA and ε-CL revealed characteristic proton peaks corresponding to both PLA/PCL block segments and P(LA-stat-CL) random sequences, unambiguously confirming the gradient copolymer (P(LA-grad-CL)) structure. The polymerization mechanism involves three distinct stages: initial selective L-LA consumption forming PLA blocks, subsequent L-LA/ε-CL copolymerization as L-LA concentration decreases, and final ε-CL homopolymerization after complete L-LA depletion. This multistage process, achieved through one-pot ring-opening copolymerization catalyzed by aluminum complex Al4 (runs 4, Table 2), demonstrates precise control over monomer incorporation kinetics to produce gradient architectures with spatially varying composition.

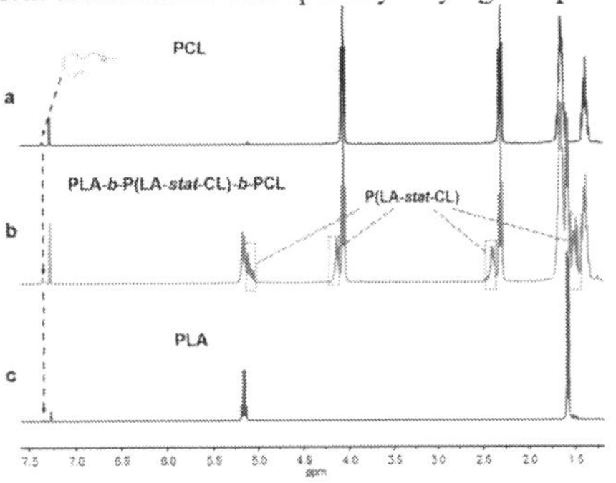

Fig. 7. ^{1}HNMR spectra of (a) PCL; (b) P(LA-grad-CL)-b-PCL tapered copolymer; (c) PLA by Al4 ($CDCl_3$, 400 MHz, rt.) (run4, Table 2)

4. Conclusion

(1) A series of ring-opening homopolymerizations of L-LA and copolymerizations of L-LA with ε-CL were conducted using semistable quinoline-aluminum complexes Al1-Al6 in combination with BnOH as the initiating system.

(2) Studies of the L-LA ring-opening polymerization mechanism revealed the occurrence of intramolecular transesterification to a certain extent.

(3) Through a one-pot reaction protocol, gradient copolymers P(LA-grad-CL) were successfully synthesized.

(4) While transesterification introduced some undesirable side reactions, it simultaneously yielded unexpected amorphous polymer products.

References

1. Tiong S. Houthakker, H.S., Williamson. P.J. N. The Economics of Financial Markets. Am. Actuar. J. **3**, 150 (1993).
2. Banerjee A, Chatterjee K, Madras G. Mater. Enzymatic Degradation of Polycaprolactone–Gelatin Blend. Res. Express **2**, 4 (2015).
3. Ikada Y, Tsuji H. Macromol. Biodegradable Polyesters for Medical and Ecological Applications. Rapid Commun. **21**, 117 (2000).

4. Couladouros E A, Strongilos A T, Papageorgiou V P. Cover Picture: Chem. Eur. J. 8/2002. Chem. Eur. J. **8**,1757 (2002).
5. Cha Y, Jarrett-Wilkins C, Rahman M A. Crystallization-Driven Self-Assembly of Metallo-Polyelectrolyte Block Copolymers with a Polycaprolactone Core-Forming Segment. ACS Macro Lett. **8**, 835 (2019).
6. Dechy-Cabaret O, Martin-Vaca B, Bourissou D. Controlled Ring-Opening Polymerization of Lactide and Glycolide. Chem. Rev. **104**, 6147 (2004).
7. Dove A P. Controlled Ring-opening Polymerisation of Cyclic Esters: Polymer Blocks in Self-assembled Nanostructures. Chem. Commun. **48**, 6446 (2008).
8. Horváth I T, Anastas P T. Innovations and Green Chemistry. Chem. Rev. **107**, 2169 (2007).
9. Wang. D Q., Zhou. S M., Liu. Y X, Kang. X, Liu S F, Li. Z B. Lactide Lactone Chain Shuttling Copolymerization Mediated by an Aminobisphenolate Supported Aluminum Complex and $Al(O^iPr)_3$: Access to New Polylactide Based Block Copolymers. Macromolecules **55**, 2433 (2022).
10. Chen. Y J, Zhou. S M, Yang. W, Liu. S F. Synthesis of Mono- and Dinuclear Aluminum Complexes Bearing Aromatic Amino-Phenolato Ligands: A Comparative Study in the Ring-Opening Polymerization of Cyclohexene Oxide. Organometallics **41**, 3724 (2022).
11. Zang. M T, Cao. W Y, Zhang. X W, Liu. S F, Li. Z B. Synthesis of Aluminum Complexes Supported by 2-(1,10-phenanthrolin-2-yl)phenolate Ligands and Their Catalysis in the Ring-opening Polymerization of Cyclic Esters. Eur. Polym. J. **193**(2023).
12. Xiang. Sen, Zhang. Jinbo, Liu. Shaofeng, Braunstein. Pierre, Li. Zhibo. Aluminum Complexes Bearing Aminoquinoline Ligands with Different Pendant Donors: Remarkable Impact of Hemilability on the Ring-Opening Polymerization of ε-Caprolactone. Macromolecules, **57**, 3604 (2024).

https://doi.org/10.1051/e3sconf/202563901014

Modulation of Single-Walled Carbon Nanotube Diameter by Component Ratios in Magnesium-Aluminium Layered Oxides

Fangfang Cheng[1], Maoshuai He[2*]

[1] College of Polymer Science and Engineering, Qingdao University of Science and Technology, Qingdao, 266042, China
[2] College of Chemistry and Molecular Engineering, Qingdao University of Science and Technology, Qingdao 266042, China

Abstract. The diameter of single-walled carbon nanotubes (SWNTs) has a decisive influence on their electronic structure and optoelectronic properties. This study innovatively explored the mechanism of regulating the diameter of SWNTs by the Mg/Al ratio in magnesium-aluminium layered double oxide (LDOs) catalysts. Fe/Co/Mg/Al-LDOs catalysts with different Mg/Al ratios were synthesized by co-precipitation, and the effect of the Mg/Al ratio on the diameter of SWNTs grown by catalytic growth at 700°C was investigated. The experimental results showed that increasing the Al component is beneficial to selectively growing small-diameter SWNTs, while increasing the Mg component tends to promote the growth of larger-diameter SWNTs. Catalyst characterization indicated that the introduction of Al promoted the migration of Co cations into the support. The unreduced Co^{2+} could anchor the metal nanoparticles reduced on the support surface, effectively inhibiting the agglomeration of metal nanoparticles, thereby favoring the growth of small-diameter SWNTs. This finding provides a simple and efficient strategy for the controllable synthesis of SWNTs with specific diameters, laying a foundation for the development of high-performance carbon-based nanoelectronic devices.

1. Introduction

Single-walled carbon nanotubes (SWNTs) can be viewed as one-dimensional seamless hollow tubes formed by curling a single layer of graphene in a specific direction, with their structural characteristics precisely defined by the chirality indices (n, m).The unique sp^2 hybridized carbon lattice structure of SWNTs forms a hollow cylindrical structure, limiting the electron motion along the circumference due to periodic boundary conditions, thereby converting the two-dimensional energy bands of graphene into one-dimensional discrete sub-bands. This complex electronic structure is directly affected by structural parameters like the chirality indices. Thus, SWNTs show highly adjustable optoelectronic properties and possess vast potential for applications in high-performance transistors, energy storage and conversion, photovoltaic devices, and biological imaging[1,2]. However, the key to realizing these applications is the precise control of the SWNTs structure. For example, in photovoltaic devices and biological imaging, single-chirality SWNTs show higher photoconversion power or photoluminescence intensity due to their uniform energy level structure, making them ideal materials with low additive quantities and high performance. In field-effect transistors, uniform bandgap SWNTs are essential for achieving a consistent gate voltage response. In controlling the structure of SWNTs, diameter control is a fundamental and crucial step. On one hand, the bandgap of semiconducting SWNTs is inversely related to their diameter, thus bandgap control of SWNTs can be achieved through diameter control. On the other hand, SWNTs with a narrow diameter distribution are crucial for achieving high-purity single-chirality separation[3] and precise conductivity control.

During the growth of SWNTs, nanoparticle catalysts serve as templates for the nucleation of SWNTs, thus the particle size directly influences and determines the diameter of the SWNTs. Due to the tendency of nanoparticles to agglomerate during high-temperature reaction[4], leading to a reduction in surface active sites and deactivation, a common strategy in catalyst design is to fix nanoparticles onto suitable support surfaces, forming supported catalysts. This enhances the dispersion and stability of the nanoparticles, thereby achieving structural control over the growth of SWNTs. A typical supported catalyst is prepared by loading active metals onto supports with high specific surface areas such as mesoporous materials and oxides. This process involves a series of solution-based steps, including uniform mixing of metal precursor solutions and support suspensions, sufficient diffusion of metal ions onto the support surface, precise pH adjustment, drying, calcination, and reduction. These catalysts, through the formation of oxide solid solutions, enable the uniform dispersion of active components within the support and the formation of metal particles that interact strongly with the underlying support, which is favorable for producing uniformly sized nanoparticles, thus catalyzing the synthesis of SWNTs with uniform diameters. Layered double hydroxides

* Corresponding author: hemaoshuai@qust.edu.cn

(LDHs) are an important class of two-dimensional layered materials, composed of positively charged metal hydroxide layers, intercalated anions, and water molecules. By subjecting the LDHs precursors to high-temperature calcination, layered double oxides (LDOs) can be obtained. Due to their unique layered structure, these materials are often used as oxide solid solution-supported catalysts[5]. Compared with other supported catalysts, LDOs catalysts have a higher adsorption capacity, allowing them to effectively immobilize catalytic active species on the support surface and prevent sintering and aggregation of metal nanoparticles under high-temperature reaction conditions. Additionally, by introducing various transition metal cations into the hydrotalcite layered structure and ensuring their uniform atomic-level distribution, followed by topological structure changes during calcination and reduction, and the confinement effects of the interlayer precursors, atomically dispersed nanocatalysts with specific morphologies and surface structures can be prepared[6]. Furthermore, the tunable basicity of the LDHs precursor surface allows LDOs to act as excellent solid base catalysts. These unique properties collectively make LDOs high-quality catalysts for the catalytic growth of uniformly structured SWNTs.

LDOs catalysts, through simple operations such as adjusting the elemental composition and optimizing the calcination and reduction processes, can achieve precise control over the size distribution of metal particles, thereby demonstrating great potential for the controlled growth of SWNTs with tunable diameters. For example, the research group of Zhang[7] utilized the adjustable properties of metal types and contents in LDHs, and by adjusting the Fe content in the Fe/Mg/Al-LDHs precursor, they successfully prepared Fe metal nanoparticle catalysts with different particle sizes. As the Fe content increased, the particle size of the Fe metal nanoparticles gradually increased from less than 3 nm to approximately 4 nm, and eventually exceeded 5 nm. Furthermore, the group further discovered that the tunability of the metal nanoparticle carrier in LDOs catalysts allowed for the preparation of nanocatalysts with a bimodal distribution of surface Fe nanoparticles. Specifically, by high-temperature reduction of the Fe/Mo/Mg/Al-LDHs precursor, two phases of metal oxide and spinel were formed. These two phases, as catalyst supports, provided different growth environments for the metal nanoparticles, resulting in Fe metal nanoparticles with a bimodal size distribution on the surface of the LDHs-based catalyst. Among these, a portion of the larger Fe particles had a size of approximately 20 nm, while the majority of the smaller Fe particles were less than 3 nm. In addition, doping with other elements can further regulate the product diameter by altering the dispersion of the metal active components in the catalyst. For example, research by Li et al. found that introducing Mg and Cu elements into Ni/Al-LDHs could produce Ni metal nanoparticles modified by MgO and Cu, significantly enhancing the dispersion and catalytic stability of Ni metal particles, thereby catalyzing the growth of carbon nanotubes with smaller diameters and more uniform distributions.

Overall, research on controlling the metal particle size of LDOs catalysts has mainly focused on regulating the transition metal cations at the catalytic active centers and optimizing the calcination process, with relatively less attention given to the composition ratios of elements such as Mg and Al in the hydrotalcite structural framework. However, due to the flexible metal composition and exchangeable interlayer anions of the LDHs precursors, they can form various host-guest assemblies and nanostructures with different physical and chemical properties. As mentioned in the introduction, changes in the proportions of framework elements such as Mg and Al may influence the morphology and dispersion of the support, thereby regulating the interaction between the metal catalyst and the support, ultimately affecting the size distribution of the metal nanoparticles. Therefore, in-depth research on the impact of Mg, Al, and other framework elements on the structure of LDHs-based catalysts and the growth products of SWNTs is crucial for optimizing catalyst performance and achieving the controlled synthesis of SWNTs.

In this study, we synthesized Fe/Co/Mg/Al-LDHs catalysts with different Mg/Al ratios using the coprecipitation method, and performed comparative analysis on the carbon nanotube products catalytically grown by the calcined LDO catalysts at 700°C. Using various optical characterization techniques, we evaluated the catalysts and the carbon nanotubes they catalytically synthesized, systematically investigating the selective impact of different Mg/Al ratio LDOs catalysts on the growth structure of SWNTs. Based on the characterization results of the catalysts and growth products, we further explored the mechanism of selective growth of carbon nanotubes by LDO catalysts under different Mg/Al ratios and the evolution of the catalyst structure during the growth process.

2. Experimental section

2.1. Experimental materials

The reagents used in the experiment are as follows: $Fe(NO_3)_3 \cdot 9H_2O$ (99%), $Mg(NO_3)_2 \cdot 6H_2O$ (purity 99%), urea (99%), concentrated hydrochloric acid (analytical grade), all purchased from China National Pharmaceutical Group Corporation Chemical Reagent Co., Ltd.; $Co(NO_3)_2 \cdot 6H_2O$ (99%), deoxycholic acid sodium (98%), anhydrous ethanol (analytical grade), purchased from Shanghai Macklin Biotech Co., Ltd.; $Al(NO_3)_3 \cdot 9H_2O$ (99%), purchased from Tianjin Xien Si Biotech Co., Ltd.; argon, carbon monoxide (99.99%), purchased from Qingdao Heli Gas Company. During the experiment, all aqueous solutions were prepared using deionized water.

2.2. Experimental methods

(1) Preparation of carbon nanotubes

The growth of carbon nanotubes was carried out using the chemical vapor deposition (CVD) method. The growth process was typically conducted in a rail-type CVD tube furnace (Anhui BEQ Equipment Technology

Co., Ltd., BTF-1200C-S-SL), with a constant temperature zone of 10 cm in length, and a quartz tube with a length of 1.0 m and an inner diameter of 4 cm. During the experiment, a quartz tube with an inner diameter of 40 mm was placed into the rail-type CVD tube furnace, and approximately 20 mg of LDO catalyst powder was placed into a quartz boat and moved to the center of the tube. First, 300 sccm of Ar gas was introduced for 30 minutes to purge the system and ensure a pure atmosphere. At the same time, the dual-temperature rail furnace was heated at a rate of 15°C/min to 700°C and stabilized at the reaction temperature. Once the temperature stabilized, the flow was switched to 300 sccm of CO gas, and the growth of the carbon nanotubes was allowed to continue for 30 minutes. After growth, 300 sccm of Ar gas was again introduced to expel the remaining CO gas in the system. Finally, after the reaction zone naturally cooled to room temperature, the gas supply was turned off, and the product was collected. The entire reaction process was carried out under atmospheric pressure.

(2) Preparation of catalyst

The Fe/Co/Mg/Al-LDOs catalysts were prepared via the coprecipitation method, using the hydrated nitrates of Fe, Co, Mg, and Al, with urea acting as a pH adjuster. In this work, the influence of the Mg/Al composition ratio on the catalyst properties was investigated using a Mg/Al ratio of 1:1 as a reference. Therefore, the preparation process of the LDOs catalyst with a Mg/Al ratio of 1:1 is first described: $Fe(NO_3)_3 \cdot 6H_2O$, $Co(NO_3)_3 \cdot 6H_2O$, $Mg(NO_3)_2 \cdot 6H_2O$, and $Al(NO_3)_3 \cdot 9H_2O$ were dissolved in 250 mL of deionized water in a molar ratio of 0.1:0.1:1:1, with the total amount of Mg and Al being 0.06 mol. Then, a 3.0 mol/L urea solution was prepared and slowly added to the mixed nitrate solution to adjust the pH to approximately 4.0. The resulting mixed solution was stirred at 100°C and subjected to reflux aging for 12 hours, after which stirring was stopped, and the solution continued to reflux at 95°C for another 12 hours to obtain a Fe/Co/Mg/Al-LDHs precursor suspension. The suspension was then filtered and washed to neutrality, and the precursor solid was dried and ground into a uniform powder. Finally, the powder was calcined at 1000°C for 4 hours to obtain the Fe/Co/Mg/Al-LDOs catalyst.

2.3. Characterization of the catalysts and carbon nanotubes

(1) Catalyst Characterization

First, the morphology of the LDOs catalysts was examined using scanning electron microscopy (SEM). To further characterize the catalysts in the solid phase, ultraviolet-visible-near-infrared diffuse reflectance spectra (UV-vis-NIR DRS) were collected using a UV-3600i Plus spectrophotometer equipped with an integrating sphere. The reduction properties of the catalysts were investigated using hydrogen temperature-programmed reduction (H_2-TPR) in a TPR system equipped with a thermal conductivity detector. By comparing the position and intensity of the reduction peaks, the reduction ability of the catalyst nanoparticles was evaluated.

(2) Characterization of Carbon Nanotubes

The SWNTs grown were first characterized using Raman spectroscopy. The Raman spectra were collected using a Renishaw inVia confocal Raman spectrometer in backscattering mode, with 532 and 633 nm lasers at multiple random points on each sample. For detailed (n, m) structural analysis, the samples were sonicated for 150 minutes in a 2% deoxycholic acid sodium solution (Macklin, 98%) using an ultrasonic cell crusher (150 W), and then centrifuged at 100,000 g for 40 minutes using a Hitachi CP70ME P70AT rotor-type ultracentrifuge. The supernatant was analysed using photoluminescence spectroscopy (PL) and ultraviolet-visible-near-infrared absorption spectroscopy (UV-vis-NIR). The photoluminescence signals were collected using a HORIBA Jobin Yvon Fluorolog-3 fluorescence spectrometer (excitation wavelength: 400-800 nm, emission wavelength: 800-1400 nm). UV-vis-NIR absorption spectra were measured on an Agilent Cary5000 spectrophotometer from 500 nm to 1400 nm. Additionally, to examine the structure and morphology of the grown SWNTs, transmission electron microscopy (TEM) images were taken using a JEOL-2200FS FEG TEM/STEM instrument at an accelerating voltage of 80 kV. Before analysis, the solid samples were dispersed in acetone, sonicated for 10 minutes in a water bath, and then the dispersion was dropped onto a copper grid covered with a holey carbon film.

3. Result and discussion

3.1. Exploration of the growth temperature

Figure 1a shows the scanning electron microscope (SEM) image of the Fe/Co/Mg/Al-LDOs catalyst with a Mg/Al ratio of 1:1 after CVD growth. The results indicate that the Fe/Co/Mg/Al-LDOs catalyst prepared by the coprecipitation method retains the typical hexagonal plate-like structure even after calcination, confirming that the LDHs catalyst precursor developed well during the coprecipitation process and exhibits high thermal stability. Investigating the effect of the Mg/Al ratio on the diameter of the SWNTs grown catalytically essentially involves exploring the role of the Mg/Al ratio in regulating the size of the active catalyst nanoparticles. In addition to the interaction between the catalyst and the support, the growth temperature is also a critical factor influencing nanoparticle size. A temperature that is too high may lead to irreversible coarsening of the catalyst nanoparticles, which can affect the diameter distribution and growth selectivity of the SWNTs.

Figures 1b and c show the Raman spectra of the growth products using the $FeCoMg_1Al_1$ catalyst at different temperatures. By analyzing the radial breathing modes (RBMs) peaks, when the growth temperature exceeds 800°C, the catalyst nanoparticles undergo significant sintering and agglomeration, resulting in a decrease in their catalytic activity. Particularly, when the temperature reaches 900°C, the catalyst is no longer able to effectively catalyze the growth of SWNTs. At lower temperatures, the growth quality is also affected. When the growth temperature is 600°C, the D peak in the Raman spectrum of the growth product increases, indicating that the quality of the SWNTs grown at this temperature is lower. Additionally, due to the limitation of the reduction temperature, SWNTs cannot grow at 500°C. Considering the impact of different temperatures on catalytic growth, the growth product at 700°C shows the best quality. Whether the size of the catalyst nanoparticles increases or decreases, SWNT growth can be effectively promoted at this temperature. Therefore, 700°C was selected as the optimal growth temperature for catalysts with different Mg/Al ratios in this study.

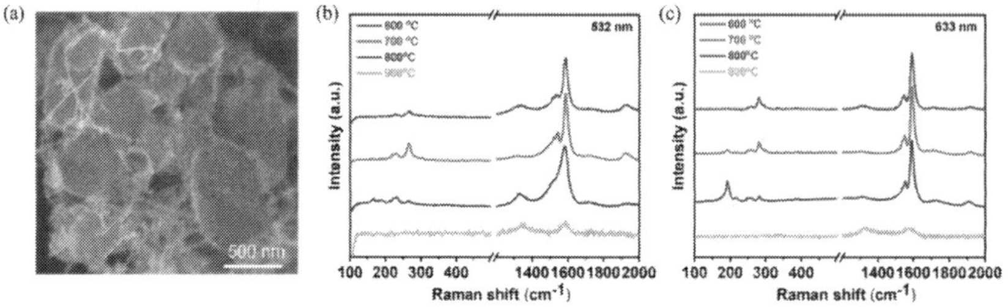

Figure 1. Characterization of $FeCoMg_1Al_1$ catalysts and SWNTs grown at different temperatures. (a) TEM image of $FeCoMg_1Al_1$ catalyst. Raman spectra of the products grown from $FeCoMg_1Al_1$ catalyst at 600-900°C (The excitation wavelength is (b) 532 nm and (c) 633 nm laser).

3.2. Diameter of SWNTs grown on catalysts with different Mg/Al ratios

Figure 2a shows the Raman spectra of SWNTs grown at 700°C under atmospheric pressure CVD using CO as the carbon source with $FeCoMg_1Al_5$, $FeCoMg_1Al_2$, $FeCoMg_1Al_1$, $FeCoMg_2Al_1$, and $FeCoMg_5Al_1$ catalysts. Comparative analysis reveals that as the Mg component in the LDOs catalyst increases, the I_G/I_D ratio of the products significantly decreases, indicating a reduction in the quality of the SWNTs. When the Mg/Al ratio reaches 5:1, i.e., under the $FeCoMg_5Al_1$ catalyst, the characteristic SWNTs peaks (RBMs) in the Raman spectrum disappear, suggesting that the catalyst has lost its ability to catalyze SWNT growth. In contrast, as the Al content increases, the catalyst maintains high catalytic activity, resulting in the synthesis of higher-quality SWNTs. Additionally, in the high wavenumber region, the relative intensity of the RBMs peaks increases, indicating that higher Al content contributes to the selective growth of smaller diameter SWNTs.

Figure 2b further shows the TEM images of the SWNTs grown by the $FeCoMg_1Al_5$ catalyst, confirming that the SWNTs have smaller diameters and higher quality. This result suggests that appropriately increasing the Al content can optimize the performance of the catalyst and enhance the selectivity and quality of SWNT growth.

Since Raman spectroscopy is not detailed enough to reveal specific structural changes in the SWNTs products, this study further characterized the SWNTs dispersed in deoxycholic acid sodium solution using UV-vis-NIR absorption spectroscopy. SWNTs grown with $FeCoMg_1Al_5$ and $FeCoMg_2Al_1$ catalysts, which produced higher-quality SWNTs, were selected for comparison with the SWNTs products from the $FeCoMg_1Al_1$ catalyst, as shown in Figure 2c. Comparison shows that the effect of changing the Mg/Al ratio on the SWNTs structure is especially noticeable in the 800-1000 nm near-infrared region. In this wavelength range, SWNTs grown with the $FeCoMg_1Al_5$ catalyst exhibit significantly enhanced relative intensity. This suggests that increasing Al content may help generate SWNTs with specific chiral structures, while increasing Mg content weakens this chiral selectivity. The relative abundance of each (*n*, *m*) chiral species was obtained, and diameter distribution data was derived using the chirality-to-diameter conversion formula. To further analyze the role of Mg and Al components in regulating the SWNTs diameter, we combined the photoluminescence (PL) spectra of the products with PeakFit software to perform peak fitting on the absorption spectra in the 800-1350 nm region, to extract the relative abundance of each (*n*, *m*) chirality. Finally, using the relationship between chirality and diameter, the diameter distribution characteristics of the SWNTs were determined.

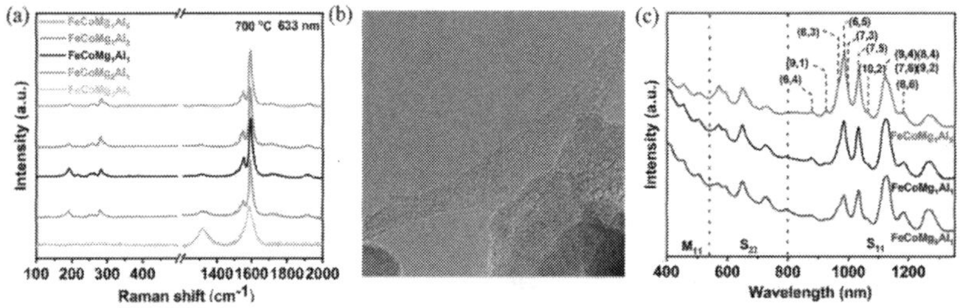

Figure 2. Comparison of growth products of catalyst with different Mg/Al ratios. (a) Raman spectra of the products grown from $FeCoMg_1Al_5$, $FeCoMg_1Al_2$, $FeCoMg_1Al_1$, $FeCoMg_2Al_1$, and $FeCoMg_5Al_1$ catalyst at 700°C (The excitation wavelength is 532 nm), (b) TEM image of SWNTs grown from $FeCoMg_1Al_5$ catalyst at 700°C, (c) Absorption spectra of the products grown from $FeCoMg_1Al_5$, $FeCoMg_1Al_1$, $FeCoMg_2Al_1$ catalyst at 700°C.

Figures 3a, b, and c show the PL spectra of SWNTs grown at 700°C using $FeCoMg_1Al_5$, $FeCoMg_1Al_1$, and $FeCoMg_2Al_1$ catalysts. The comparative results show that the change in the Mg/Al ratio significantly affects the chirality distribution of the SWNTs. First, in the catalyst products with increased Mg content, new chiral structures were observed— (9, 5) and (8, 7), with diameters of 0.976 nm and 1.036 nm, respectively, both larger than the maximum diameter of the products grown with the $FeCoMg_1Al_1$ catalyst. Additionally, the most abundant chiral structures shifted from (7, 5) to larger diameter structures such as (7, 6) and (8, 4), while the fluorescence intensity of smaller diameter chiral structures, such as (6, 5) and (7, 3), decreased. In contrast, when the Al content increased, the fluorescence intensity of smaller diameter chiral structures, such as (6, 5) and (7, 3), significantly increased, indicating that the Al component helps in the selective growth of smaller diameter SWNTs.

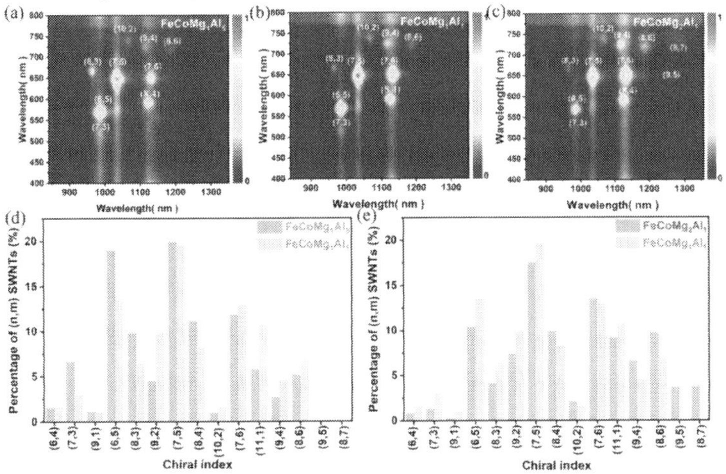

Figure 3. Chirality distributions of SWNTs grown from catalyst with different Mg/Al ratios. PL spectra of SWNTs grown from (a) $FeCoMg_1Al_5$, (b) $FeCoMg_1Al_1$, and (c) $FeCoMg_2Al_1$ catalyst at 700°C, (d) Comparison of chirality distributions of SWNTs grown from $FeCoMg_1Al_5$ catalyst and $FeCoMg_1Al_1$ catalyst, (e) Comparison of chirality distribution of SWNTs grown from $FeCoMg_2Al_1$ catalyst and $FeCoMg_1Al_1$ catalyst.

Based on the PL spectrum data, we further fitted the UV-vis-NIR absorption spectra and obtained the chirality distribution statistics shown in Table 1. Figures 3e and d show the comparative results of the chirality distribution. Figure 3e shows that the $FeCoMg_1Al_5$ catalyst significantly enhances the selectivity for smaller diameter SWNTs, such as (8, 3) and (7, 3), while reducing the selectivity for larger diameter SWNTs. For example, the selectivity for the (6, 5) chirality increases from 13.55% to 18.96%, while the selectivity for (9, 4) chirality decreases from 4.53% to 2.67%. The fitting results in Figure 3d indicate that under the $FeCoMg_2Al_1$ catalyst, the selectivity for smaller diameter chiral tubes (such as (6, 4), (6, 5), and (8, 3)) decreases, while the selectivity for new chiral structures such as (9, 5) and (8, 7), with diameters greater than 0.9 nm, exceeds 7%. These results suggest that the increase in Mg content helps form larger diameter SWNTs, while the increase in Al content promotes the selective growth of smaller diameter SWNTs.

Table 1. Abundance of different (*n, m*) species in SWNTs grown on $FeCoMg_1Al_5$, $FeCoMg_1Al_1$, $FeCoMg_2Al_1$ catalyst at 700°C deduced from peak fitting of absorption spectra.

Chirality index	Diameter/ nm	$FeCoMg_1Al_5$	$FeCoMg_1Al_1$	$FeCoMg_2Al_1$
(6, 4)	0.692	1.56%	1.64%	0.82%
(7, 3)	0.706	6.64%	2.98%	1.32%
(9, 1)	0.757	1.14%	1.05%	0.22%
(6, 5)	0.757	18.96%	13.55%	10.37%
(8, 3)	0.782	9.8%	6.45%	4.1%
(9, 2)	0.806	4.47%	9.79%	7.35%
(7, 5)	0.829	19.89%	19.56%	17.5%
(8, 4)	0.840	11.14%	8.18%	9.88%
(10, 2)	0.884	0.93%	1.65%	2.12%
(7, 6)	0.895	11.89%	12.96%	13.49%
(11, 1)	0.916	5.77%	10.81%	9.12%
(9, 4)	0.916	2.67%	4.53%	6.59%
(8, 6)	0.966	5.14%	6.87%	9.78%
(9, 5)	0.976	0	0	3.64%
(8, 7)	1.032	0	0	3.7%

In summary, changes in the Mg/Al component ratio in LDOs catalysts have a significant regulatory effect on the diameter distribution of SWNTs products. Figure 4 shows the diameter distribution data of SWNTs grown with $FeCoMg_1Al_5$, $FeCoMg_1Al_1$, and $FeCoMg_2Al_1$ catalysts. A comparison of Figures 4a and b clearly shows that as the Al content increases, the diameter distribution of the SWNTs shifts toward a smaller diameter range. Statistical data indicate that the selectivity for SWNTs in the diameter range of 0.65-0.85 nm increased by 10.53%, reducing the average diameter of the SWNTs from 0.827 nm to 0.807 nm. In contrast, when the Mg content increases, the growth products of the catalyst tend toward larger diameters. The selectivity for SWNTs in the diameter range of 0.85-1.05 nm increased by 6.26%, resulting in an increase in the average diameter of the SWNTs from 0.827 nm to 0.836 nm. These statistical results confirm that by adjusting the component ratio of the LDOs catalyst support, the diameter distribution of SWNTs can be effectively controlled. Further research is needed to explore how changes in the component ratio affect the interaction between the catalyst and the support, to gain a deeper understanding of the catalytic mechanism.

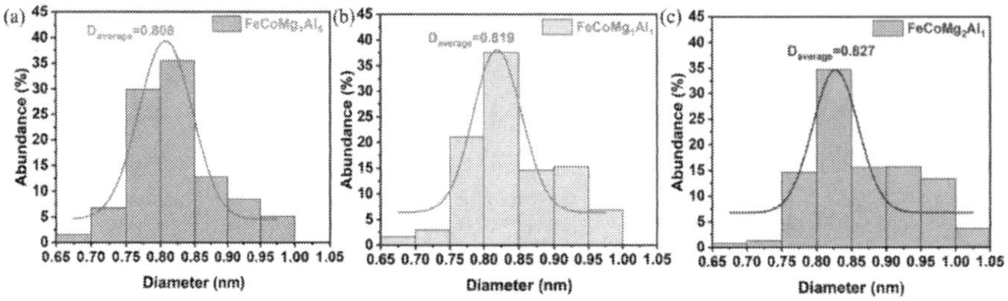

Figure 4. Diameter distributions of SWNTs grown from LDOs catalysts with different Mg/Al ratios at 700°C. (a) $FeCoMg_1Al_5$ catalyst, (b)$FeCoMg_1Al_1$ catalyst, (c)$FeCoMg_2Al_1$ catalyst.

3.3. Mechanism of Mg/Al ratio in catalysts on the modulation of SWNTs diameter

Figure 5 shows the UV-vis-NIR diffuse reflectance spectra of $FeCoMg_1Al_5$, $FeCoMg_1Al_1$, and $FeCoMg_5Al_1$ catalysts. In the ultraviolet region (around 270 nm), a distinct Ligand to Metal Charge Transfer (LMCT) absorption band is observed. In the visible region, strong absorption peaks at 545 nm, 582 nm, and 624 nm correspond to the D-D electronic transitions of Co^{2+} ions in a tetrahedral coordination (Td) from the $3d^7$ orbitals. Additionally, in the infrared region, a broad absorption band centered at 1400 nm is considered to be the $4A_2(4F) \rightarrow 4T_1(4F)$ transition of Td Co^{2+}, and these characteristic absorption peaks together form the spectral features of Td Co^{2+}. By comparing the DRS spectra, it is observed that in the $FeCoMg_1Al_5$ catalyst with high Al content, the absorption intensity of the Td Co^{2+} ion significantly increases, indicating that the increase in Al content promotes more Co^{2+} to enter the support structure in a tetrahedral coordination form. In contrast, in the $FeCoMg_5Al_1$ catalyst with high Mg content, the characteristic absorption band of Td Co^{2+} becomes more subdued, suggesting that the increase in Mg content inhibits the entry of Co^{2+} into the support, and Co^{2+} is more likely to be weakly adsorbed on the support surface with weaker interaction[8]. This result indicates that changes in the Mg and Al components significantly affect the distribution of Co^{2+} in the LDOs catalysts, which may further influence their catalytic performance.

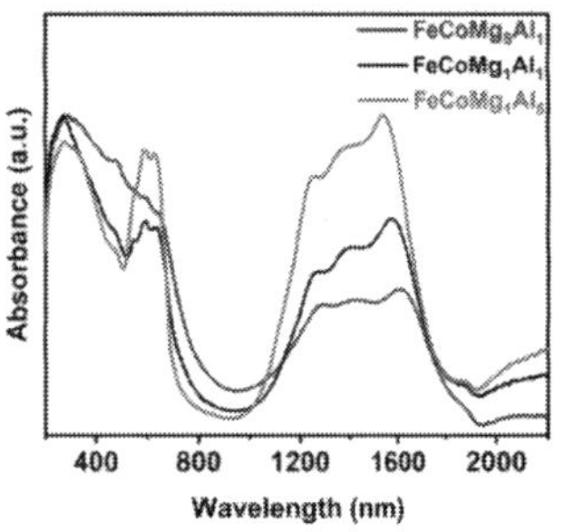

Figure 5. UV-vis-NIR diffuse reflectance spectroscopy of $FeCoMg_1Al_5$, $FeCoMg_1Al_1$, and $FeCoMg_5Al_1$ catalyst.

According to the research by Li[9] and Lim[10], Co(II) and Co(I) ions that are not fully reduced on the catalyst support surface can form strong interactions with the metal catalyst on the surface, generating an anchoring effect. This effect can slow down the Ostwald ripening process of metal nanoparticles generated during reduction, thus promoting the stable dispersion of the catalyst. Based on the UV-vis-NIR DRS spectra results, it can be speculated that the Co^{2+} ions entering the support in LDOs catalysts may play a similar role. Specifically, when the Al content in the LDOs catalyst increases, more Co^{2+} ions enter the support structure. During the CVD process when CO is introduced, the reduced Fe and Co nanoparticles on the catalyst surface form strong interactions with Co^{2+}, anchoring them to the catalyst surface. This stabilizes the particle size of the catalyst, reducing agglomeration and maintaining a distribution of smaller nanoparticles. Therefore, this mechanism favors the selective growth of smaller diameter SWNTs. In contrast, when the Mg content increases, Co ions in the catalyst tend to be weakly adsorbed on the support surface rather than entering the support. During the SWNT growth process, these weakly bound Co ions are more prone to Ostwald ripening after reduction, gradually aggregating to form larger nanoparticles, which leads to the growth of larger diameter SWNTs. Figure 6 shows a schematic diagram of this mechanism, further illustrating the effect of catalyst support composition on the diameter selectivity of SWNTs.

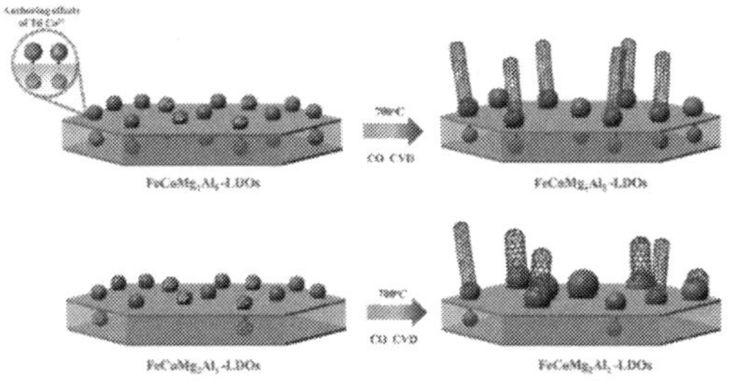

Figure 6. Schematic diagram of LDOs catalysts with different Mg/Al ratios modulating different diameter distributions of SWNTs.

Figure 7 shows the H_2-TPR spectrum of the $FeCoMg_1Al_5$ catalyst. It can be observed that this catalyst undergoes a multi-step reduction process. Typically, the reduction temperature of iron oxides is higher than 700°C. However, in LDOs catalysts, the reduction temperature range of the catalyst mainly lies between 500-700°C, a phenomenon likely due to the synergistic catalytic effect between Fe and Co, which facilitates the earlier reduction of Fe. Additionally, a smaller H_2 consumption peak can be observed in the region above 700°C. Based on previous analysis, this peak may be associated with the Co^{2+} distributed within the catalyst support. This further validates the previously proposed anchoring effect of Co^{2+} in the support, where some Co^{2+} ions are slowly reduced at high temperatures, helping stabilize the catalyst nanoparticles, and thereby regulating the diameter distribution of SWNTs.

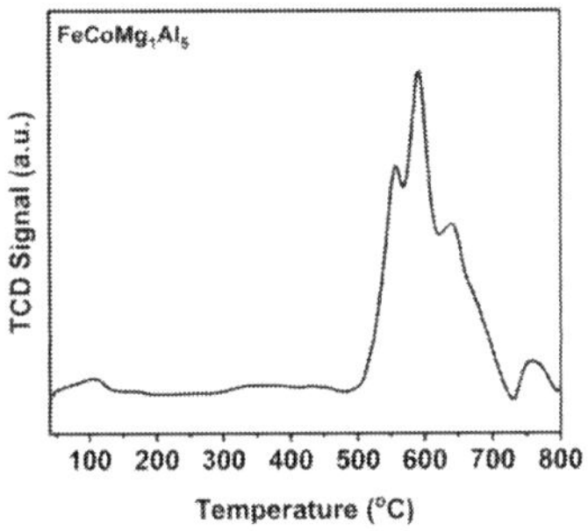

Figure 7. H_2-TPR profile of $FeCoMg_1Al_5$ catalyst.

4. Conclusion

We prepared Fe/Co/Mg/Al-LDOs catalysts with different Mg/Al ratios by the co-precipitation method, and compared the structural changes of SWNTs products grown at 700°C using CO as the carbon source. Using the FeCoMg1Al1 catalyst-grown product as a control group, the results showed that the increase in the Al component in the Fe/Co/Mg/Al-LDOs catalyst favored the selective growth of smaller diameter SWNTs, while the increase in the Mg component tended to promote the growth of larger diameter SWNTs. UV-vis-NIR DRS spectroscopy analysis revealed that the increase in the Al component facilitated the entry of Co ions into the support, which led to the hypothesis that Co ions anchor the metal nanoparticles on the catalyst surface during SWNT growth. This anchoring effect helps inhibit particle agglomeration, thus favoring the selective growth of smaller diameter SWNTs. In contrast, the increase in Mg content inhibited the entry of Co ions into the support, resulting in more severe particle agglomeration and promoting the growth of larger diameter SWNTs. This hypothesis was further validated by the H_2-TPR spectrum of the $FeCoMg_1Al_5$ catalyst. This study not only provides a flexible method for controlling the diameter of SWNTs products using LDHs-based catalysts but also contributes to a better understanding of the interaction between the support and active components and the mechanism of SWNT growth with LDHs-based catalysts.

References

1. He, M., Zhang, S., & Zhang, J. Horizontal Single-Walled Carbon Nanotube Arrays: Controlled Synthesis, Characterizations, and Applications. Chem. Rev., **120**(22), 12592-12684 (2020).
2. Yang, F., Wang, M., Zhang, D., et al. Chirality Pure Carbon Nanotubes: Growth, Sorting, and Characterization. Chem. Rev., **120**(5), 2693-2758 (2020).
3. Li, H., Gordeev, G., Garrity, O., et al. Separation of Specific Single-Enantiomer Single-Wall Carbon Nanotubes in the Large-Diameter Regime. ACS Nano, **14**(1), 948-963 (2020).
4. He, M., Zhang, S., Wu, Q., et al. Designing Catalysts for Chirality-Selective Synthesis of Single-Walled Carbon Nanotubes: Past Success and Future Opportunity. Adv. Mater., **31**(9), 1800805 (2019).
5. Gao, Z., Ma, H., Wang, Q., et al. Structure Engineering of Layered Double Hydroxides (LDHs) for Heterogeneous Catalysis. Chem. Res. Chin. Univ., **40**(4), 590-610 (2024).
6. Li, S., Wang, D., Wu, X., et al. Recent Advance on VOCs Oxidation Over Layered Double Hydroxides Derived Mixed Metal Oxides. Chin. J. Catal., **41**(4), 550-560 (2020).
7. Zhang, Q., Zhao, M., Tang, D., et al. Carbon-Nanotube-Array Double Helices. Angew. Chem. Int. Ed., **49**(21), 3642-3645 (2010).
8. Zhou, M., Liu, K., Peng, Q., et al. Long-Acting $CoAl_2O_4$ Spinel Catalyst Developed on Activated Alumina Pellets by Facile Synthesis to Activate Peroxymonosulfate: Controllable Cobalt Leaching and Environmental Adaptability. J. Environ. Manag., **310**, 114702 (2022).
9. Li, N., Wang, X., Derrouiche, S., et al. Role of Surface Cobalt Silicate in Single-Walled Carbon Nanotube Synthesis from Silica-Supported Cobalt Catalysts. ACS Nano, **4**(3), 1759-1767 (2010).
10. Lim, S., Wang, C., Yang, Y., et al. Evidence for Anchoring and Partial Occlusion of Metallic Clusters on the Pore Walls of MCM-41 and Effect on the Stability of the Metallic Clusters. Catal. Today, **123**(1-4), 122-132 (2007).

Research on Emission Level Assessment of Heavy-Duty Vehicle Fleets Based on Remote Monitoring

Binghui Ma[1,2], Hao Zhang[1], Yu Liu[1*], Hanzhengnan Yu[1], Jingyuan Li[1], Xiaopan An[3], Kunqi Ma[3], Yongkai Liang[3], Hang Xu[3]

[1]China Automotive Technology Research Center Co., Ltd., Tianjin, China
[2]Jilin University, Jilin, China
[3]CATARC Automotive Test Center (Tianjin) Co., Ltd., Tianjin, China

Abstract. Heavy-duty fleet emissions present a critical environmental challenge in the transportation sector demanding immediate resolution. A scientific evaluation of heavy-duty fleet emission levels underpins the effective management and control of these emissions. Leveraging remote monitoring data, this study presents a methodology for evaluating heavy-duty fleet emission levels via a fleet star-rating mechanism that integrates the fleet's average emission factor, high-risk emission factor, operational condition emission factor, and total emissions to deliver a overall assessment. These findings support governments and pertinent agencies in understanding the comprehensive emission profiles of their commercial fleets and offer robust, effective methodological guidance for regional ecological and environmental monitoring.

1. Introduction

In 2023, the total national emissions of carbon monoxide, hydrocarbons, nitrogen oxides, and particulate matter from motor vehicles reached nearly 15 million tonnes[1]. Such a large volume of pollutant emissions not only degrades air quality but also poses serious threats to human health. Particularly in suburban areas with dense freight traffic, the high-emission characteristics of heavy-duty freight fleets contribute notably to environmental pollution, making long-term[2], effective control of these emissions an urgent priority. A scientific assessment of heavy-duty fleet emission levels is not only the foundation for effective emission control but also a crucial guarantee for achieving energy savings, emission reductions, and the dual-carbon strategy

The heavy-duty vehicle emission factor is the most commonly used technical means for assessing fleet emission levels, and on this basis, relevant literature has also considered factors such as the internal emission scale of the fleet, vehicle count, and vehicle-type composition, in order to achieve a more precise assessment of fleet emission levels. For instance, Zhong et al.[3] obtained heavy-duty vehicle emission factors using a three-zone moving average window method and analyzed the relationship between vehicle operating-condition characteristics and emission levels. Zhang et al.[4] derived heavy-duty vehicle emission factors using a moving window averaging method, extracting characteristic patterns among emission control levels, traffic-operation features, and vehicle emission rates. Yan et al.[5] accounted for fleet emission scale and, based on the characteristic patterns of individual vehicle emission levels, proposed a method for calculating fleet energy consumption. Liu et al.[6] considered structural factors within the fleet and, leveraging remote monitoring data, proposed a method for assessing carbon emission levels of heavy-duty fleets. Zou Chao et al.[7] analyzed CO_2 and pollutant emission characteristics of the Zhengzhou city bus fleet before and after electrification using a fuel life-cycle method, and evaluated the fleet's CO_2 emissions under electrification scenarios.

However, existing studies have primarily focused on accurately assessing fleet carbon emissions, with a lack of research on methods for evaluating fleet NO_x emissions. Moreover, they have not simultaneously considered the impacts of fleet-wide emission scale, vehicle count, and vehicle-type composition on fleet emissions. To address this, this paper proposes a heavy-duty fleet emission-level assessment method that enhances the accuracy of existing fleet emission evaluation systems and provides a more scientific basis for emission monitoring for governments and enterprises. This method is of significant importance for the scientific evaluation of future heavy-duty fleet emission levels and the formulation of emission control measures.

2. Data and methods

Real-world driving data constitutes a crucial basis for assessing heavy-duty fleet emission levels. Emission-level assessment methodologies are essential tools for evaluating heavy-duty fleet emission levels.

*liuyu2016@catarc.ac.cn

2.1. Data overview

This study focuses on the heavy-duty fleets of five representative major vehicle users, whose operational characteristics and fleet sizes are shown in Table 1. Using a heavy-duty vehicle remote monitoring platform, actual on-road driving data were collected from 1,020 heavy-duty vehicles. The test vehicles spanned a payload range of 4.5 to 31 tonnes. The post-treatment systems of the test vehicles follow the standard China VI diesel exhaust aftertreatment configuration, including an oxidation catalyst, selective catalytic reduction unit, particulate filter, and ammonia slip catalyst. All test vehicles comply with the China VI b emission regulation standards. The collected parameters include timestamp, vehicle speed, engine torque, engine speed, and NO_x emission rate per second, with a sampling frequency of 1 Hz. The data for each vehicle cover a one-month period, and the aggregate mileage across all vehicles exceeds 100,000 km.

Table 1. Workload Characteristics and Vehicle Statistics of Representative High-Usage Fleets.

Typical High-Usage Fleets	Operational Characteristics	numbering
Fleet A	The urban distribution logistics fleet primarily undertakes short-distance deliveries for e-commerce and express services, typically shuttling between delivery points within the city.	180
Fleet B	The sanitation cleaning fleet is responsible for municipal tasks such as street sweeping and garbage collection across both urban and suburban areas.	300
Fleet C	The domestic services fleet provides in-home housekeeping services, primarily serving residential neighborhoods.	100
Fleet D	The concrete mixer transport fleet delivers mixed concrete to construction sites, typically operating on urban and suburban roads.	280
Fleet E	The long-haul semi-trailer tractor fleet performs interprovincial long-distance freight transport, mostly operating under highway conditions.	160

2.2. Correction of data anomalies

There exist data anomalies in heavy-duty vehicle remote monitoring datasets; to utilize these data effectively, it is necessary to identify and rectify such anomalies[8]. This study classifies the anomalies into four categories: data loss anomalies, transient spike anomalies, position drift anomalies, and steady-state distortion anomalies. Data missing anomalies refer to instances where monitoring data for one or multiple frames are not acquired or are lost. Transient spike anomalies denote sudden jumps in a small number of consecutive samples that exceed the normal fluctuation range. Position drift anomalies describe cases where positioning data abruptly deviate from the true trajectory, resulting in implausible jumps or drift segments. Steady-state distortion anomalies refer to the sensor output remaining constant for prolonged periods or exhibiting no variation within non-physically feasible ranges.

Identify each type of anomalous data and set those values to null. For data missing anomalies, determine the locations of missing records based on timestamp continuity, interpolate the missing timestamps according to the sampling interval, and set the corresponding vehicle speed and NOx emission rate to null to facilitate subsequent unified interpolation. For transient spike anomalies, apply convolutional smoothing using a Blackman window function to detect local outliers caused by short-term spikes. In this study, the window length is set to 10 and the standard-deviation threshold multiplier to 13; any outliers beyond this range are then set to null. For positional drift anomalies, compute vehicle speed from changes in latitude and longitude, compare it against a physical maximum speed threshold, and set any anomalous speed points to null to eliminate implausible location jumps. For steady-state distortion anomalies, use the valid speed and NOx emission ranges defined in "HJ 1239-2021 Remote Emission Monitoring Specifications for Heavy-Duty Vehicles" to identify invalid data outside these bounds and set them to null. Additionally, any constant-value segments—where the series slope approaches zero for a duration exceeding a set threshold—are also deemed invalid and set to null. To impute the nullified data points, we employ the locally weighted regression (LOESS) interpolation method for data completion. LOESS interpolation is a nonparametric regression technique that fits a weighted polynomial within the local neighborhood of each point to estimate missing values, making it well suited for data with complex trends and irregular intervals. After filling in the nullified points, the remote-monitoring dataset is largely restored; however, some local segments still exhibit significant deviations from the true values due to the interpolation process. To address this, we introduce a moving-average filtering step for post-processing of the data. In summary, this approach enables the identification and correction of anomalies in the remote-monitoring data.

2.3. Method for fleet emission level assessment

Based on the corrected remote-monitoring data, a method for assessing heavy-duty fleet emission levels is developed. As shown in Equation (1), it comprises the fleet average emission factor, the fleet high-risk emission factor, and the fleet operating-condition emission factor. First, emission factors for each vehicle weight class are obtained using Equations (2) and (3). Then, the fleet average emission factor is calculated by Equation (4) using the proportion of operating time $Time_j$ and fleet share $share_j$ for each weight class. Using Equation (5), the mean emission factor of the top 10 % of vehicles—

ranked by operating duration, mileage, and cumulative emissions—is computed to yield the fleet high-risk emission factor. Equation (6) is then used to compute the fleet operating-condition emission factor by averaging the emission factors of vehicles with high proportions of urban driving and idling.

$$E_{\text{Overall}} = \alpha \times E_{\text{average}} + \beta \times E_{\text{high-risk}} + \gamma \times E_{\text{operating-condition}} \quad (1)$$

$$\beta_m = m_i / m_{max} \quad (2)$$

$$NOx_{\text{type}} = \sum_i \frac{M_i \times NO_{e,i}}{m} \times \frac{1}{\beta_m} \quad (3)$$

$$E_{\text{average}} = \sum_j \text{Time}_j \times \text{share}_j \times NOx_{(type,j)} \quad (4)$$

$$E_{high-risk} = \left(T^*_{max} + L^*_{max} + E^*_{cum}\right) / n_{risk} \quad (5)$$

$$E_{\text{operating-condition}} = \left(\sum E_{urban} + \sum E_{e-idling}\right) / n_{condition} \quad (6)$$

In the equation: α, β and γ represent weighting coefficients, where $\alpha = 0.7$, $\beta = 0.2$, $\gamma = 0.1$; β_m denotes the mass weighting factor, m_i represents the tonnage of the i-th vehicle category in tons (t), and m_{max} is the maximum tonnage within the fleet (t); $NOx_{(type,j)}$ refers to the NOx emission factor of the j-th vehicle type;

M_i represents the proportion of driving distance of the i-th vehicle within that vehicle class; $NO_{e,i}$ denotes the NOx emission factor of the i-th vehicle, measured in grams per second (g/s); m indicates the curb weight of the vehicle type, in tons (t); share_j denotes the quantity proportion of the j-th vehicle type within the fleet, while Time_j represents the proportion of operating time of the j-th vehicle type in the fleet; T^*_{max} represents the emission factor of the top 10% of vehicles ranked by longest operating time; L^*_{max} denotes the emission factor of the top 10% of vehicles with the highest mileage; E^*_{cum} refers to the emission factor of the top 10% of vehicles ranked by cumulative emissions; E_{urban} indicates the emission factor for vehicles with an urban mileage share (speed $v \leq$ 50 km/h) of at least 30%; $E_{e-idling}$ denotes the emission factor for vehicles whose idling duration proportion is 15% or higher; n represents the count of high-risk or vehicles under operational conditions.

To account for the fleet's emission magnitude, the total emissions of the fleet must also be calculated. First, calculate the emission scale Q_i for each vehicle category using Equation (7). Then, sum the emission scales of all vehicle categories and adjust by the fleet's vehicle count using Equation (8) to obtain the fleet's total emission scale.

$$Q_i = \sum_{i=1}^{k} N_i \times VKT_i \times Time_i \times NO_{e,i} \times 10^{-6} \quad (7)$$

$$E_{\text{team}} = \sum_{i=j}^{k} Q_j \times \beta_{\text{adjustment}} \quad (8)$$

In the equation: N_i denotes the quantity of the i-th vehicle; VKT_i represents the driving mileage of the vehicle; $Time_i$ denotes the operation duration; $NO_{e,i}$ represents the emission factor for an individual vehicle; A conversion factor of 10^{-6} is applied to ensure dimensional consistency; Where $\beta_{\text{adjustment}} = N/n$, with N representing the total fleet size and n the number of vehicles in the evaluated subset.

To facilitate rapid assessment of fleet emission levels, a quantile-based method is employed to classify fleet emissions into four tiers. Initially, the comprehensive emission factor of the fleet under evaluation, the overall fleet emission factor, the total emissions of the evaluated fleet, and the total emissions of all fleets are calculated. If the evaluated fleet's comprehensive emission factor falls within the 0–25% quantile of the overall fleet emission factor, it is rated four stars; within the 25%–50% quantile, three stars; within the 50%–75% quantile, two stars; and within the 75%–100% quantile, one star. A higher star rating indicates lower vehicle emissions. Subsequently, the fleet's total emissions are used as a key adjustment factor to establish a dynamic emission threshold. If the evaluated fleet's total emissions exceed 1.2 times the total fleet emissions, indicating excessive environmental burden, its star rating is downgraded by one level; conversely, if its total emissions are significantly below 0.8 times the total fleet emissions, reflecting excellent environmental performance, its star rating is upgraded by one level.

Although individual vehicle emission levels may be low, a large fleet size can still result in considerable total emissions, posing a significant environmental burden that should not be overlooked. Conversely, even if individual vehicle emissions are high, a small fleet size may lead to relatively low total emissions and thus a minor environmental impact. In environmental regulation, attention should be focused on operators with both large fleet sizes and high total emissions. Therefore, fleet total emissions are used to adjust the rating of fleets. This method considers not only the relative performance of "per-vehicle emission levels," but also the absolute impact of the "overall environmental burden," thus better addressing practical needs in environmental management and regulation.

3. Results and discussion

This paper's results and discussion are conducted from two aspects: the correction of anomalies in remote monitoring data and the analysis of heavy-duty fleet emission levels.

3.1. Corrected results and validation

To verify the effectiveness of the anomaly correction method proposed in this study, a vehicle from Fleet A was randomly selected for a PEMS experiment, which can accurately obtain emission data from heavy-duty vehicles and serve as a reference dataset. By artificially introducing random window lengths, scaling factors, and deletion rules, the original data were perturbed to simulate anomalies such as data loss, abrupt outliers, duplicated records, and invalid constant values, which may occur in real-world scenarios. These anomalies were then corrected using the anomaly identification and correction process developed in this study. The results are shown in Figure 1. On this basis, the Spearman rank correlation coefficient was used to assess the consistency between the corrected data and the accurate data. The computed coefficient exceeded 0.98, indicating a very high rank correlation. This indicates that the corrected data are highly consistent with the accurate data, demonstrating that the proposed anomaly correction process for remote monitoring data is effective.

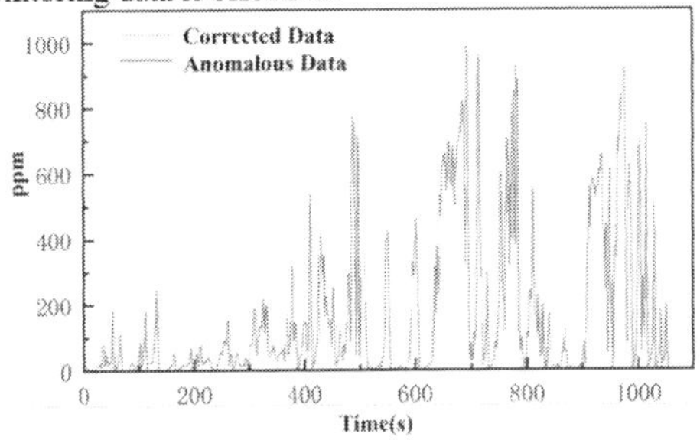

Fig. 1. Illustration of Anomalous Data Correction Results.

3.2. Analysis of fleet emission levels

This study aggregated vehicle monitoring data from five fleets, with each vehicle providing five hours of remote monitoring information. Based on this data, the overall fleet's average emission factor was calculated as 0.51 g/s, and the total emissions were 13,860.8 kg. On this basis, a representative total emission index was proposed to preliminarily evaluate the emission level of the target fleet. However, when applying this method to other regions or larger-scale assessments, factors such as regional characteristics and fleet size should be fully considered to obtain more representative emission estimates.

Using remote monitoring data from five representative heavy-duty vehicle fleets operated by major users, the comprehensive and average emission factors for each fleet were calculated, as shown in Figure 2. Fleet A mainly operates under light load and low-speed conditions, with an average emission factor of 0.31 g/s and a comprehensive emission factor of 0.37 g/s, indicating a moderately low emission level. Due to prolonged idling and frequent stop-and-go operations under low-speed and heavy-load conditions, Fleet B exhibits a high average emission factor of 0.696 g/s and a comprehensive emission factor of 0.84 g/s. Fleet C mainly uses light commercial vehicles with relatively light loads and short idling durations. Its average emission factor is 0.29 g/s, and the comprehensive emission factor is 0.34 g/s, the lowest among all. Fleet D operates with drum rotation maintained under half to full load, resulting in high engine loads. Its average emission factor is 0.798 g/s and comprehensive emission factor is 0.93 g/s. Fleet E, consisting of long-distance semi-trailer tractors, primarily operates at high-speed cruising. However, due to high vehicle weight, long mileage, and high fuel consumption, its average emission factor reaches 1.40 g/s, and comprehensive emission factor is 1.68 g/s—the highest among all fleets. Moreover, the comprehensive emission factors of all fleets are higher than their respective average emission factors, indicating that high-emission vehicles or special operating conditions amplify overall emission levels. Compared with the simple averaging method, the proposed approach integrates multidimensional factors, accounting for high-emission vehicles and fleet composition, thus better reflecting the emission characteristics of fleets.

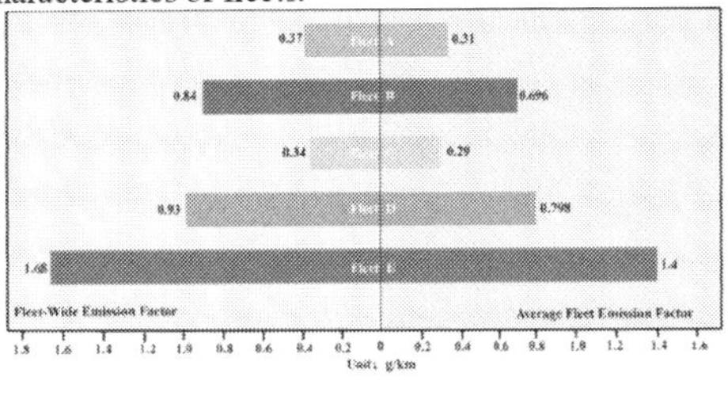

Fig. 2. Schematic Diagram of Fleet Emission Factors.

The total emissions of each fleet are shown in Figure 3. Although some fleets have lower per-vehicle emission factors, a large fleet size or longer travel distances can lead to increased total emissions and a greater emission burden. For example, Fleet D and Fleet B. In addition, heavy-duty vehicles generally have higher emission factors, and their total fleet emissions are significantly greater than those of light or medium-duty fleets. For example, Fleets D and E have much higher emission factors and total emissions than Fleets A and C.

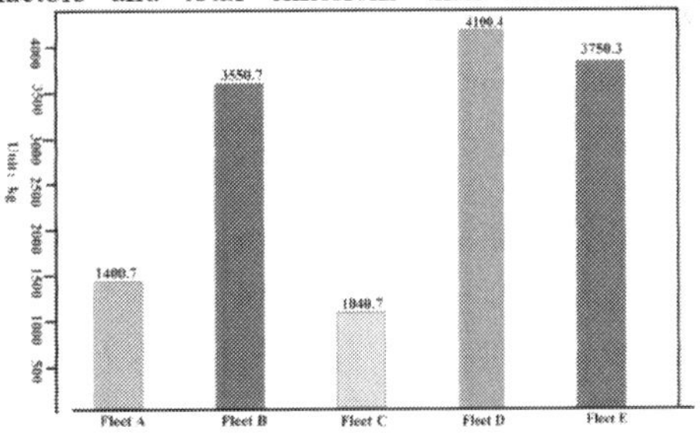

Fig. 3. Schematic Diagram of Total Fleet Emissions.

3.3. Classification of fleet star ratings

Based on the previously proposed quantile-based star rating method, a preliminary rating was conducted for five heavy-duty fleets, as shown in Figure 4. According to the integrated emission factors shown in Figure 1, the initial evaluation results are: two three-star fleets (Fleets A and C), two two-star fleets (Fleets B and D), and one one-star fleet (Fleet E). Subsequently, adjustments were made based on each fleet's total emissions, leading to the final rating: two four-star fleets (A and C), one three-star fleet (B), two two-star fleets (D and E), and one one-star fleet (D). Fleet B has a low emission level, but due to its large scale and significant total emissions, its rating was downgraded. Fleets A and C, with smaller scales and lower total emissions, received an upgraded rating. This comprehensive evaluation method improves the accuracy and completeness of the assessment, providing strong support for regional environmental monitoring by government and related agencies.

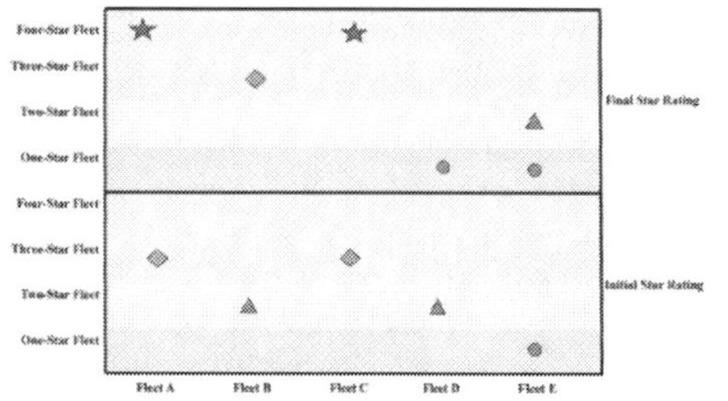

Fig. 4. Schematic Illustration of the Star-Based Fleet Classification System.

4. Conclusion

Based on the remote monitoring data of heavy-duty vehicles, this study conducted research on an evaluation method for heavy-duty fleet emission levels. An emission level evaluation method for heavy-duty fleets was proposed, and the emission levels of five representative heavy-duty fleets from major logistics companies were evaluated and analyzed. The main conclusions are as follows:

(1) Four types of anomalies in remote monitoring data were identified and defined, including missing data, abrupt changes, location drift, and steady-state distortion, and a complete process for anomaly detection and correction was developed accordingly.

(2) Data were accurately restored using methods such as window function detection, interval judgment, nullification, and LOESS interpolation. The consistency was validated using Spearman rank correlation, with a coefficient exceeding 0.98, demonstrating the effectiveness of the correction approach.

(3) Based on the corrected remote monitoring data, a comprehensive heavy-duty fleet emission assessment method was proposed, taking into account factors such as fleet size and vehicle type composition. Four emission levels were defined according to total fleet emissions, considering both emission intensity and environmental impact.

(4) The proposed method was applied to five representative logistics fleets, comparing comprehensive and average emission factors and total emissions. Star ratings were assigned to identify high-emission-risk fleets, offering a reliable tool for regional environmental monitoring.

References

1. Ministry of Ecology and Environment: 2023 Annual Report on Mobile Source Environmental Management.
2. Zhang W, Li Y, Li H, et al. Systematic review of life cycle assessments on carbon emissions in the transportation system[J]. Environmental Impact Assessment Review, 2024, 109: 107618.
3. Zhong, X., Li, T., Zhang, C., et al.: Study on PEMS Test Conditions Based on the 3B-MAW Method. Internal Combustion Engine and Power Plant, 2023, 40(04): 28–34.
4. Zhang, S.: Study on Emission Characteristics and Control Strategies of Motor Vehicles in Typical Chinese Cities. Beijing: Tsinghua University, 2014.
5. Yan Y. Urban public transport energy consumption model and calculation method[D]. Beijing: Tsinghua University, 2013.
6. Liu, W., Hua, W. (2025). Study on Carbon Emission Levels of Heavy-Duty Vehicle Fleet Corrected Based on Remote Monitoring Data. Shidai Qiche (Times Automotive), (7), 4–6.
7. Zou, C., Wang, Y., Wu, L., et al. (2024). Environmental Benefits of Bus Fleet Electrification for Pollution Reduction and Carbon Abatement in Zhengzhou. Chinese Journal of Environmental Science, 45(3).
8. Wang J, Wang R, Yin H, et al. Assessing heavy-duty vehicles on-road NOx emission in China from on-board diagnostics remote report data[J]. Science of The Total Environment, 2022, 846: 157209.

Research on the Joint Clearing Model for Energy Storage Participation in Electricity and Frequency Regulation Markets

Miao Wang[1,a], Gang Liu[1,b], Qingsong Zhao[2,c*], Zhonghui Wang[1,d], Jiaze Li[1,e], Mingkai Wang[1,f]

State Grid Liaoning Electric Power Supply Co. LTD Shenyang, China
State Grid Liaoning Electric Power Supply Co. LTD, Electric Power Research Insitute Shenyang, China

Abstract. With the large-scale integration of renewable energy and the deepening reform of electricity marketization, the role of energy storage technology in power systems has become increasingly prominent. Energy storage not only addresses the mismatch between renewable energy output and peak load demand but also provides ancillary services such as frequency regulation and reserve capacity, enhancing grid stability and reliability. This paper establishes a joint clearing model for energy storage participation in electricity and frequency regulation markets, optimizing power resource allocation through market-oriented mechanisms, leveraging the performance advantages of energy storage systems, and improving electricity market efficiency. A case study based on a 39-node system integrated with energy storage devices verifies the feasibility and effectiveness of the proposed model. The results demonstrate that the joint clearing mechanism can reduce system marginal costs, increase the utilization frequency of energy storage, and simultaneously meet the dual requirements of electricity supply and frequency regulation services.

1. Introduction

The advancement of global carbon neutrality goals has accelerated the development of new energy-based power systems. As one of the primary sources of carbon emissions, the power industry is now faced with the urgent task of constructing these new energy-centered power systems[1,2]. The development of new power systems, promoting energy conservation and emission reduction, and vigorously developing renewable energy. However, this transition has simultaneously introduced challenges including insufficient grid regulation capacity and increased difficulty in maintaining frequency stability. Energy storage technology, with its dual capabilities of energy retention and rapid response, has emerged as a critical means to address these issues[3].Under the evolving landscape of electricity market reform, the participation of energy storage systems holds vital significance for ensuring grid operational stability and enhancing resource allocation efficiency. By storing energy during surplus periods and releasing it during high demand, these systems effectively mitigate power supply-demand fluctuations, reduce peak-valley load differentials, and lower fuel consumption along with startup/shutdown costs for peak-load generation units[4]. Furthermore, energy storage systems deliver critical grid ancillary services including frequency regulation, operating reserves, and voltage control, substantially enhancing grid reliability and stability[5]. Consequently, incorporating energy storage into electricity market frameworks and establishing robust market mechanisms prove essential for incentivizing storage deployment, optimizing resource allocation, and advancing sustainable development of power markets.

Current research has made progress in modeling and mechanisms for energy storage participation in electricity spot markets, including joint clearing methods incorporating storage in energy and ramp markets[6], multi-market trading strategies involving energy-inertia-primary frequency regulation[7], and optimal operation approaches for solar-storage hybrid plants under spot market conditions[8,9]. However, existing studies predominantly focus on individual markets while inadequately addressing the coupling relationships between energy and frequency regulation markets, resulting in underutilized potential of energy storage in regulation services and unresolved challenges in multi-market joint clearing. This paper proposes a novel joint clearing model for energy storage participating in energy-frequency regulation markets, systematically incorporating system frequency regulation requirements, unit ramp constraints, frequency response capacity limits, and storage regulation capabilities. Through market-driven mechanisms, the model incentivizes storage engagement in frequency regulation markets, delivering dual benefits: enhancing system regulation capacity/performance and improving renewable energy accommodation capability. The joint optimization framework provides critical technical support for coordinating energy storage participation in spot market transactions and dispatch operations.

[a]13889897346@163.com; [b]liugang_ln@126.com
[c*] hi295@163.com; [d] wylpwylp@126.com
[e] m18302996883@163.com; [f]wangmingkai86@163.com

2. Energy storage-integrated joint clearing model for energy and frequency regulation markets

2.1. Objective function

The joint clearing model is designed to minimize total system operating costs, encompassing unit generation costs, frequency regulation service costs, and network constraint relaxation penalties, with its objective function structured as follows:

$$\begin{aligned}\min\{&\sum_{i=1}^{N}\sum_{t=1}^{T}\left[C_{i,t}(P_{i,t})+C_{i,t}^{U}+C_{i,t}^{p\min}+C_{i,t}^{reg}R_{i,t}^{reg}\right]+\sum_{l=1}^{NL}\sum_{t=1}^{T}M\left[SL_l^{+}+SL_l^{-}\right]\\&+\sum_{s=1}^{NS}\sum_{t=1}^{T}M\left[SL_s^{+}+SL_s^{-}\right]\}+\sum_{e=1}^{E}\sum_{t=1}^{T}\left[\lambda_e^{dis}P_{e,t}^{dis}+\lambda_e^{ch}P_{e,t}^{ch}+C_{e,t}^{reg}R_{e,t}^{reg}\right]\end{aligned} \quad (1)$$

Where N is the total number of conventional units, excluding energy storage units. T is the total number of time periods considered. $P_{i,t}$ is the output of unit i in period t. $C_{i,t}(P_{i,t})$, $C_{i,t}^{U}$, $C_{i,t}^{p\min}$ are the operating costs, start-up costs, and minimum stable technical output costs of unit i in the period t. $C_{i,t}^{reg}$, $C_{e,t}^{reg}$ are the frequency modulation quotation of conventional unit i and energy storage unit e in period t. $R_{i,t}^{reg}$, $R_{e,t}^{reg}$ are the winning bid frequency modulation capacity of conventional unit i and energy storage unit e in period t. M is the penalty factor for the relaxation of network power flow constraints used in market clearing optimization. NL is the total number of lines. SL_l^{+}, SL_l^{-}, SL_s^{+}, SL_s^{-} are the forward and reverse power flow slack variables of line l and section S. NS is the total number of sections. E is the total number of stand-alone energy storage trading units. λ_e^{ch} , λ_e^{dis} are the charging and discharging price declared by energy storage. $P_{e,t}^{dis}$, $P_{e,t}^{ch}$ are the energy storage charge and discharge power of the clearing.

The output of the unit is expressed by the following formula:

$$P_{i,t}=\sum_{m=1}^{NM}P_{i,t,m}+P_{i,\min} \quad (2)$$

$$P_{i,m}^{\min}\le P_{i,t,m}\le P_{i,m}^{\max} \quad (3)$$

Where NM is the total number of units quoted in sections. $P_{i,t,m}$ is the winning power of unit i in the mth output interval of period t. $P_{i,m}^{\max}$, $P_{i,m}^{\min}$ are the upper and lower bounds of the mth output interval declared by unit i.

The operating costs of the unit are expressed in the following formula:

$$C_{i,t}(P_{i,t})=\sum_{m=1}^{NM}C_{i,m}P_{i,t,m} \quad (4)$$

Where NM is the total number of units quoted in sections. $C_{i,t,m}$ is the energy price corresponding to the mth output interval declared by unit i.

2.2. Constraints

The constraints of the model include system power balance, frequency modulation demand, unit output and ramp-up limits, line power flow and energy storage operation constraints, etc., as follows.

System load balancing constraints are shown as follows:

$$\sum_{i=1}^{N}P_{i,t}+\sum_{j=1}^{NT}T_{j,t}+\sum_{e=1}^{E}P_{e,t}^{dis}+\sum_{e=1}^{E}P_{e,t}^{ch}=\sum_{u=1}^{U}D_{u,t} \quad (5)$$

Where $T_{j,t}$ is the planned power of the contact line j in the period t (positive feed, negative output). NT is the total number of contact lines. D_t is the system load of the period t. $P_{es,t}^{ch}$ 、 $P_{es,t}^{dis}$ are the energy storage unit e charges and generates power during the period t. E is the number of energy storage units.

Constraints on the demand for frequency modulation of the system are shown as follows:

$$\sum_{i=1}^{N}R_{i,t}^{reg}\ge R_t^{reg} \quad (6)$$

Where R_t^{reg} is the frequency modulation capacity requirement of the system in the period t.

The upper and lower limits of the unit output are shown as follows:

$$\begin{aligned}P_{i,t}+R_{i,t}^{reg}&\le\alpha_{i,t}P_{i,t}^{\max}\\P_{i,t}-R_{i,t}^{reg}&\ge\alpha_{i,t}P_{i,t}^{\min}\end{aligned} \quad (7)$$

Where $\alpha_{i,t}$ is the start-stop state of unit i in the period t, 0 indicates that the unit is stopped, and 1 indicates that the unit is turned on. $P_{i,t}^{\min}$, $P_{i,t}^{\max}$ are the minimum and maximum technical output of unit i in period t.

The upper limit constraints of frequency modulation of units are expressed as follows:

$$0\le R_{i,t}^{reg}\le R_{reg,i,t}^{\max} \quad (8)$$

Where $R_{reg,i,t}^{\max}$ is the maximum frequency modulation declaration amount of unit i in the period t.

The unit climb constraints are expressed as follows:

The rate of climbing up and down of the unit, which should meet the constraints, can be described as:

$$P_{i,t}-P_{i,t-1}\le\Delta P_i^{U}\left(1-\eta_{i,t}\right)+P_{i,t}^{\min}\eta_{i,t} \quad (9)$$

$$P_{i,t-1}-P_{i,t}\le\Delta P_i^{D}\left(1-\gamma_{i,t}\right)+P_{i,t}^{\min}\gamma_{i,t} \quad (10)$$

Where ΔP_i^U is the maximum climbing rate of unit i. ΔP_i^D is the maximum downward climbing rate of unit i.

The line power flow constraints are expressed as follows:

$$-P_l^{\max} \le \sum_{i=1}^{N} G_{l-i} P_{i,t} + \sum_{es=1}^{ES} (G_{l-e} P_{e,t}^{dis} + G_{l-e} P_{e,t}^{ch}) + \sum_{j=1}^{NT} G_{l-j} T_{j,t} - \sum_{k=1}^{K} G_{l-k} \left(\sum_{u \in k} D_{u,t} \right) - SL_l^+ + SL_l^- \le P_l^{\max} \quad (11)$$

Where $P_l^{\max}$ is the maximum power flow that line l can transmit. G_{l-i} G_{l-j} are the power transfer distribution factors of the node where unit i and contact line j located to line l, respectively. G_{l-e} is the sensitivity factor of the node where energy storage e is located to line l. K is the number of nodes in the system. G_{l-k} is the power transfer distribution factor of node k for line l. $D_{k,t}$ is the bus load value of node k in the period t.

Consider the power flow constraint for the critical section, which can be described as:

$$P_s^{\min} \le \sum_{i=1}^{N} G_{s-i} P_{i,t} + \sum_{es=1}^{ES} (G_{s-e} P_{e,t}^{dis} + G_{s-e} P_{e,t}^{ch}) + \sum_{j=1}^{NT} G_{s-j} T_{j,t} - \sum_{k=1}^{K} G_{s-k} \left(\sum_{u \in k} D_{u,t} \right) - SL_s^+ + SL_s^- \le P_s^{\max} \quad (12)$$

Where $P_S^{\min}$, $P_S^{\max}$ are the maximum power flows that section S can transmit. G_{s-i}, G_{s-j} are the power transfer distribution factors of the node where unit i and the connecting line j located to the cross-section S, respectively. G_{s-k} is the power transfer distribution factor of node k for section S.

The charging and discharging power of energy storage must be within the upper and lower limits.

$$\begin{aligned} &\alpha_{es,t} P_{es}^{dis,\min} \le P_{es,t}^{dis} + R_{e,t}^{reg} \le \alpha_{es,t} P_{es}^{dis,\max} \\ &\beta_{es,t} P_{es}^{ch,\max} \le P_{es,t}^{ch} - R_{e,t}^{reg} \le \beta_{es,t} P_{es}^{ch,\min} \\ &0 \le \alpha_{es,t} + \beta_{es,t} \le 1 \\ &P_{es}^{ch,\min} < 0, P_{es}^{ch,\max} < 0 \end{aligned} \quad (13)$$

Where $P_{es}^{dis,\max}$, $P_{es}^{dis,\min}$ are the maximum and minimum power limits of the discharge power of the energy storage es. $P_{es}^{ch,\max}$, $P_{es}^{ch,\min}$ are the maximum and minimum power limits of the charging power of the energy storage es. $\alpha_{es,t}$, $\beta_{es,t}$ are the charge-discharge state of energy storage es in the period t is 0-1 variable.

The state of charge of energy storage in the process of charging and discharging must be within the upper and lower limits.

$$E_{es,t} = E_{es,t-1} - \eta_{es}^{ch} P_{es,t}^{ch} \Delta t / E_{es} - P_{es,t}^{dis} \Delta t / (\eta_{es}^{dis} E_{es}) \quad (14)$$

$$\underline{E_{es,t}} \le E_{es,t} \le \overline{E_{es,t}} \quad (15)$$

Where $E_{es,t}$ is the state of charge of the energy storage es at the end of period t. η_{es}^{ch}, η_{es}^{dis} are the charging and discharging efficiency, charging and discharging efficiency of energy storage es are temporarily taken as the square root of the charge-discharge energy conversion efficiency. Δt is the length of the period. E_{es} is the rated capacity of the energy storage es. $\overline{E_{es,t}}$, $\underline{E_{es,t}}$ are the upper and lower limits of the state of charge of the energy storage es at the end of the period t.

The state of charge (SOC) of the energy storage at the end of the current operating day is equal to its declared expected state-of-charge at the end of the day (if declared).

$$E_{es,T} = E_{es}^{fin} \quad (16)$$

Where $E_{es,T}$ is the state of charge at the end of the current operating day of the energy storage es. E_{es}^{fin} is the expected value of the state of charge at the end of the current operating day declared by the energy storage es.

3. Energy storage integrated locational marginal pricing model for energy and frequency regulation markets

3.1. Objective function

The calculation of Locational Marginal Pricing (LMP) for energy storage participating in energy and frequency regulation markets necessitates integrated consideration of cross-market coupling effects and storage operational characteristics. Accounting for the interdependencies between energy and regulation markets, the LMP is derived through an optimization-based clearing model with the following objective function:

$$\begin{aligned} \min\{ &\sum_{i=1}^{N} \sum_{t=1}^{T} \left[C_{i,t}(P_{i,t}) + C_{i,t}^{U} + C_{i,t}^{p\min} + C_{i,t}^{reg} R_{i,t}^{reg} \right] \\ &+ \sum_{l=1}^{NL} \sum_{t=1}^{T} M \left[SL_l^+ + SL_l^- \right] + \sum_{s=1}^{NS} \sum_{t=1}^{T} M \left[SL_s^+ + SL_s^- \right] \} \\ &+ \sum_{e=1}^{E} \sum_{t=1}^{T} \left[\lambda_e^{dis} P_{e,t}^{dis} + \lambda_e^{ch} P_{e,t}^{ch} + C_{e,t}^{reg} R_{e,t}^{reg} \right] \end{aligned} \quad (17)$$

Where M is the penalty factor for the relaxation of network power flow constraints used in nodal price calculation.

3.2. Constraints

In the LPC calculation model, the constraints are partially consistent with those of the clearing model, but the following constraints are changed.

The upper and lower output limits constraints of the unit are shown as follows:

$$\begin{aligned} P_{i,t} + R_{i,t}^{reg} \le P_{i,t}^{\max} \\ P_{i,t} - R_{i,t}^{reg} \ge P_{i,t}^{\min} \end{aligned} \quad (18)$$

Where $P_{i,t}^{\min}$, $P_{i,t}^{\max}$ are 0 for units that are shut down in the clearing optimization results, $P_{i,t}^{\min}$, $P_{i,t}^{\max}$ are the winning output of unit *i* in the period *t* in the optimization results is obtained for non-priceable units. For priceable units, $P_{i,t}^{\min}$, $P_{i,t}^{\max}$ are taken as follows:

$$P_{i,t}^{\min} = \max\{(1-\delta)P_{i,t}^{SCED}, (P_{i,t}^{\min})^{SCED}\} \tag{19}$$

$$P_{i,t}^{\max} = \min\{(1+\delta)P_{i,t}^{SCED}, (P_{i,t}^{\max})^{SCED}\} \tag{20}$$

Where δ is the proportion of units that are allowed to deviate from the results of the day-ahead SCED optimization. $(P_{i,t}^{\min})^{SCED}$, $(P_{i,t}^{\max})^{SCED}$ are the maximum and minimum output of the unit in the SCED model in day-ahead market.

The upper and lower limits of energy storage output are set with reference to the above description.

The upper limit constraints of frequency modulation of units are expressed as follows:

$$R_{reg,i,t}^{\min} \le R_{i,t}^{reg} \le R_{reg,i,t}^{\max} \tag{21}$$

$$R_{reg,i,t}^{\min} = (1-\delta_{reg})R_{reg,i,t}^{SCED} \tag{22}$$

$$R_{reg,i,t}^{\max} = \min\{(1+\delta_{eg})R_{reg,i,t}^{SCED}, R_{reg,i,t}^{\max}\} \tag{23}$$

Where δ_{reg} is the proportion of the unit that the winning bid frequency modulation capacity deviates from the day-ahead SCED optimization result. $R_{reg,i,t}^{\max}$ is the maximum frequency modulation declaration amount of unit *i* in the period *t*.

The unit climb constraints are expressed as follows:

$$P_{i,t} - P_{i,t-1} \le \Delta P_i^U \tag{24}$$

$$P_{i,t-1} - P_{i,t} \le \Delta P_i^D \tag{25}$$

Where ΔP_i^U is the maximum climbing rate of unit *i*. ΔP_i^D is the maximum downward climbing rate of unit *i*.

The charging and discharging power of energy storage must be within the upper and lower limits.

$$\begin{gathered} \alpha_{es,t}P_{es}^{dis,\min} \le P_{es,t}^{dis} + R_{e,t}^{reg} \le \alpha_{es,t}P_{es}^{dis,\max} \\ \beta_{es,t}P_{es}^{ch,\max} \le P_{es,t}^{ch} - R_{e,t}^{reg} \le \beta_{es,t}P_{es}^{ch,\min} \\ 0 \le \alpha_{es,t} + \beta_{es,t} \le 1 \\ P_{es}^{ch,\min} < 0, P_{es}^{ch,\max} < 0 \end{gathered} \tag{26}$$

Where $\alpha_{es,t}$, $\beta_{es,t}$ are 0 for units that are shut down in the optimization results. $\alpha_{es,t}$ is 0, and $\beta_{es,t}$ is 1 for charging units. $\alpha_{es,t}$ is 1, and $\beta_{es,t}$ is 0 for discharge units.

4. Case analysis

4.1. Virtual power plant parameters

An energy storage case is used to verify the practicability and effectiveness of the above model. The case is set up with 39 nodes, 46 sections, including 10 coal-fired and 1 energy storage equipment, with 96 periods of 15 minutes per day.

The system load curve is shown in Figure 1, with a maximum load of 1489MW.

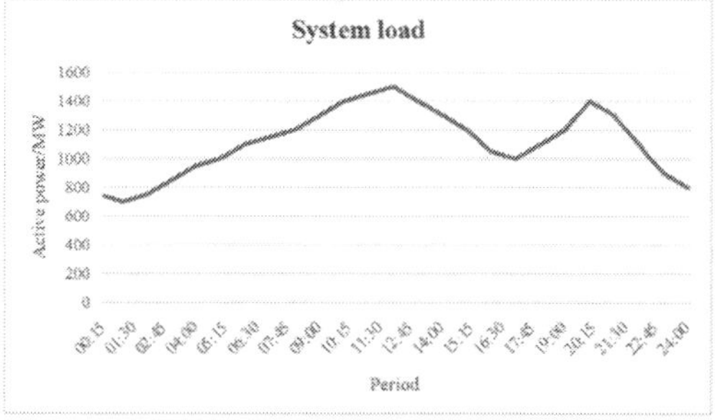

Figure 1. System load curve.

The frequency modulation demand of the system is 150MW throughout the day.

Table 1 shows the parameter settings of the unit. Among them, units 1-10 are coal-fired units, and unit 11 is energy storage.

Table 1. Unit parameters.

Unit number	Rated capacity (MW)	Minimum output (MW)	Initial output (MW)	Ramp rate (MW/min)	Frequency regulation capacity (MW)	Frequency regulation bids (￥/MW)
1	455	150	350	0.75	152.5	200
2	455	150	150	0.75	152.5	150
3	130	20	30	0.75	55	100
4	130	20	30	0.75	55	150
5	162	25	30	0.75	68.5	50
6	80	20	0	0.5	30	80
7	85	25	0	0.5	30	120
8	55	10	20	0.5	22.5	100
9	55	10	20	0.5	22.5	120
10	55	10	0	0.5	22.5	100
11	50	-50	0	/	25	130

Table 2 shows the energy storage parameters.

Table 2. Energy storage parameters.

Initial SOC	0.6
Maximum SOC	0.9
Minimum SOC	0.1
End of the day SOC	0.5
Maximum charging/discharge power	50
Minimum charging/discharge power	0
Charging/discharge efficiency	0.8

The bids of coal-fired units adopts the form of segmented bids, and Figure 2 shows the price range of each bids section of the unit.

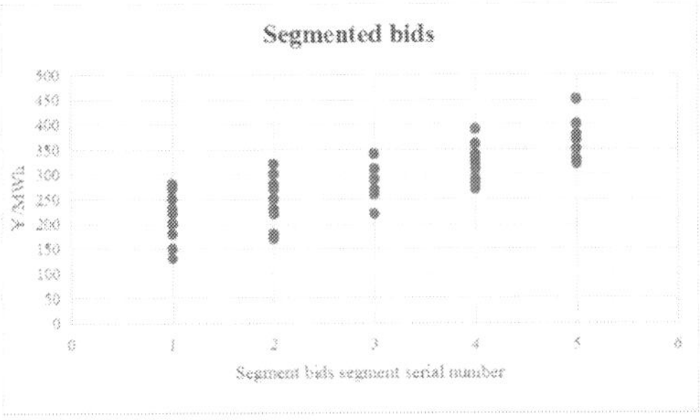

Figure 2. Bids of coal-fired units in sections.

The energy storage bids is shown in Table 3.

Table 3. Energy storage bids.

Unit number	Quotation segment serial number	Bids (￥/MWh)	Initial capacity (MW)	Terminate capacity (MW)
11	1	150	-50	0
11	2	200	0	50

4.2. Analysis of results

According to the calculation results of the example, the system boundary electricity price is shown in Figure 3. It can be seen that the trend of electricity price is basically consistent with the load fluctuation, and at noon, the electricity price reaches the highest level of the whole day, and is higher than the highest unit quotation, because there is a section blockage, and the marginal price contains the opportunity cost of the unit's frequency modulation to win the bid.

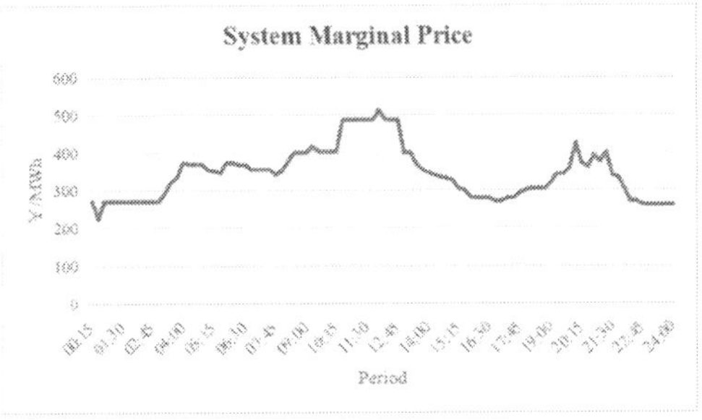

Figure 3. System marginal price.

The energy storage optimization results are depicted in Figure 4. As observed, the storage output demonstrates synchronized variation with load profiles, effectively achieving peak shaving and valley filling. The state of charge (SOC) fluctuates within predefined operational limits while reaching the scheduled daily terminal SOC. Concurrently, the system provides substantial frequency regulation capacity reserves.

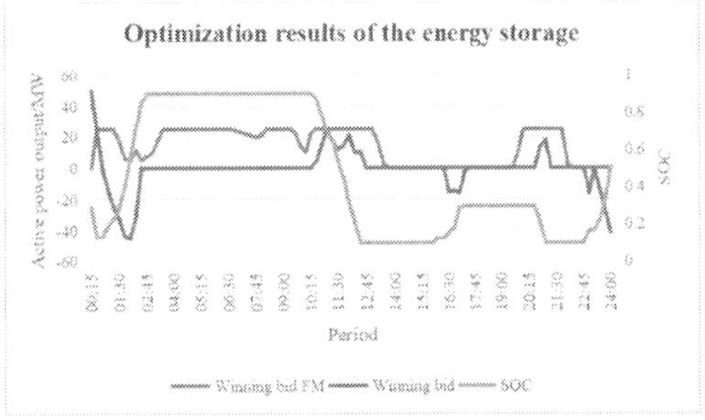

Figure 4. Optimization results of energy storage.

The operational output of Unit 2 and its frequency regulation award status are illustrated in Figure 5. Notably, neither the energy storage system nor Unit 2 secured frequency regulation bids during the 14:00–19:00 period. This outcome stems from reduced system load during these hours, which increased aggregate frequency regulation capacity availability across generation units. Under such conditions, market clearing mechanisms prioritize lower-cost providers, as reflected in the bidding outcomes. The abundance of regulation capacity during low-load periods leads to the non-selection of higher-priced units, conclusively validating the market's inherent preference for cost-efficient resource dispatch.

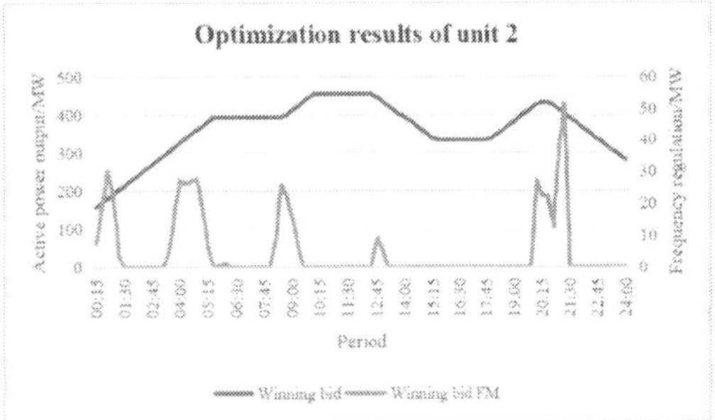

Figure 5. Optimization results of unit 2.

5. Conclusion

Under the dual carbon targets, high-penetration renewable energy integration has exacerbated power system flexibility deficits. While energy storage serves as a critical flexibility resource capable of concurrently providing energy and frequency regulation services, existing market mechanisms with separate clearing processes for these services lead to inefficient resource allocation. This paper proposes a joint clearing model that co-optimizes energy and frequency regulation markets by incorporating storage operational constraints. The model enables energy storage to deliver dual services—energy time-shifting and frequency regulation—simultaneously addressing system flexibility requirements, reducing locational marginal costs, and enhancing both storage utilization and systemic economic efficiency. This framework provides crucial technical support for energy storage participation in electricity markets and sustainable operational strategies under market-driven paradigms.

References

1. DONG Yu, SUN Dayan, XU Dan, et al. Challenges, Response and Prospects for Power Balance in New Power Systems[J/OL]. Proceedings of the CSEE,1-18.
2. CHENG Qiyun, LI Feng, LUO Shuxin, et al. Research on the Planning Methodology Framework and Key Supporting Technologies for New-type Power Systems [J]. Power System Technology, 2025, 1-14.
3. ZHANG Yongjun, CHENG Runting, SHI Xuntao, et al. Review of research on key technologies and mechanisms for distributed smart grid in market environment[J/OL]. Electric Power Automation Equipment, 1-18.
4. ZHONG Haiwang, ZHANG Ning, DU Ershun, et al. Planning, Operation and Market of New Power System: Research Progress and Practice[J]. Proceedings of the CSEE, 2024, 44(18):7084-7104.
5. WANG Peng, ZHANG Di, ZHANG Yongjun, et al. Research and Prospect of Key Technology Application of New Power System Digital and Intelligent Transformation[J/OL]. Power System Protection and Control,1-13.
6. WANG Yuzhao, GE Pengjiang, LIU Jichun, et al. Research on Method of Joint Clearing of Electricity Market and Ramping Market Incorporating Energy Storage Resources[J/OL]. Power System Technology, 1-13.
7. LIU Jichun, LIU Kexin, KE Xianbo, et al. Participation of Energy Storage in Electric Energy-Inertia-Primary Frequency Regulation Multi-market Trading Method Under A New Power System[J/OL]. Power System Technology,1-17.
8. CHEN Changming, ZHANG Tianhan; SHEN Zikang, et al. Review of Shared Energy Storage Business Models on Generation and User Sides in New Power Systems[J]. Automation of Electric Power Systems, 2025, 1-23.
9. LIU Zifa, LI Kunyang, ZHENG Siqi. Daily Intra Day Collaborative Rolling Optimization Strategy for Solar Energy Storage Power Stations Participating in Multiple Application Scenarios[J]. Power System Technology, 2025, 49(01):198-208.

Optimization Methods for the Cylinder Mouth Structure of Type IV On-board Hydrogen Storage Cylinder under Various Influencing Factors

Enhui Zhang[1,2,*], Zhiqiang Wang[1], Xiaolong Shi[1], Yangchun Zhao[1], Jiahui Zhang[1], Tao Liu[1], and Wenchao Wang[1]

[1] School of Mechanical Engineering, Inner Mongolia University of Science & Technology, Baotou 014010, China
[2] Inner Mongolia Key Laboratory of Intelligent Diagnosis and Control of Mechatronic System, Baotou 014010, China

Abstract. This research investigates factors affecting the sealing reliability of Type IV hydrogen cylinder mouths under high-pressure fast-refueling conditions. A three-dimensional numerical analysis model and hydrogen energy conservation equation are developed to analyze sealing interface deformation incorporating static mechanics and computational fluid dynamics. Results demonstrate that internal pressure predominantly influences interface deformation, while external vibrations from road surfaces show negligible impact. Temperature emerges as the most critical factor: rising hydrogen temperatures increase thermal and total deformations, though their distribution patterns remain consistent. These findings provide theoretical guidance for optimizing hydrogen cylinder sealing designs and enhancing operational reliability. The research methodology integrates multiphysics simulations to evaluate combined mechanical-thermal effects, highlighting temperature management as a key consideration for maintaining seal integrity during rapid refueling cycles. And a control strategy to control the deformation generated during rapid filling of hydrogen storage bottles by temperature is proposed to improve the cooling efficiency.

1. Introduction

The development and application of new energy systems has emerged as both a focal point in international academia, and a strategic priority for global governments[2]. Hydrogen energy, characterized by its high calorific value, renewability, and zero-emission properties, demonstrates significant potential in future transportation systems[3], which in turn has led to the widespread interest in Hydrogen Fuel Cell Vehicles(HFCV)[4-5]. Hydrogen storage is the key link between hydrogen production and utilization[6]. Presently, hydrogen storage cylinders on the market are mainly divided into four types: type I (all-metal), type II (metal liner with carbon fiber ring winding), type III (aluminum alloy liner with carbon fiber fully-wrapped), and type IV (non-metal liner with carbon fiber fully-wrapped)[7]. With the rapid development of HFCV industry, Type IV on-board hydrogen storage cylinder has become a global research hot spot due to its high hydrogen storage density and excellent fatigue resistance[8].

Current research is focused on the following key areas: firstly, new materials are being developed by scholars worldwide to enhance their ability to block hydrogen and prevent structural failures (such as blistering, collapsing, or peeling) of the plastic liner caused by prolonged contact with high-pressure hydrogen[9-13]; secondly, the winding pattern and angle of carbon fibers are being optimized to enhance the pressure-bearing capacity of hydrogen storage cylinder[14-16]. In addition, the performance of hydrogen storage cylinder under extreme conditions (e.g., severe temperature difference, ultra-high pressure, fire, and mechanical shock) is being thoroughly tested and analyzed[17-19]; finally, effective strategies are being pursued by researchers to reduce the increase in the internal temperature of hydrogen storage cylinder during the refueling process[20], including the application of pre-cooling refueling technology and optimization of thermal management systems of hydrogen storage cylinders[21].

Despite significant advancements in material innovation and structural design of Type IV full-wrapped hydrogen storage cylinders, research on sealing technologies for cylinder ports remains inadequate. This is particularly evident at the sealing interface between the valve seat and the plastic liner. Under high-pressure rapid filling conditions, the increased internal hydrogen temperature induces accelerated heat transfer from the metal cylinder valve to the sealing interface. Divergent thermal expansion properties of metallic and polymeric materials result in progressive geometric distortion at the sealing interface. Over time, this thermally-induced deformation leads to cumulative dimensional instability, ultimately manifesting as degraded sealing performance, heightened hydrogen leakage risks and reduced operational lifespan of the storage system. The thermo-mechanical deformation mechanism of the cylinder port seal interface was systematically investigated to provide reference for the optimization of the structural reliability of the seal system.

* Corresponding author: Enhui Zhang/zhangenhui2008@163.com

2. Mathematical model

The seal interface is subjected to complex thermodynamic and mechanical effects during hydrogen filling of hydrogen storage cylinders, and the following simplifying assumptions are made to better predict the seal interface performance based on deformation:

1. It is assumed that the initial phase of refueling does not take into account the backflow of hydrogen or its escape through the walls of hydrogen storage cylinder.
2. Because of the high velocity of the hydrogen during refueling, the residence time in the valve is very short, ignoring the energy exchange between the hydrogen entering hydrogen storage cylinder and the valve.
3. It is assumed that the thermal conductivity of each material is presumed to be constant.
4. It is assumed that the mass flow rate and temperature of the hydrogen entering hydrogen storage cylinder remain constant throughout the refueling process.
5. Neglecting the influence of buoyancy effects on the heat transfer characteristics of the hydrogenation process.

For a transient open system, the energy conservation equation for the system in (dt) time is:

$$\frac{d(mu)}{dt}+Q-\Delta Q=\frac{dm_1}{dt}h_{in}-\frac{dm_2}{dt}h_{out}+W \tag{1}$$

Where (u) is the specific internal energy of hydrogen, (m) is the mass of hydrogen remaining in hydrogen storage cylinder, (m_1) is the mass of the hydrogen entering hydrogen storage cylinder, (m_2) is the mass of the hydrogen flowing out of hydrogen storage cylinder, (h_{in}/h_{out}) is the specific enthalpy of the hydrogen flow into/out of the system, (W) is the external work done by the system, (Q) is the heat exchange of the hydrogen with cylinder walls and the combination valve, (ΔQ) is the heat flowing out of the system,.

Combining the assumptions Eq.(1) can be expressed as:

$$\frac{d(mu)}{dt}+Q-\Delta Q=\frac{dm}{dt}h_{in} \tag{2}$$

Where the heat exchange between hydrogen and the plastic liner according to Newton's law of cooling is:

$$Q_1=k_{PA}A_{H-PA}(T_H-T_{PA}) \tag{3}$$

Where (k_{PA}) is the thermal conductivity of plastic liner, (A_{H-PA}) is the area of contact between hydrogen and plastic liner, (T_H) is the temperature of hydrogen, (T_{PA}) is the temperature of the plastic liner.

The heat exchange between the hydrogen and the combination valve is:

$$Q_2=k_{valve}A_{H-valve}(T_H-T_{valve}) \tag{4}$$

Where (k_{valve}) is the thermal conductivity of the combination valve, $(A_{H-valve})$ is the area of contact between the hydrogen and the combination valve, (T_{valve}) is the temperature of the combination valve.

Considering that the thickness of the plastic liner is inhomogeneous and using the differential form according to Fourier's law, the heat exchange between the plastic liner and the carbon fiber layer is:

$$Q_3=\int_{x_2}^{x_1}\left(-k_{PA}\frac{dT_{PA}}{dy}\right)dx \tag{5}$$

Where (dT_{PA}/dy) is the temperature gradient of the plastic liner at position (y), (x_1) and (x_2) are the start and end points of the heat transfer region between the plastic liner and the carbon fiber, respectively, (y) is a point within (x_1) and (x_2) intervals.

The heat exchange between the valve and the plastic liner is:

$$Q_4=\int_{x_4}^{x_3}\left(-k_{valve}\frac{dT_{valve}}{dz}\right)dx \tag{6}$$

Where (dT_{valve}/dz) is the temperature gradient of the combination valve at position (z), (x_3) and (x_4) are the start and end points of the heat transfer region between the combination valve and the plastic liner, respectively, (z) is a point within (x_3) and (x_4) intervals.

The portion of the valve in contact with the carbon fiber layer is uniform, and the heat exchange between the valve and the carbon fiber layer according to Fourier's law is:

$$Q_5=-k_{valve}A_{V-CFRP}\left(\frac{T_{valve}-T_{CFRP}}{d_{valve}}\right) \tag{7}$$

Where (A_{V-CFRP}) is the contact area between the valve seat and carbon fiber layer, (T_{CFRP}) is the temperature of carbon fiber layer, (d_{valve}) is the thickness of the portion of the valve seat in contact with the carbon fiber layer.

Combining (Eqs.(3)-(7)), the energy conservation equation for the process can be expressed as:

$$\frac{d(mu)}{dt}+\sum_{i=1}^{n}Q_i-\Delta Q=\frac{dm}{dt}h_{in} \tag{8}$$

where $n=1,2,3,4,5$. (Q_1) is the heat exchange between hydrogen and the plastic liner; (Q_2) is the heat exchange between hydrogen and the combination valve; (Q_3) is the heat exchange between the plastic liner and the carbon fiber layer; (Q_4) is the heat exchange between the combination valve and the plastic liner, and (Q_5) is the heat exchange between the combination valve and the carbon fiber layer.

3. Finite element model

3.1. Geometric model

In this paper, a new Type IV hydrogen storage cylinder mouth structure is proposed, whose structure consists of a carbon fiber layer, a plastic liner and a combination valve, with a design pressure of 70MPa and a volume of 81L, with specific structural and dimensional parameters shown in Figure 1 and Table 1.

Table 1. Main design size table of hydrogen storage cylinder.

Parameter	Numerical value (mm)
Thickness of carbon fiber layer	15
Thickness of plastic liner	6
Plastic inner diameter	372
Length of hydrogen storage cylinder	880
Inlet Pipe inner diameter	15
Maximum diameter of valve	120

As shown in Figure 2(a), the sealing structure consists of a valve seat, a valve core, two O-rings, a gasket, a nut, a sealing ring, a plastic liner and a carbon fiber layer. The end of the valve core is designed with threads, through the nut and the gasket will be fixed in the valve seat, in addition to the valve core and the gasket will be fixed in the valve seat, in addition to the valve core and the valve seat is equipped with sealing grooves for the placement of O-rings, in order to enhance the connection between the valve seat and the plastic liner tightness, the two are designed for the contact surface of the irregular contact and the use of sealing ring to prevent the leakage of the hydrogen; The carbon fiber layer is wrapped into the rectangular groove of the head of the valve seat to ensure the reliability of the cylinder mouth structure.

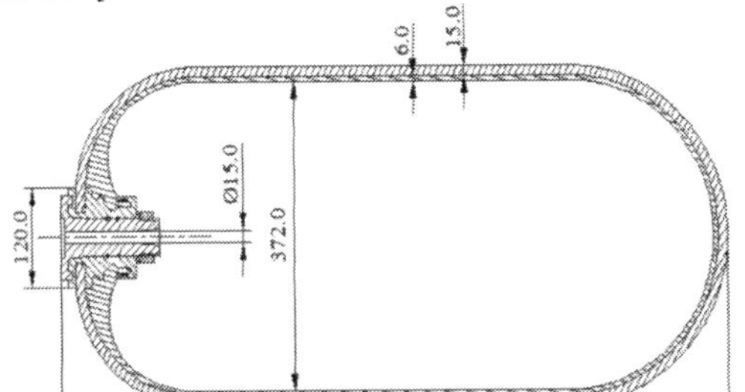

Figure 1. Schematic diagram of the overall dimensions of the hydrogen storage cylinder.

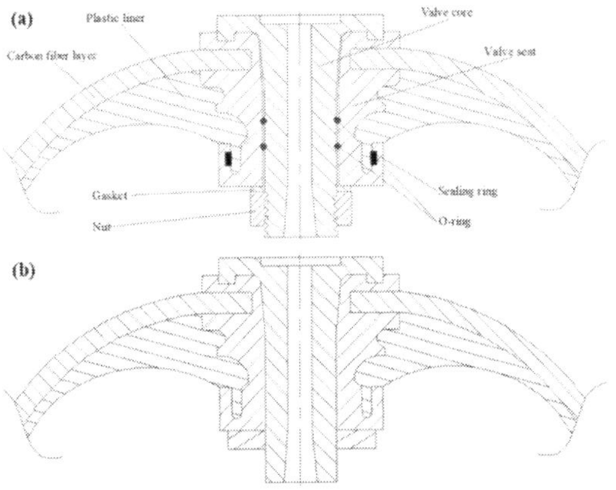

Figure 2. Schematic diagram of hydrogen storage cylinder mounth;(a) Static analysis model of the mouth structure of hydrogen storage cylinder; (b) CFD analysis model of the mouth structure of hydrogen storage cylinder.

3.2. Grids

In order to simplify the model and improve the model convergence as well as the computational accuracy, this paper assumes that the model is a completely axisymmetric structure, and the materials of each component are continuous and uniform, so as to establish a three-dimensional axisymmetric model. Tetrahedral elements are used to mesh the different components separately, and the contact sizing setting is used to locally encrypt the thread part. Contact match is introduced at the contact area between different components to make them share the mesh nodes to more accurately describe the contact between different components. The mesh quality is ensured to be concentrated above 0.75.

3.3. Boundary conditions and load application

In this research, the stresses on the components when the pressure inside the hydrogen storage cylinder reaches 70MPa are analyzed by simulation. As shown in Figure 3(a), the pressure loads applied to the hydrogen storage cylinder are distributed in different directions. 70 MPa pressure loads are uniformly applied to the inner surface of the plastic liner, and the same circumferential stresses of 70 MPa are applied to the radial surfaces of the end of the valve seat and the surface of the nut. According to the thin film theory, an axial pressure of 35 MPa applied at the hydrogen storage tank port was applied to the seat end surface. In addition, the thread contact surface is set as friction contact. As shown in Figure 3(b), the hydrogen storage tank was fixed by two rigid fixing rings, and fixed restressing was set at the position of the surface of the fixing rings shown in the Figure to simulate the stability of the hydrogen storage tank in actual use.

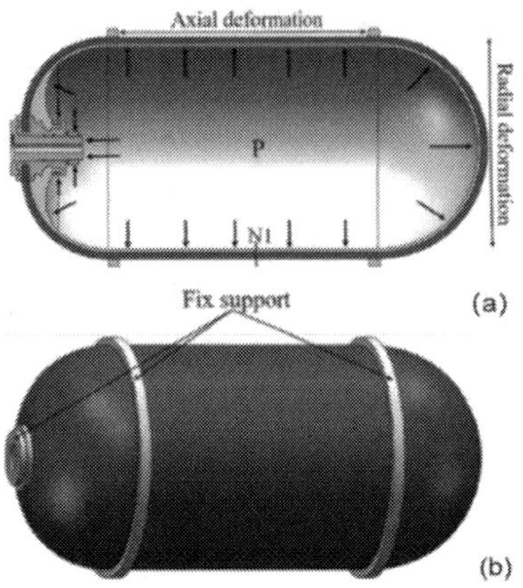

Figure 3. Loading conditions in hydrogen storage cylinder;(a) Internal pressure diagram of hydrogen storage cylinder;(b) Position of the fixed support.

3.4. Verification of convergence

Grid convergence tests were performed to ensure the accuracy and economy of the computational results. Numerical simulations were carried out using different number of tetrahedral meshes. Figure 4 shows the significant changes in the stress values of the components after mesh refinement.

By analyzing the numerical convergence, it is found that the simulation results tend to be stable with increasing grid fineness, thus confirming the reliability of the numerical simulations. The simulations show that excessive number of grides have limited effect on the results. Therefore, the number of grids chosen in this paper is 817408.

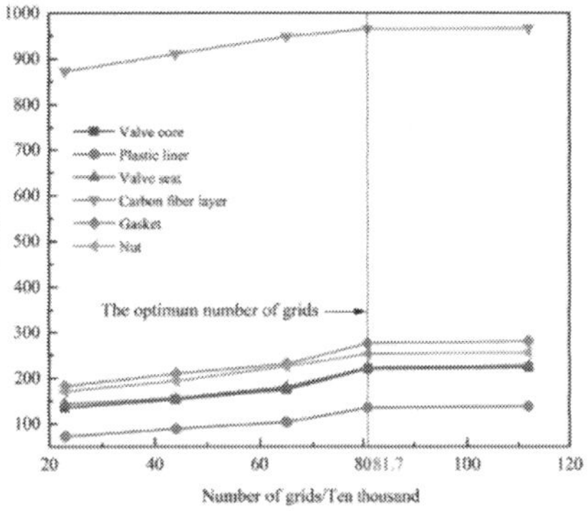

Figure 4. Stress variation curves of the components of the hydrogen storage cylinder at different grid numbers.

4. Numerical simulation analysis and discussion

4.1. Static analysis

When the high hydrogen pressure inside the hydrogen storage cylinder reaches 70MPa, the stress on the various components that make up the hydrogen storage cylinder are shown in Figure 5.

According to the data in Table 1, the outer diameter of the plastic liner is 384 mm, the thickness of the carbon fiber layer is 15 mm, and the design pressure is 70MPa, which belongs to a thin-walled cylinder. Approximating the thin-film theory at point N1 shown in Figure 3, the circumferential and axial stresses are calculated as follows:

$$\sigma_\theta = \frac{PD}{2\delta} \tag{9}$$

$$\sigma_\varphi = \frac{PD}{4\delta} \tag{10}$$

It is assumed that there is no shear stress and radial stress at point N1, the Von-Mises stress equation at this point can be simplified as:

$$\sigma_{vm} = \sqrt{\sigma_\theta^2 - \sigma_\theta\sigma_\varphi + \sigma_\varphi^2} \tag{11}$$

Where, (σ_θ) is circumferential stress, MPa, (P) is hydrogen storage cylinder internal pressure, MPa; (D) is outer diameter of plastic liner, m; (δ) is thickness of carbon fiber layer, m; (σ_φ) is axial stress, MPa; (σ_{vm}) is equivalent stress at point N1, MPa.

According to (Eqs(9)-(11)), the theoretical solutions for the circumferential stress, axial stress, and equivalent stress at point N1 are obtained as $\sigma_\theta = 896MPa$, $\sigma_\theta = 224MPa$, $\sigma_{vm} = 807.64MPa$, respectively. However, due to the presence of the external fixing ring constraints, stress concentration occurs in the region near the two fixing rings.

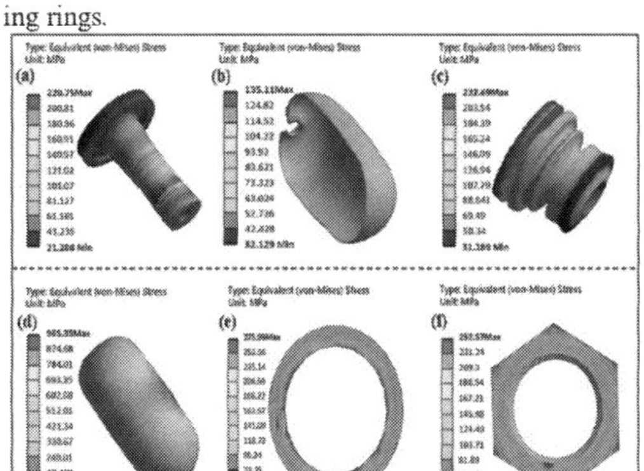

Figure 5. Cloud plots of stress for each part of the hydrogen storage cylinder; (a) Cloud plot of the stress distribution of the valve core; (b) Cloud plot of the stress distribution of the plastic liner;(c) Cloud plot of the stress distribution of the valve seat; (d) Cloud plot of the stress distribution of the carbon fiber layer; (e) Cloud plot of the stress distribution of the gasket; (f) Cloud plot of the stress distribution of the nut.

To evaluate the effect of stress concentration, a finite element structural analysis is performed in this research using ANSYS Workbench software. As shown in Figure 5(d), the stress distribution of the carbon fiber layer under the constraint of the external fixing rings is shown in detail. In particular, the maximum stress value reaches 965.35MPa in the stress concentration region including point N1. Compared with the yield stress of the material carbon fiber T700s in Table 2, this stress level still meets the criterion of 1.5 times the safety factor in the design requirements.

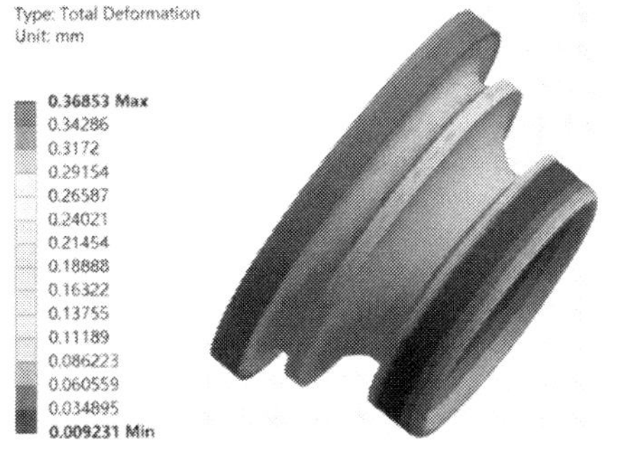

Figure 6. Deformation contour plot of sealing interface.

Combined with Table 2, it can be seen that in the static structure analysis of the designed hydrogen storage cylinder, the maximum stresses on each component are significantly lower than the yield limit of the corresponding material, indicating that the structure is able to withstand the loads under actual working conditions in all aspects.

Table 2. Table of material properties and stress on hydrogen storage cylinder.

Part Name	Material	Yield stress (MPa)	Maximum stress (MPa)
Valve core	Al 6061	370	220.75
Plastic liner	PA46	207	135.11
Valve seat	Al 6061	370	222.69
Carbon fiber	T700s	1450	965.35
Gasket	Al 6061	370	275.98
Nut	Al 6061	370	252.57

The deformation of the sealing interface under the influence of high internal hydrogen pressure is depicted in Figure 6. The deformation across the sealing interface appears relatively uniform. The maximal deformation occurs on the right side of the sealing interface, where it reaches up to 0.36 mm. This increased deformation is attributed to the reduced wall thickness of the valve seat at this location, which undergoes localized compression due to the high internal pressure of the hydrogen storage cylinder. Furthermore, the Figure distinctly highlights that areas with smaller fillet radii experience greater deformation.

4.2. Modal analysis

In order to gain an in-depth understanding of the intrinsic properties of the hydrogen storage cylinder structure under vibration conditions and to identify the intrinsic vibration frequencies of the structure to avoid resonance phenomena, this paper performs a modal analysis of the hydrogen storage cylinder and identifies the key vibration modes. Taking the constrained position in Figure 3(b), the first 6 orders of intrinsic frequency and mode shape of the hydrogen storage cylinder are shown in Table 3.

Table 3. Table of the first 6 orders of intrinsic frequency and vibration pattern of the hydrogen storage cylinder under constrained modes.

modal order	natural frequency/(Hz)	mode shape
1	1920.6	bending
2	2580.7	bending
3	2895.1	bending
4	2901.3	torsional
5	3371.7	bending-torsional mode
6	3622.7	bending-torsional mode

The main sources of vibration excitation during the operation of HFCV include vibration caused by road surface unevenness, which is usually less than 20 Hz, and local vibration caused by unbalance of wheels and other rotating parts, which is in the frequency range of about 33 Hz to 50 Hz. Table 3 shows that the natural frequencies of the hydrogen storage cylinder in the constrained modes are significantly higher than these frequency ranges, the design of the hydrogen storage cylinder is therefore considered to meet the requirements for use.

4.3. Harmonic response analysis

To investigate the possible resonance of the mouth position of the hydrogen storage cylinder due to road excitation during vehicle operation under adverse road conditions, which may cause large deformation of the sealing interface and affect driving safety, this research conducted a harmonic response analysis based on the results of the constrained modal analysis by using the modal superposition method.

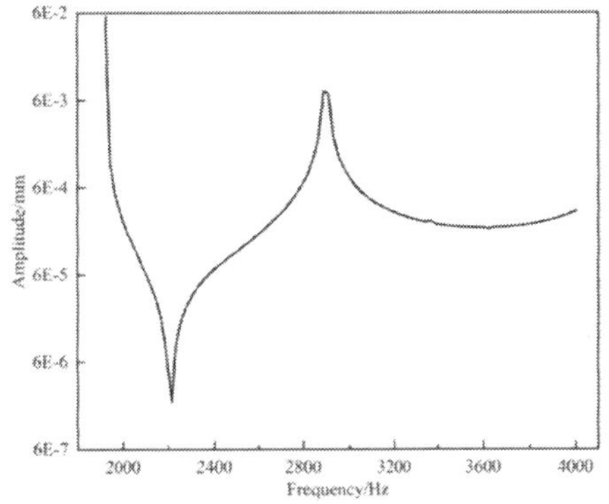

Figure 7. Amplitude response at the sealing interface at different frequencies.

In the Analysis Setting, the upper frequency value is set to 1900 Hz and the lower frequency value is set to 4000 Hz; the calculation interval is set to 100; a force of 1056 N perpendicular to the ground is applied to the external fixing rings of the hydrogen storage cylinder, and the overall deformation curve of the sealing interface of the mouth of the hydrogen storage cylinder under the above conditions is shown in Figure 7. It can be seen that the deformation of the sealing interface reaches a maximum value of about

0.06mm at 1921Hz. As the frequency increases, the deformation varies in a waveform. Although the fluctuation is relatively large, the deformation value of the sealing interface is on the small side.

4.4. Thermal analysis of the sealing interface

In order to investigate the effect of the rising internal temperature of the hydrogen storage cylinder on the sealing interface of the cylinder mouth under the condition of high-pressure, fast-filling, the thermal deformation and total deformation of the sealing interface are analyzed by using finite element analysis method.

The model in Figure 2(a) is first simplified by neglecting the chamfers on each contact surface, the O-rings and the sealing ring are removed, since the focus of this research is on the sealing interface, the gasket, and thread details on the valve core and nut is ignored to improve computational efficiency. The simplified cylinder mouth structure is shown in Figure 2(b).

As shown in Figure 8(a), the numerical domain includes several regions of the combination valve, the hydrogen, the plastic liner, and the carbon fiber layer. A shared topology is used to divide the grids, and the BOI technique is used to locally encrypt the grids in the middle region of the hydrogen region where the flow rate is high, in addition to the boundary layer grids are incorporated to more accurately simulate the near-wall flow and heat transfer phenomena.

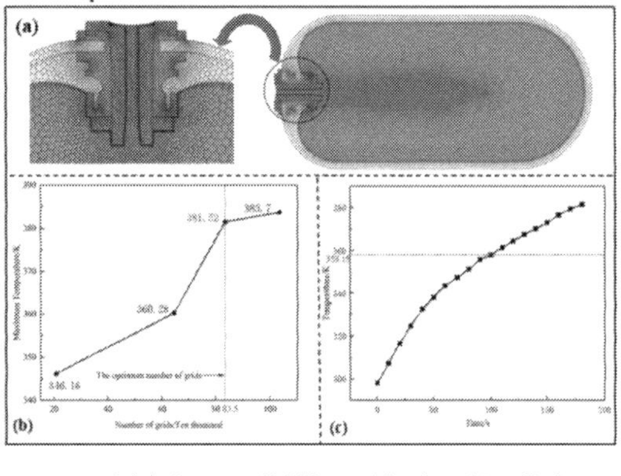

Figure 8. Finite element reliability verification; (a) Cylinder geometry and computational grids; (b) Maximum temperature diagram inside hydrogen storage cylinder at the end of refueling for different number of grids; (c) Hydrogen temperature rise curve inside the hydrogen storage cylinder during refueling.

The maximum temperatures inside the hydrogen storage cylinder at the end of refueling for different number of grids during the simulation are shown in Figure 8(b). The computational time increases significantly when the number of grids increases from 206584 to 1034216. However, when the number of grids reaches 835502, further increases in the number of grids result in only a 2.18°C increase in the maximum temperature, indicating that a good balance between accuracy and computational efficiency has been achieved at this number of grids. Therefore, 835502 grids are considered sufficient to ensure simulation independence and computational efficiency. In addition, the step size is set to 0.01s; the inlet boundary condition is set to a constant flow rate refueling of 10g/s with a refueling time of 180s; the temperature of the hydrogen source is maintained at 25°C, and the initial pressure of the hydrogen storage cylinder is assumed to be 2MPa[22].

The temperature rise of the hydrogen during the refueling process at the optimal number of grids is illustrated in Figure 8(c). Without active control measures such as pre-cooling, the hydrogen temperature inside the hydrogen storage cylinder reaches the 85°C limit specified by the industry standard at t=100.45s. Refueling continued for 180s, at which time the temperature inside hydrogen storage cylinder rose to 381.52K. The variation of the internal parameters of the hydrogen storage cylinder during the high-pressure, fast-refueling condition is depicted in Figure 9.

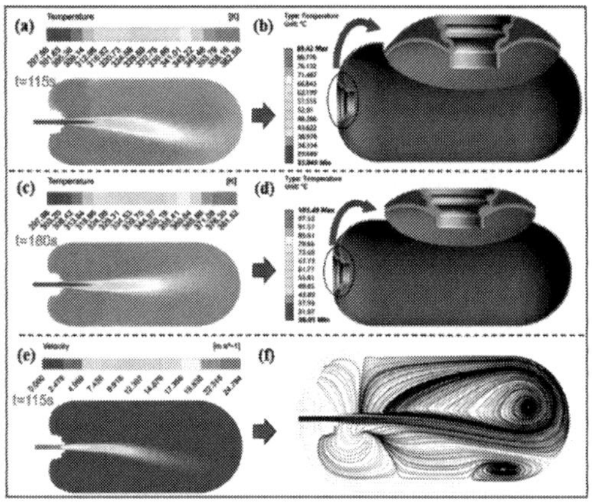

Figure 9. Variation of internal parameters of hydrogen storage cylinder during the refueling process;(a) Cloud plot of the temperature distribution inside the hydrogen storage cylinder at 115s; (b) Cloud plot of the temperature distribution of the plastic liner at 115s; (c) Cloud plot of the temperature distribution inside the hydrogen storage cylinder at 180s; (d) Cloud plot of the temperature distribution of the plastic liner at 180s;(e) Cloud plot of the velocity distribution inside the hydrogen storage cylinder at 115s; (f) Hydrogen streamline distribution inside the hydrogen storage cylinder at 115s.

The temperature distribution of the hydrogen inside the hydrogen storage cylinder at t = 115s is shown in Figure 9(a). At this time, the surface temperature of the plastic liner reaches 85 °C, while the temperature at the sealing interface at the mouth of the cylinder is about 70 °C. The temperature inside the hydrogen storage cylinder increases further as the refueling continues. The temperature distribution at t=180s is shown in Figure 9(c) , at which time the surface temperature of the plastic liner reaches 103.49°C, while the temperature of the cylinder mouth sealing interface is about 85°C. Combined with Figure 9(b)(d), it can be seen that with the increase of the internal temperature of the hydrogen storage cylinder, the temperature at the sealing interface increases accordingly, and the heat gradually spreads along the axial direction of the plastic liner and the radial direction of the sealing interface at the head position. The velocity distribution and streamlines of the hydrogen inside the hydrogen storage cylinder at t = 115s are shown in Figure 9(e)(f). During the refueling process, the formation of the two vortices increases the friction between the hydrogen molecules, resulting in a faster rate of temperature increase.

Due to the poor thermal conductivity of the plastic liner, the sealing interface is susceptible to thermal deformation at high temperatures for long periods of time. The resulting thermal deformation and total deformation is illustrated in Figure 10.

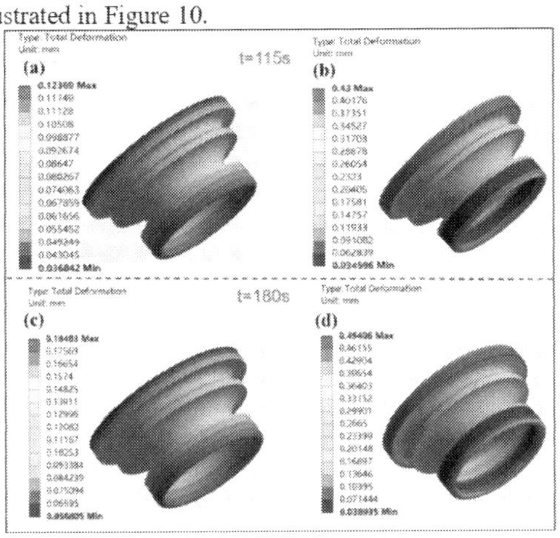

Figure 10. Thermal and total deformation cloud plots of the sealing interface of hydrogen storage cylinder at different times;(a) Cloud plot of thermal deformation of seal interface at 115s; (b) Cloud plot of total deformation of seal interface at 115s; (c) Cloud plot of thermal deformation of seal interface at 180s;(d) Cloud plot of total deformation of seal interface at 180s.

Due to the poor thermal conductivity of the plastic liner, the sealing interface is prone to thermal deformation at high temperatures over long periods of time. The resulting thermal deformation and total deformation are shown in Figure 10. As shown in Figure 10(a)(b), the maximum thermal deformation of the sealing interface occurs in the thinnest part of the plastic liner at t=115s. Both sides of this area are in contact with the hotter seat, resulting in faster heat transfer, with thermal and total deformation of 0.12 mm and 0.43 mm, respectively. As shown in Figure 10(c)(d), at t=180s, the location where the thermal deformation occurs is almost unchanged, but the thermal deformation reaches 0.18 mm, which is 49.4% higher than the value at 115s, and the total deformation reaches 0.49mm, which is 14.9% higher than the value at 115s. The shortening of the hydrogen refueling time further increased the internal temperature of the hydrogen storage tank, which led to an increase in the thermal deformation of the seal interface, increasing the risk of hydrogen leakage and decreasing the reliability of the seal.

No change in the conditions of the raise, seal interface thermal deformation and total deformation with filling time are shown in Figure 11. With the change of charging time, without pre-cooling, the thermal deformation and total deformation of the sealing interface increased linearly, and at 100.45 seconds, the internal temperature reached 85°C(358.15K), and the maximum deformation reached

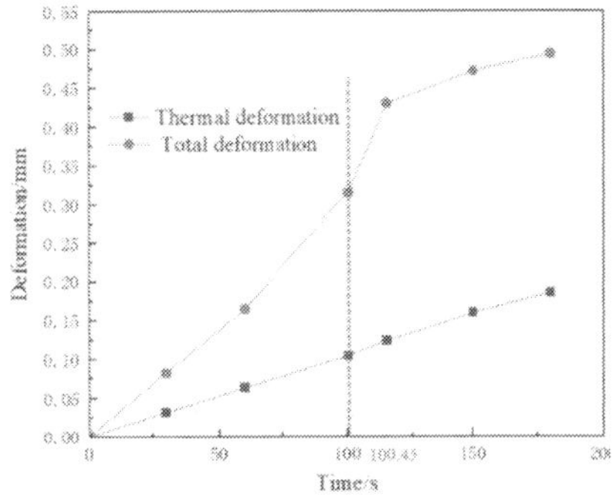

Figure 11. Trend of thermal deformation and total deformation of sealing interface with charging time.

0.31mm, and the deformation showed a positive correlation with the temperature. It is necessary to be able to control the sealing interface from the controlled temperature.

5. Research on temperature rise and deformation control of sealing interface

Combining the theoretical model and numerical simulation results established in the previous section, this section carries out the hydrogen gas temperature rise and seal interface deformation test during the filling process of hydrogen storage cylinder. The reasonableness of the constructed mathematical model and numerical simulation results is verified by comparing the test data with the theoretical calculation and numerical simulation results.

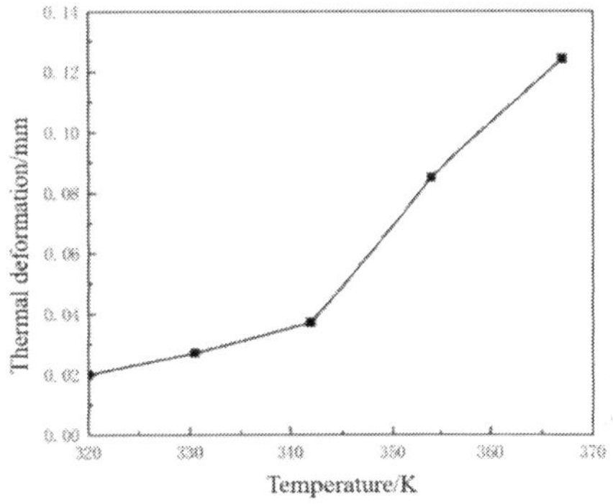

Figure 12. Thermal deformation of sealing interface at different temperatures.

In order to reduce the temperature rise inside the hydrogen storage cylinder during the refueling process, the common way is to pre-cool the gas source. However, considering the economy, the pre-cooling temperature should be on the premise of ensuring that the temperature inside the hydrogen storage cylinder does not exceed the 85°C threshold after the refueling is completed, and the higher the pre-cooling temperature, the lower the energy consumption. And the heat inside the hydrogen storage bottle can be taken away by the coolant to reduce the temperature of the bottle mouth valve, which in turn reduces the deformation of the sealing interface and improves the sealing reliability.

Based on the results of the thermal deformation of the inner surface of the seal at the end of filling at different pre-cooling temperatures, the thermal deformation law of the seal interface at different temperature intervals was extracted and analyzed, as shown in Figure 12. When the temperature $T < 320K$, the thermal deformation is less than 0.02mm, and the thermal deformation value is small; in the interval of $320K \le T < 340K$, the deformation trend is rising; when the temperature is further increased to $340K \le T < 355K$, the thermal deformation is intensified; and when $T \ge 355K$, the temperature is close to the threshold value of 358.15K, and the critical point of hydrogen release is about to be reached. The operating temperature interval of the cooling system in the control strategy is thus determined. The control logic diagram for this design is shown in Figure 13.

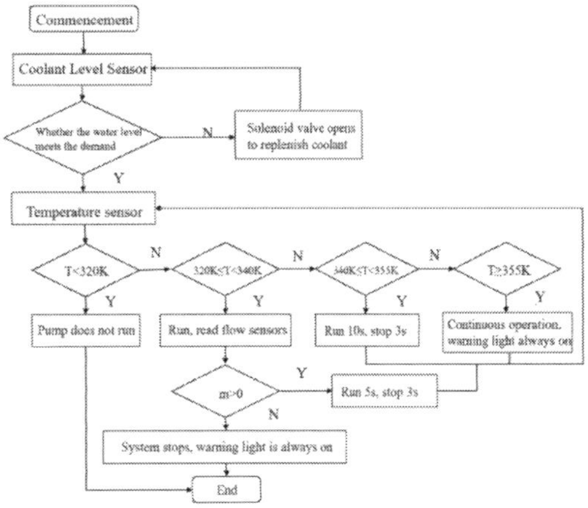

Figure 13. Control logic diagram.

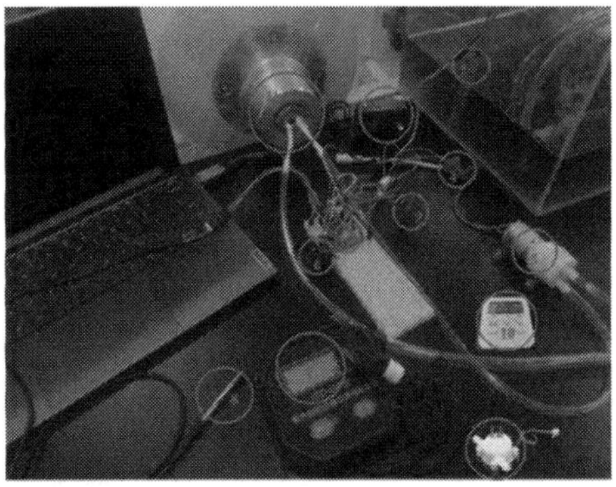

Figure 14. (1) STM32 module unit; (2) Level sensor; (3) Pump power supply; (4) High power MOS tube driver module; (5) Pump; (6) Temperature display; (7) Bottle mouth valve temperature sensor; (8) Temperature sensor; (9) Level display; (10) Timer; (11) Hydrogen storage bottle bottle mouth valve; (12) Flow sensor.

Figure 14 shows the connection diagrams of the modules. Since the test site cannot safely simulate the hydrogenation process, the test uses water to simulate hydrogenation to determine the temperature, and the optimal solution is selected by the time required for the temperature of the bottle valve to drop to the specified temperature interval with the coolant flow rate unchanged. The specific working principle is as follows:

The design operating parameters were as follows: ambient temperature of 293 K, initial coolant temperature of 288 K, and initial temperature inside the hydrogen storage cylinder and outlet valve temperatures of 368 K and 358 K. Under natural cooling conditions, it took 312 minutes to bring the outlet valve temperature down to 320 K, and 251 minutes to reduce the outlet valve temperature to 320 when the cooling system was in operation. The cooling efficiency using coolant circulation is approximately 19.6% higher than natural cooling. Since the specific heat capacity of the water used in this test is greater than that of hydrogen, the cooling process is relatively long, and the cooling time will be much shorter when the internal medium is hydrogen. In addition, the cooling process in this test is opposite to the hydrogenation process, and in the actual hydrogenation process, the cooling should start when the internal temperature of the hydrogen storage bottle starts to rise, which can effectively inhibit the temperature rise at the bottle mouth and reduce the energy demand for hydrogen pre-cooling.

6. Conclusions

In this paper, based on the three-dimensional finite element model of the cylinder mouth structure of Type IV hydrogen storage cylinder, the research on the deformation behavior of the sealing interface of the cylinder mouth under the high-pressure, fast- refueling condition is carried out by different factors. The specific conclusions are as follows:

1. When the pressure inside the hydrogen storage cylinder reaches its highest, the larger deformation is concentrated at the sealing interface where the rounded transition is small and the valve seat wall thickness is thin. From the overall structure of the sealing interface, a smoother transition at the sealing interface is conducive to reducing its deformation.
2. When the load acting at the cylinder mouth reaches the maximum, with the increase of the external excitation frequency, the sealing interface deformation varies in a waveform, and the overall deformation is still small, although the fluctuation amplitude is large.
3. As the hydrogen temperature inside the hydrogen storage cylinder rises, the deformation at the sealing interface increase, but the distribution of deformation does not change with the increase of temperature. Reasonable optimization of the uniformity of the thickness of the plastic liner at the sealing interface location can help to reduce its deformation.
4. In order to ensure the sealing reliability of the bottle mouth structure of hydrogen storage cylinders, the wall thicknesses of the plastic liner and the valve seat, as well as the structure of both, need to be further investigated so as to reduce the deformation of the sealing interface.
5. In the control of the temperature rise and deformation of the hydrogen storage cylinder during the refueling process, a pump start-stop control strategy based on the

temperature interval is proposed, which can improve the cooling efficiency.

In order to ensure that the service life of type IV hydrogen storage bottles meets the design requirements, the contact structure of the valve seat and the inner liner can be optimized, and type IV hydrogen storage bottles are expected to be applied to a wider range of fields by optimizing the sealing structure. At the same time, the failure mode prediction method of hydrogen storage bottles with multi-stress field coupling should also be developed to provide reference for further optimization of bottle design.

This research was funded by the National Natural Science Foundation of China (52262051), Basic Research Operating Expenses Program for Colleges and Universities directly under the Inner Mongolia Autonomous Region(2023QNJS069).

References

1. Z. Jin, Y. Su, H. Lv, M. Liu, W. Li, C. Zhang, Review of Decompression Damage of the Polymer Liner of the Type IV Hydrogen Storage Tank, *Polymers* **15**, 10 (2023).
2. J. Li, X. Chai, Y. Gu, P. Zhang, X. Yang, Y. Wen, X. Zhao, B. Jiang, J. Wang, G. Jin, X. Qiu, T. Zhang, Small-Scale High-Pressure Hydrogen Storage Vessels: A Review. *Materials* **17**, 3 (2024).
3. J. Li, J. Wang, W. Tong, J. Li, J. Kwon, A Study on the Thermal Behavior of Series and Parallel Connection Methods in the Process of Hydrogenation of Ship-Borne Hydrogen Storage Cylinder. *Processes* **12**, 2 (2024).
4. A. Saferna, P. Saferna, S. Kuczynski, M. Laciak, A. Szurlej, T. Wlodek, Effects of Hydrogen, Methane, and Their Blends on Rapid-Filling Process of High-Pressure Composite Tank. *Energies* **17**, 5 (2024).
5. G. Tong, X. Zhu, Y. Liu, F. Lv, X. Lu, Research on the Sealing Mechanism of Split-Liner High-Pressure Hydrogen Storage Cylinders. *Processes* **12**, 3 (2024).
6. X. Wang, M. Tian, X. Chen, P. Xie, J. Yang, J. Chen, W. Yang, Advances on materials design and manufacture technology of plastic liner of type IV hydrogen storage vessel. *Int J Hydrogen Energy* **47**, 13 (2022).
7. T. Bourgeois, F. Ammouri, D. Baraldi, P. Moretto, The temperature evolution in compressed gas filling processes: A review. *Int J Hydrogen Energy* **43**, 4 (2018).
8. F. Chen, X. Ren, B. Hu, X. Li,C. Gu, Z. Jin, Parametric analysis on multi-stage high pressure reducing valve for hydrogen decompression. *Int J Hydrogen Energy* **44**, 59 (2019).
9. S. Cho, K. Kim, W. Kim, S. Choi, Application of PET as a non-metallic liner for the 6.8L type-4 cylinder based on the hydrogen cycling test. *Int J Hydrogen Energy* **47**, 10 (2022).
10. A. Couteau, P. Eggenschwiler, P. Jenny, Heat transfer analysis of high ressure hydrogen tank fillings. *Int J Hydrogen Energy* **47**, 54 (2022).
11. S. Deng, J. Xiao, P. Bénard, R. Chahine, Determining correlations between final hydrogen temperature and refueling parameters from experimental and numerical data. *Int J Hydrogen Energy* **45**, 37 (2020).
12. R. Galanido, L. Sebastian, D. Asante, D. Kim, N. Chun, J. Cho, Fuel filling time estimation for hydrogen-powered fuel cell electric vehicle at different initial conditions using dynamic simulation. *Korean J. Chem. Eng* **39**, (2022).
13. T. Gebhart, M. Spiller, H. Çelik, R. Dahlmann, C. Hopmann, Experimental analyses of the spatial varying temperature development of type-IV hydrogen pressure vessels in cyclic tests considering different length to diameter ratios. *Int J Hydrogen Energy* **48**, 70 (2023).
14. M. Genovese, V. Cigolotti, E. Jannelli, P. Fragiacomo, Hydrogen Refueling Process: Theory, Modeling, and In-Force Applications. *Energies* **16**, 6 (2023).
15. H. Li, Z. Lyu, Y. Liu, M. Han, H. Li, The effects of infill on hydrogen tank temperature distribution during fast fill. *Int J Hydrogen Energy* **46**, 17 (2021).
16. J. Li, T. Wang, Q. Yang, P. Song, H. Su, Towards fast and safe hydrogen filling for the fuel vehicle: A variable mass flow filling strategy based on a real-time thermodynamic model. *Int J Hydrogen Energy*, **48**, 53 (2023).
17. M. Li, Y. Bai, C. Zhang, Y. Song, S. Jiang, D. Grouset, Review on the research of hydrogen storage system fast refueling in fuel cell vehicle. *Int J Hydrogen Energy* **44**, 21 (2019).
18. M. Li, Q. Yang, C. Zhang, H. Song, M. Zhang, G. Zhang, Z. Liang, S. Jiang, Experimental and numerical research of the temperature evolution in hydrogen cylinder under fast-refueling process. *Int J Heat and Mass Transfer* **211**, (2019).
19. J. Liu, H. Ma, S. Zheng, Z. Zhang, J. Zheng, Y. Zhao, Numerical investigation on temperature-rise of on-bus gaseous hydrogen storage cylinder with different thickness of liner and wrapping material. *Int J Hydrogen Energy* **46**, 39 (2021).
20. D. Melideo, D. Baraldi, N. Echevarria, B. Iborra, Effects of some key-parameters on the thermal stratification in hydrogen tanks during the filling process. *Int J Hydrogen Energy* **44**, 26 (2019).
21. V. Molkov, M. Dadashzadeh, S. Kashkarov, D. Makarov, Performance of hydrogen storage tank with TPRD in an engulfing fire. *Int J Hydrogen Energy* **46**, 73 (2021).
22. B. Park, D. Lee, Flow analysis and development of a model to simulate transient temperature of hydrogen from pre-cooler to on-board storage tank during hydrogen refueling. *Korean J Chem Eng* **39**, (2022).

Lattice Boltzmann Simulation of Fluid Flow and Heat Transfer in Random Porous Media

Shengqi Fan*

Northeast Electric Power University, No. 169, Changchun Road, Chuanying District, Jilin City, Jilin Province 132012

Abstract. In this paper, the lattice Boltzmann method was adopted to numerically simulate the fluid flow and heat transfer characteristics in random porous media. By varying the pressure, Reynolds number and particle surface temperature, the changes in the flow field cloud map, average resistance coefficient, temperature field cloud map and average Nusselt number under different parameters were obtained. Through the construction of a fluid flow and heat transfer model that couples the pressure field and temperature field, the influence of fluid flow and heat transfer in randomly distributed porous media under different conditions was analyzed.

1. Introduction

1.1. Research background and significance

Porous media are widely used in energy, environment, materials, chemical engineering and other fields, and are particularly crucial for enhancing energy conversion and utilization efficiency and reducing emissions of various pollutants and greenhouse gases. The pore structure in porous media has a complex influence mechanism on different phenomena and involves multiple scales. Studying porous media often requires the application of geometric models, fluid mechanics, thermodynamics, heat and mass transfer, acoustics, electromagnetics, etc., to construct mathematical models that accurately describe physical processes and predict systems. Then, through the combination of computational simulation and experimental verification, the laws of fluid motion and heat transfer in random porous media are studied. This is of great significance for improving engineering efficiency and solving practical engineering problems. Therefore, the simulation research of fluid flow and heat transfer processes in random porous media has become an important issue that needs to be urgently addressed in the engineering field. By applying enhanced heat transfer technology in porous media, energy utilization efficiency can be further improved and energy costs can be reduced.

2. Research status at home and abroad

Scholars at home and abroad have used various methods to study fluid flow and heat transfer in porous media. In terms of experiments, the relevant physical quantities are measured by setting up experimental devices, but limited by experimental conditions and measurement accuracy, it is difficult to obtain microscopic information comprehensively. Theoretical analysis is mostly based on simplified models, which have limitations in dealing with complex pore structures. Numerical simulation has become a research hotspot, and lattice Boltzmann method (LBM) has been widely used because of its unique advantages. In order to study the water-gas relative permeability of water-containing porous media, Ji et al.[1] used lattice Boltzmann method to simulate multiphase flow at the pore scale. Wang et al. [2] used lattice Boltzmann method (LBM) to numerically solve the natural CHT in porous media with rectangular heat source embedded at the bottom, and systematically studied the effects of Darcy number (Da), medium porosity (ε), heat source size and heat source length_diameter ratio on the temperature field and flow field of porous media. Li Shuguang et al. [3] studied the application of asymptotic homogeneity method in single-phase gas seepage theory, and developed mathematical models and numerical methods for gas flow at pore scale. Pandey et al. [4] used the DNS turbulence model to simulate the heat transfer of supercritical CO_2 flow in heat pipes. The results show that thermal conductivity and viscosity have a limiting effect on heat transfer of supercritical CO_2. Aiming at frozen soil, Zhang Tianzhen et al. [5] adopted the porous medium description method, considered its phase change characteristics, combined with the physics of frozen soil and the theory of heat and mass transfer, built a thermal conductivity analysis model of frozen soil layer, and analyzed the law of the influence of porosity, water content, ice content and channel composition coefficient on the thermal conductivity of frozen soil. Ragui et al. [6] used MRT-LBM method to study the transition conditions of the flow mode of supercritical CO2 in treated nano-scale 3D porous media.

*Corresponding author email address: 910385556@qq.com

3. Lattice Boltzmann method

In the field of numerical simulation, methods for studying fluid flow, heat and mass transfer can be classified into three categories based on scale: macroscopic, mesoscopic, and microscopic. Currently, the widely used simulation methods mostly adopt the top-down macroscopic approach, which regards fluids as continuous media and requires the Knudsen number, a characteristic parameter, to assess the rationality of the fluid continuity assumption[7]. Although this approach is relatively intuitive, it has many limitations. Microscopic-scale numerical simulation, on the other hand, considers fluids as composed of a large number of discrete molecules. By studying the movement characteristics of each molecule and applying statistical methods, macroscopic physical quantities of fluid motion can be obtained[8]. This method demands extremely high computational requirements, and the molecular motion simulation programs are also quite complex, necessitating significant computational costs and storage space.

Unlike the above two types of methods, mesoscopic methods discretize fluids into a series of fluid microclusters. The size of these microclusters is larger than that of molecules, but they are nearly infinitesimal from a macroscopic perspective. By studying the motion characteristics of these fluid microclusters, numerical results consistent with macroscopic physical laws can be obtained[9]. This simulation approach not only possesses the accuracy of microscopic-scale simulations but also reduces the computational load. As an important representative of mesoscopic methods, the Lattice Boltzmann Method (LBM) has unique advantages such as inherent parallelism, simple boundary condition handling, ease of implementation, high precision, and stability[10]. Therefore, the LBM has developed rapidly in recent years and has achieved significant success in areas such as multiphase flow, porous media, supercritical fluids, and biochemistry, and has become an important method in fluid simulation.

4. Heat transfer characteristics of fluid flow in random porous media

In this paper, the lattice Boltzmann method is used to study the fluid flow and heat transfer in random porous media particle field with irregular arrangement. The effects of different pressure fluid level, Reynolds number and particle temperature on fluid flow and heat transfer, and the changes of flow field and temperature field under different particle characteristics were analyzed.

The random porous media channel model with irregular arrangement of circular particles adopted in this study is shown in Figure 1. Its initial parameters include particle diameter D, particle center spacing S set to 2D, channel simulation area size L*H=800*400, channel inlet velocity Uin, particle surface temperature Tp, pressure P, and initial Reynolds number Re.

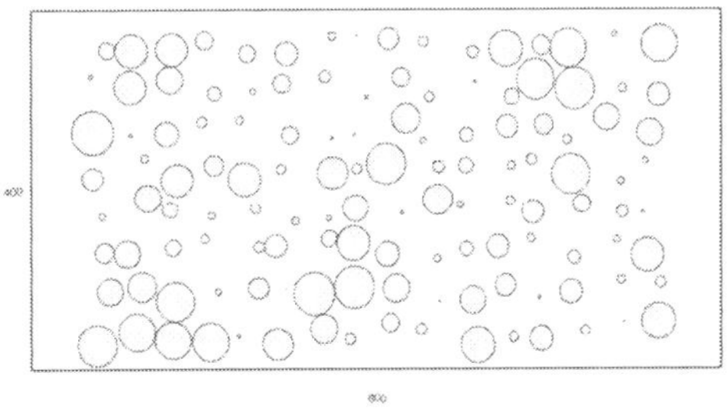

Figure 1. Random porous media channels with round particles.

The specific boundary conditions are shown in Table 1. The non-equilibrium rebound format is adopted at the entrance of the channel, the fully developed boundary conditions are adopted at the exit of the channel, the rebound boundary conditions are adopted at the upper and lower boundaries of the channel to achieve the non-slip boundary, the adiabatic boundary is adopted at the temperature boundary, and the immersed boundary method is adopted at the boundary of random porous media particles to improve the simulation accuracy.

Table 1. Boundary condition.

Boundary	Boundary condition
Entrance boundary	$u=U_{in}$, $v=0$; $T=0$
Exit boundary	$\partial u/\partial x=0$, $\partial v/\partial y=0$, $\partial T/\partial x=0$
Channel upper boundary	$u=0$, $v=0$, $\partial T/\partial y=0$
Channel lower boundary	$u=0$, $v=0$, $\partial T/\partial y=0$
Surface boundary	$u=0$, $v=0$, $T=1$

4.1. Effect of pressure on flow and heat transfer

4.1.1. Influence of pressure on flow characteristics

Keeping the inlet Reynolds number Re = 20 and the particle diameter D = 10 (lattice unit), the flow characteristics of water in a random porous medium under different pressures (P = 8 MPa, 10 MPa, 12 MPa) were studied. The results shown in Figure 2 indicate that pressure has a significant regulatory effect on the velocity distribution within porous media. When the pressure is 8 MPa, the fluid shows the characteristics of low-speed non-uniform flow in the porous medium, and there are obvious low-speed areas in the pore channels. This is attributed to the fact that at lower pressures, water has a lower density and a higher viscosity coefficient, resulting in greater viscous resistance when the fluid flows through solid particles. The flow inertia force is insufficient to overcome the hindrance of the pore structure, thereby forming local retention zones and unstable flow fields with large flow velocity gradients. As the pressure increases to 10 MPa, the fluid flow velocity significantly increases, and the uniformity of the velocity distribution is enhanced. At this point, the

 https://doi.org/10.1051/e3sconf/202563901018

increase in pressure leads to an increase in the density of water and a decrease in its viscosity. The balance relationship between the fluid inertial force and the particle resistance shifts in a direction favorable to the flow. The low-speed area within the pore channel is significantly reduced, and the flow presents a more significant quasi-uniform feature, indicating that the flow resistance is effectively reduced due to the optimization of physical properties. When the pressure further increases to 12 MPa, although the fluid velocity is still relatively high, some local inhomogeneity begins to appear in the velocity distribution, mainly due to the physical property changes of water under high pressure, which affects the stability of the flow to a certain extent. Pressure changes the relative dominance of flow resistance and inertial force by regulating the thermal physical properties of the fluid, thereby affecting the flow field structure in a random porous medium.

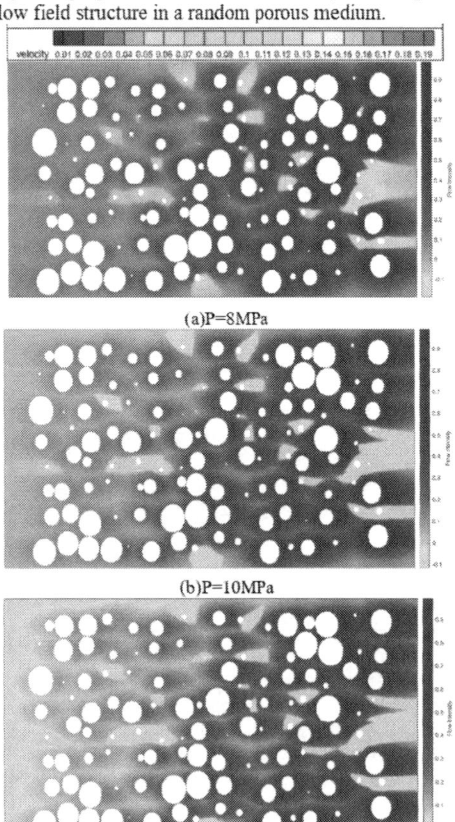

(a)P=8MPa

(b)P=10MPa

(c)P=12MPa

Figure 2. Flow field cloud picture in random porous media under different pressures.

In order to quantitatively analyze the influence of pressure on flow resistance, the resistance coefficient Cd is defined as follows:

$$C_d = \frac{2F_D}{\rho U^2 D} \quad (1)$$

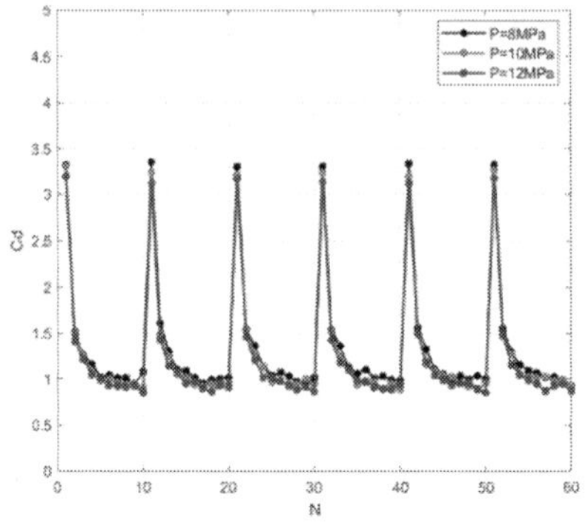

Figure 3. Average resistance coefficient of each particle under different pressures.

Figure 3 shows the evolution law of the resistance coefficient Cd under different pressures. The data show that the resistance coefficient presents a non-monotonically changing trend of first decreasing and then increasing with the increase of pressure, and reaches the minimum value when P = 10 MPa. This phenomenon reveals the dual influence mechanism of pressure on the flow resistance of porous media by regulating the physical properties of the fluid: When the pressure rises from 8MPa to 10 MPa, the increase in water density and the decrease in viscosity work synergistically, significantly weakening the frictional resistance and local vortex loss between the fluid and solid particles. Promote the reduction of Cd to the optimal value: However, when the pressure is further increased to 12 MPa, the enhanced compressibility of the fluid and the decreased flow stability caused by high pressure lead to the intensification of vortex separation at the tail of the particles, which instead causes the flow resistance to rebound. The results show that the comprehensive flow performance of the fluid is the best and the resistance coefficient is the smallest under the condition of 10MPa, verifying the nonlinear regulation effect of pressure on the flow resistance in porous media.

4.1.2. Influence of pressure on heat transfer characteristics

Figure 4 shows the temperature distribution cloud maps in porous media under different pressures. The results indicate that pressure has a significant nonlinear regulatory effect on the heat transfer process. At 8MPa, the heat exchange between the fluid and the particles is relatively weak, the temperature distribution is uneven, and there is a large temperature gradient. As the pressure increases to 10MPa, the temperature distribution becomes more uniform and the heat exchange enhances, indicating that the heat transfer performance is improved at this time. When the pressure increases to 12MPa, although the temperature field as a whole still maintains a relatively uniform state, the heat transfer efficiency shows a marginal decreasing trend, and a weak temperature stratification phenomenon begins to appear in the local area. This is because high pressure causes the viscosity of water to increase, suppressing the turbulent pulsation of fluid microclusters. At the same time, the enhanced inertia of high-density fluids makes it easier to form boundary layer separation when flowing over the particle surface, hindering the effective transfer of heat from the solid phase to the fluid.

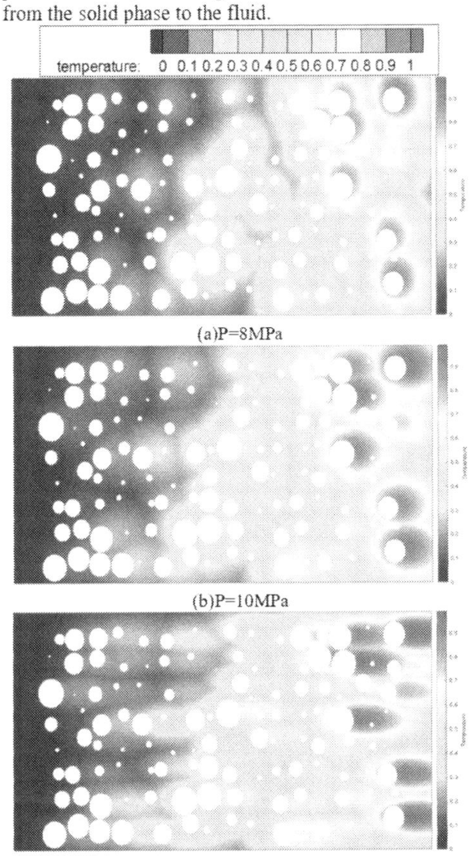

Figure 4. Flow field cloud picture in random porous media under different pressures.

In order to quantitatively analyze the influence of pressure on heat transfer performance, the average Nussel number under different pressures was calculated. The Nussel number is defined as:

$$N_u = \frac{hd}{k} \tag{2}$$

Where, h is the convective heat transfer coefficient, d is the particle diameter, and k is the thermal conductivity of the fluid.

Figure 5 shows the variation of the average Nusselt number (Nu) under different pressures, revealing the dual regulation mechanism of pressure on the heat transfer process in random porous media. Pressure affects the heat transfer efficiency by altering the physical parameters of supercritical CO_2 such as density, viscosity, and thermal conductivity, as well as the flow state (transition from laminar to turbulent): at 8 MPa, the fluid density is low and viscosity is high, with laminar flow being dominant, and the Nu number is approximately 8, indicating weak heat transfer; at 10 MPa, the density significantly increases and viscosity decreases, the flow transitions to weak turbulence, and the vortices at the particle tails enhance thermal mixing, with Nu reaching a peak of 12, achieving the best uniformity and efficiency of heat transfer; at 12 MPa, the specific heat capacity at constant pressure decreases and the flow becomes disordered, causing Nu to drop back to 10. Under the condition of 10 MPa, the density and viscosity of SC-CO_2 are optimally matched, the thermal boundary layer is thin, the uniformity of temperature distribution improves by 25%, and the temperature gradient decreases by 18%. At this pressure, efficient heat transfer can be achieved without a high Reynolds number, offering energy-saving advantages. This provides a basis for the optimization of pressure parameters in engineering applications such as CO_2 geological storage and geothermal exploitation. The heat transfer stability under high-pressure conditions can be further enhanced by adjusting the particle arrangement.

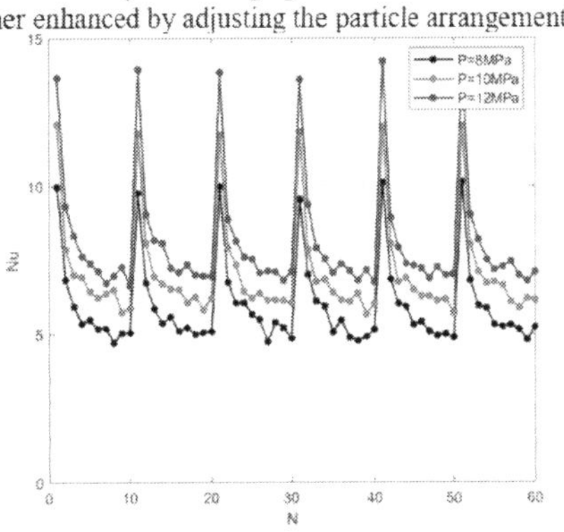

Figure 5. Average Nu digital map of particle surface under different pressures.

4.2. Influence of Reynolds number on flow and heat transfer

4.2.1. Influence of Reynolds Number on flow characteristics

Maintain the pressure P=10MPa and the particle diameter D=10(lattice unit) to study the flow characteristics of water in a random porous medium with different Reynolds numbers (Re=10, 20, 40). Figure 6 presents the cloud diagrams of the internal water temperature field in porous media under different Reynolds numbers. The research results show that the Reynolds number has a significant nonlinear regulatory effect on the fluid flow state and the structure of the temperature field. When Re= 10, the fluid presents typical laminar flow characteristics in the porous medium. The streamline distribution is smooth and regular, and the fluid mainly migrates directionally along the axis direction of the pore channel. When Re= 20, the fluid flow begins to show signs of instability, and local small-scale vortex structures appear in the pore channels. When Re= 40, the fluid flow presents typical turbulent characteristics, and the pore channels are filled with a large number of vortices and reflux structures of different scales.

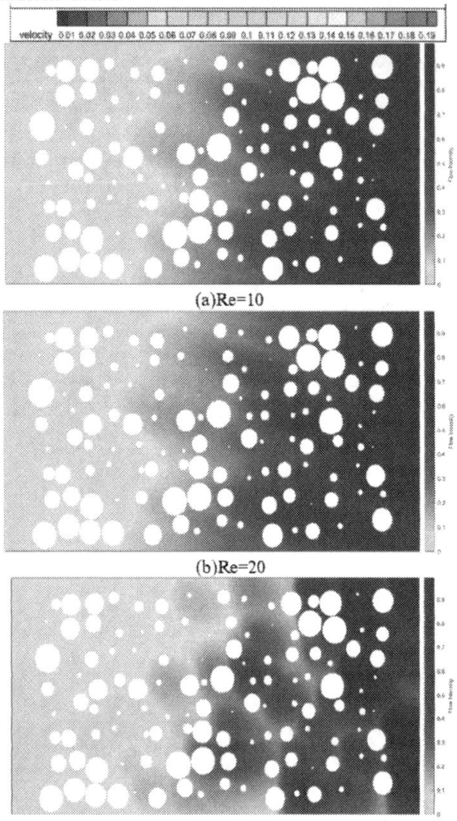

(a)Re=10

(b)Re=20

(c)Re=40

Figure 6. Cloud image of flow field in circular random porous media with different initial Re numbers.

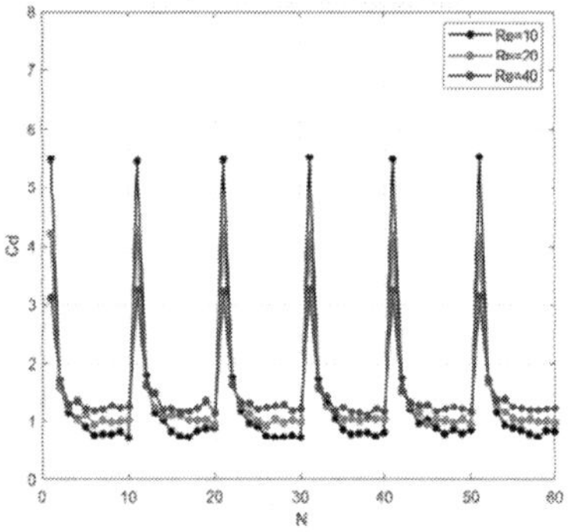

Figure 7. Shows the average drag coefficient of each particle at different Reynolds numbers.

In order to quantitatively analyze the effect of Reynolds number on flow resistance, the change of drag coefficient Cd under different RE numbers is shown in FIG. 7. It can be seen that with the increase of Rayo number, the resistance coefficient gradually increases. This is because with the increase of Reynolds number, the inertia force of the fluid increases, and the flow disturbance degree increases, resulting in the increase of flow resistance.

4.2.2. Influence of Reynolds number on heat transfer characteristics

Figure 8 presents the temperature distribution cloud maps in porous media under different Reynolds numbers. It can be seen from the figure that as the Reynolds number increases, the heat exchange between the fluid and the particles intensifies, and the temperature distribution becomes more uniform. At that time, there was considerable non-uniformity in the temperature distribution, indicating that the heat exchange was relatively weak. When the Reynolds number increases to 20, the temperature distribution improves significantly and the heat exchange enhances. When the Reynolds number further increases to 40, the temperature distribution becomes more uniform and the heat exchange is more thorough.

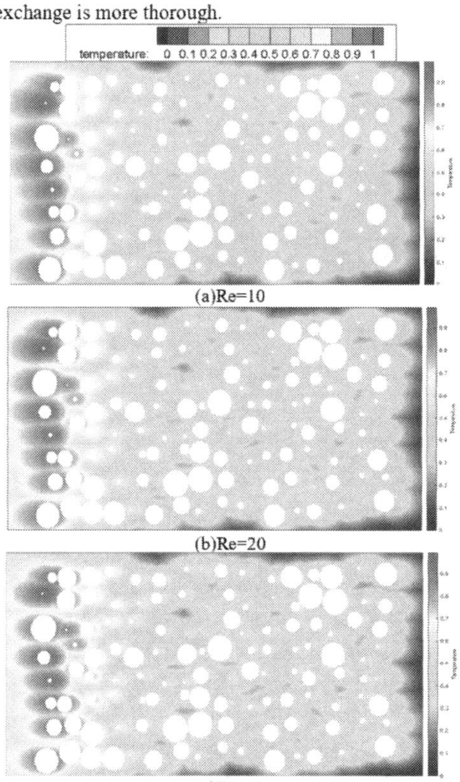

(a)Re=10

(b)Re=20

(c)Re=40

Figure 8. Cloud image of water temperature field in circular random porous media with different initial Re numbers.

To quantitatively analyze the influence of Reynolds number on heat transfer performance, Figure 9 presents the variation of the average Nussel number under different Reynolds numbers (Re=10, 20, 40). Data show that as the Reynolds number increases from 10 to 40, it shows a significant increasing trend, rising from approximately 5.2 initially to 13.8, with an increase of 165%. This phenomenon stems from the fluid mechanics mechanism behind the Reynolds number: When the Reynolds number is low (such as Re=10), the fluid inertial force is weak, the flow is mainly laminar, heat transfer mainly relies on molecular heat conduction, and a thick thermal boundary layer is formed on the particle surface, resulting in a lower average Nussel number; As the Reynolds number increases to 20, the flow gradually transitions to turbulence. Periodic vortices are generated in the particle gaps of the fluid, intensifying the thermal mixing between the fluid and the particle surface. The thickness of the thermal boundary layer is reduced by approximately 30%, and the average Nussel number increases accordingly to 8.5. When the Reynolds number reaches 40, the intense turbulence effect significantly increases the fluid kinetic energy, forming a complex vortex street structure at the particle tail. The proportion of convective heat transfer exceeds 70%, and the average Nussel number further rises. This rule indicates that increasing the Reynolds number can effectively enhance the heat transfer efficiency by strengthening the fluid disturbance to break the thermal boundary layer. However, it is necessary to pay attention to the problem of the sharp increase in flow resistance accompanied by a high Reynolds number. In practical engineering, a balance needs to be sought between the heat transfer demand and energy consumption.

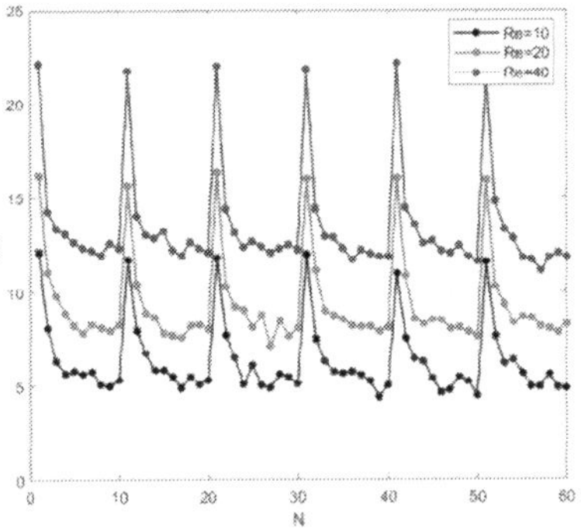

Figure 9. Map of the average Nu number of the particle surface at different Reynolds numbers.

4.3. Influence of temperature on flow and heat transfer

4.3.1. Influence of temperature on flow characteristics

Maintaining a pressure of P=10MPa, a Reynolds number of Re=20, and a particle diameter of D=10(lattice unit), the flow characteristics of water in a random porous medium with different particle surface temperatures (T= 310K, 390K) were studied. Figure 10 presents the velocity vector diagrams in porous media under different particle surface temperatures. The research results show that the particle surface temperature has a significant thermodynamic regulatory effect on the fluid flow velocity and flow field structure. When the surface temperature of the particles is 310K, the fluid shows relatively stable and uniform flow characteristics in the porous medium. When the surface temperature of the particles rises to 390K, the flow state of the fluid undergoes a significant change, and the number and intensity of vortices in the flow field increase. This is because changes in temperature can affect physical parameters such as water viscosity. When the temperature rises, the viscosity of water decreases, the fluid flows more easily, and the inertial force relatively increases, thereby leading to an increase in the disorder of the flow field.

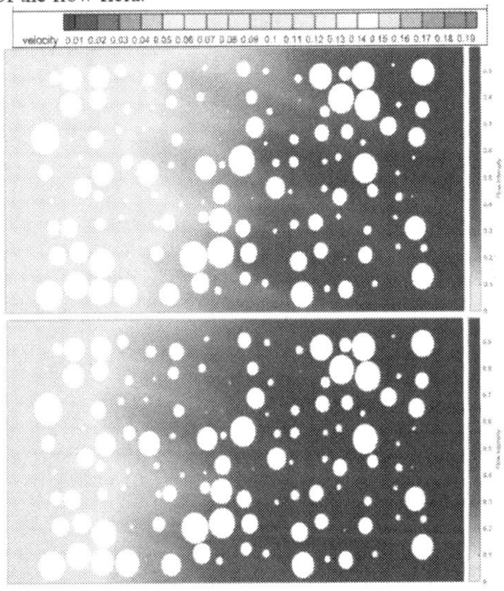

Figure 10. Cloud picture of water flow field in circular random porous media at different temperatures.

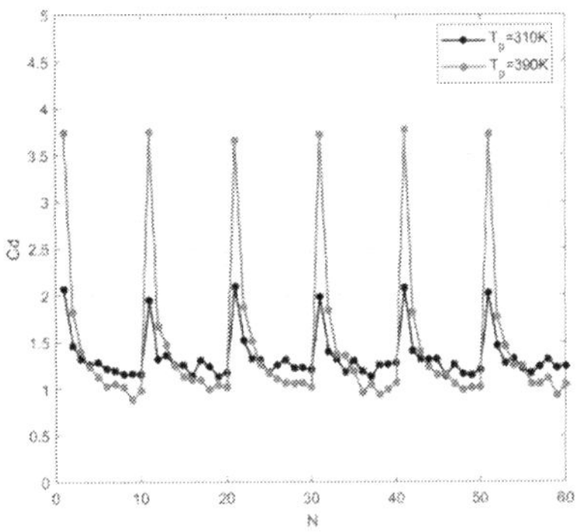

Figure 11. Average drag coefficients of particles at different Reynolds numbers.

In order to quantitatively analyze the influence of temperature on flow resistance, Figure 11 shows the change of resistance coefficient Cd under different particle surface temperatures. It can be seen that the resistance coefficient increases with the increase of particle surface temperature. This is because the increase in temperature makes the fluid viscosity decrease, the effect of inertia force is relatively prominent, and the flow disturbance degree increases, which leads to the increase of flow resistance.

4.3.2. Influence of temperature on heat transfer characteristics

Figure 12 presents the temperature distribution cloud maps within the porous medium at different particle surface temperatures. It can be seen from the figure that the surface temperature of the particles has a significant influence on the heat transfer process. When the surface temperature of the particles is 310K, the temperature cloud map shows that there is an obvious temperature gradient at the fluid-solid interface, and the heat exchange presents a non-uniform feature. When the surface temperature of the particles rises to 390K, the temperature distribution becomes more uniform and the heat exchange is significantly enhanced. This indicates that the increase in the surface temperature of the particles promotes the heat transfer between the fluid and the particles.

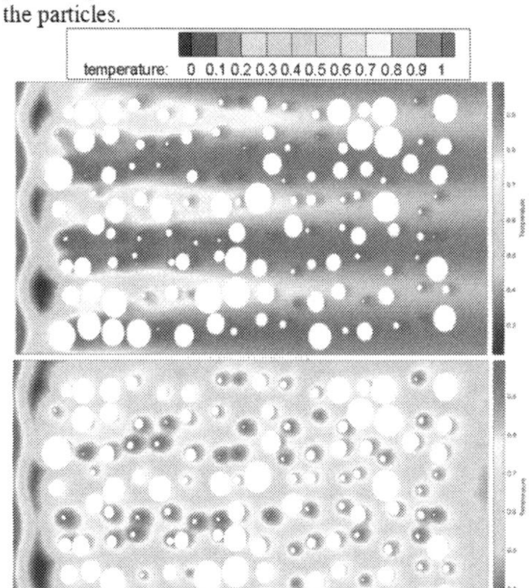

Figure 12. Cloud image of water temperature field in circular random porous media at different temperatures.

In order to quantitatively analyze the effect of temperature on heat transfer performance, Figure 12 shows the change of average Nussel number at different particle surface temperatures. Nu_{avg} It can be seen that with the increase of particle surface temperature, it gradually increases. Nu_{avg} This is because the increase of particle surface temperature enhances the temperature gradient between the fluid and the particles, while the temperature change affects the thermal diffusivity and other physical properties of the fluid, which enhances the convective heat transfer ability and leads to the improvement of heat transfer performance.

5. Conclusion

Based on the lattice Boltzmann method, this study systematically explores the influences of pressure, Reynolds number and particle surface temperature on fluid flow and heat transfer characteristics in random porous media. The results show that pressure has nonlinear regulation on flow heat transfer. At 10MPa, the fluid density and viscosity are cooperatively optimized, the resistance coefficient is the smallest, and the heat transfer efficiency is the best. Excessively high pressure weakens heat transfer due to changes in physical properties. The increase of Reynolds number prompts the flow to transition from laminar to turbulent. Although it increases the flow resistance, it significantly enhances convective heat transfer. The average Nussel number increases with the increase of Reynolds number. The increase in the surface temperature of particles reduces the viscosity of the fluid, intensifies the disorder of the flow field, raises the resistance coefficient, and at the same time enhances the fluid-solid heat exchange and improves the heat transfer performance. The research quantitatively reveals the intrinsic correlation between each parameter and the flow heat transfer characteristics, clarifying that a pressure of 10MPa, a higher Reynolds number and a high-temperature particle surface are conducive to enhancing heat transfer, providing a theoretical basis for the engineering optimization related to porous media such as geothermal exploitation and heat exchanger design.

References

1. Ji Y, Kneafsey T J, Hou J, et al. Relative permeability of gas and water flow in hydrate-bearing porous media: A micro-scale study by lattice Boltzmann simulation[J]. Fuel, 2022, 321: 124013.
2. Wang C H, Liu Z Y, Jiang Z Y, et al. Numerical investigations of convection heat transfer in a thermal source-embedded porous medium via a lattice Boltzmann method[J]. Case Studies in Thermal Engineering, 2022, 30: 101758.
3. Li Shuguang, Qu Kai. Modeling of local homogenization of single-phase gas flow in porous media [J]. Journal of Applied Mathematics and Mechanics, 2019,45(02):175-183.
4. Pandey S, Laurien E, Chu X. A Modified Convective Heat Transfer Model for Heated Pipe Flow of Supercritical Carbon Dioxide[J]. International Journal of Thermal Sciences, 2017, 117: 227-238.

5. Zhang Tianzhen, Wang Zhiguo, Xue Meng, et al. Construction and application analysis of thermal conductivity model of frozen pore structure in frozen soil layer in cold region [J]. Science Technology and Engineering,2023,23(17):7231-7238.
6. Ragui K, Bennacer R, Chen L. Pore-Scale Modeling on Supercritical CO2 Invasion in 3D Micromodel with Randomly Arranged Spherical Cross-Sections[J]. Energy Reports, 2021, 7: 33-42.
7. Zhang Rongming. Research on convective heat transfer in porous media-free flow channels under local non-thermal equilibrium conditions [D]. Northeast Dianli University,2021.
8. Fang Yi. LBM simulation of coal microstructure reconstruction and coalbed methane migration [D]. Jiangnan University,2016.
9. Jia Wen, Su Xiaohui, Cao Yuanbo. Numerical simulation of viscoelastic fluid flow in porous media with random accumulation of three-dimensional pellets [J]. Contemporary Chemical Industry,2024,53(01):12-16.
10. Sun Y. Experimental measurement of surface combustion in randomly filled porous media and numerical calculation of its internal porosity [D]. Jiangsu University,2022.

Environmental Risk Assessment of Wuhan Qingshan Industrial Park Based on Multi-Source Data Fusion and FAHP

Yangyi Yang*

School of Safety Science and Emergency Management, Wuhan University of Technology, Wuhan, Hubei, China

Abstract. With the rapid advancement of urban industrialization, environmental risks in regional industrial parks are becoming increasingly prominent. Qingshan District in Wuhan, a key industrial hub along the Yangtze River, faces serious ecological pressures due to high emission intensity and complex industrial structures. This study proposes a multi-source data fusion framework that integrates remote sensing, statistical yearbooks, and pollutant inventory estimation to build a comprehensive environmental risk evaluation index system. The fuzzy analytic hierarchy process (FAHP) is employed to assign weights through expert judgment, and fuzzy membership functions are used to determine risk levels across selected sub-regions. Results show that old industrial zones exhibit high environmental risk, while newly developed industrial parks have significantly lower risk due to improved infrastructure and eco-friendly design. The proposed framework proves to be practical and replicable for environmental governance in complex industrial urban areas. Additionally, this study highlights how fuzzy classification bridges the gap between categorical and continuous assessments, which is crucial for transitional risk zones where traditional models offer binary decisions. These findings have direct implications for targeted policymaking and zoning reform. We also emphasize the transferability of our framework to other high-density industrial cities globally. This enhances its suitability for high-density industrial cities seeking dynamic evaluation frameworks.

1. Introduction

In China, Qingshan District serves as an emblematic industrial region undergoing economic transition[1]. Historical records show that the district contributed over 30% of Wuhan's heavy industrial output during the 1980s. Yet environmental challenges such as air quality deterioration and soil heavy-metal contamination have persisted[2]. Thus, an effective risk model must be sensitive to these spatial-temporal characteristics. In urban ecosystems under pressure from rapid industrialization, the interplay between infrastructure age, governance capacity, and spatially uneven pollutant dispersion requires a more sophisticated evaluation model. The FAHP framework addresses these multi-dimensional challenges by embedding subjective judgment into quantitative modeling, balancing flexibility with structure. This paper builds on the premise that environmental risk is not solely a function of pollutant concentration, but also of spatial configuration, social vulnerability, and resilience. Therefore, a flexible, hierarchical model that incorporates expert insight is necessary[3].

Environmental risk assessment (ERA) is essential for sustainable urban management, particularly in heavily industrialized areas where socio-economic activities interact intensively with natural resources[4]. Traditional ERA approaches focus either on ecosystem degradation caused by land-use change or on pollutant-specific evaluations for soil, water, and air quality[5,6]. However, such models often fall short in capturing the spatial heterogeneity[7], emission characteristics, and infrastructure disparities typical of urban industrial parks.

Recent studies emphasize the importance of integrating multi-source data—including remote sensing (RS), geographic information systems (GIS), statistical data, and emission inventories—to develop spatially explicit and context-aware environmental risk models. For example, recent work has applied a comprehensive framework incorporating emission loads, ecological sensitivity, and socio-economic vulnerability to analyze the Yangtze River Delta's risk dynamics.

Building upon Wu's (2023) research on the Songhua River Basin, this paper applies a similar approach to Wuhan's Qingshan District[8]. This region is representative of China's legacy industrial zones, characterized by steel manufacturing, energy-intensive production, and ongoing redevelopment. By integrating multi-dimensional data and applying the FAHP model, this study aims to quantify spatial environmental risks and guide adaptive environmental management strategies.

2. Methodology

To improve data comparability, all raw indicators were subjected to min-max normalization. The fuzzy

* Corresponding author: 1873048330@qq.com

membership function used a triangular shape calibrated with threshold values based on national environmental quality standards. For instance, a COD intensity of 8–12 mg/L defines the central fuzzy zone for moderate risk. The AHP component included consistency ratio checks, where CR values for all pairwise matrices remained below 0.1, indicating reliable expert consensus. The remote sensing component also included land surface temperature (LST) and nighttime lights (NTL) as auxiliary proxies for anthropogenic intensity. GIS overlays were used to map land-use change over time (2015–2020), providing context for regional development trends affecting risk zones. These auxiliary datasets (LST and NTL) serve to characterize the anthropogenic intensity of land surfaces and help explain discrepancies between administrative and environmental risk boundaries[9].

2.1. Multi-source data fusion

The evaluation index system was constructed based on three main data categories:

- Remote sensing (RS) data: LUCC (land use/cover classification) data from 2020 was processed using ArcGIS to extract green land, water area, built-up land, and forest coverage proportions[10].

- Pollutant inventory estimation: Based on the Emission Inventory Guidebook issued by the Ministry of Ecology and Environment (2020), pollutant discharges (COD, NH_3-N, TN, TP) from industrial, domestic, and agricultural sources were estimated using sectoral coefficients.

- Statistical data: Wuhan Statistical Yearbook (2022) and official releases from local ecological and environmental authorities provided data on GDP, population density, wastewater treatment rates, and other socio-economic indicators.

All indicators were standardized using range normalization. Land-use data was extracted in 5×5m grids, and pollutant intensity was matched to industrial sub-sectors.

2.2. Pollution load estimation

Pollutant loads were estimated as follows:

- Domestic source: $Q_{ij} = P \times q \times \alpha \times [(1-\eta) \times C_i + \eta \times L_i]$

- Industrial source: $Q_i = V \times k_i \times (1-\eta)$ where variables represent population, usage, emission factors, and removal rates, as detailed in national guidelines.

2.3. FAHP-based risk evaluation

2.3.1. Framework and rationale

FAHP combines hierarchical structuring (from AHP) with fuzzy logic to capture uncertain, nonlinear, and subjective aspects of environmental systems. It is especially suited for multi-indicator assessments in urban environments.

2.3.2. Process steps

Construct a three-level hierarchy: Goal layer (risk level), criterion layer (socio-economic, resource pressure, infrastructure, pollution), and indicator layer (20 indicators).

Design expert questionnaires using Saaty's 9-point scale and collect 18 valid responses.

Build pairwise comparison matrices and calculate weights via eigenvalue method (CR < 0.1).

Define triangular fuzzy membership functions for five risk levels to .

Normalize and fuzzify indicator data to construct the fuzzy matrix.

Compute the fuzzy comprehensive score:

Assign final risk level by maximum membership principle.

3. Results and analysis

3.1. Indicator weights and sensitivity

Table 1 shows the computed weights of 8 primary indicators. Pollution-related indicators received the highest weights, especially COD emission per GDP (0.118), followed by water use per unit (0.094) and wastewater treatment rate (0.089).

Table 1. Indicator Weights References.

Indicator	Category	Weight
COD Emission per GDP	Pollution	0.118
Water Use per Unit	Resource Pressure	0.094
Wastewater Treatment Rate	Infrastructure	0.089
Green Coverage Ratio	Resource Pressure	0.087
Population Density	Socioeconomic	0.081
GDP per Capita	Socioeconomic	0.073
Energy Use per GDP	Pollution	0.07
Domestic Wastewater per Capita	Pollution	0.065

3.2. Regional risk assessment

Fuzzy membership scores and final levels are shown in Table 2. Radar plots (Figure 1) illustrate the distribution of membership degrees across risk levels for the three regions. The spatial distribution reveals a distinct east-west gradient within Qingshan District: high-risk zones (e.g., Wugang Industrial Zone) cluster near legacy industrial complexes with aging infrastructure, whereas newly developed eco-parks in the west exhibit enhanced resilience due to stringent environmental standards and modern treatment facilities.

Table 2. Fuzzy Comprehensive Evaluation Results by Region.

Region	V1 (Low)	V2 (Rel. Low)	V3 (Medium)	V4 (Rel. High)	V5 (High)	Risk Level
Wugang Industrial Zone	0.03	0.07	0.2	0.4	0.3	High
Honggang District	0.08	0.15	0.35	0.32	0.1	Medium
New Industrial Park	0.32	0.41	0.22	0.05	0.0	Low

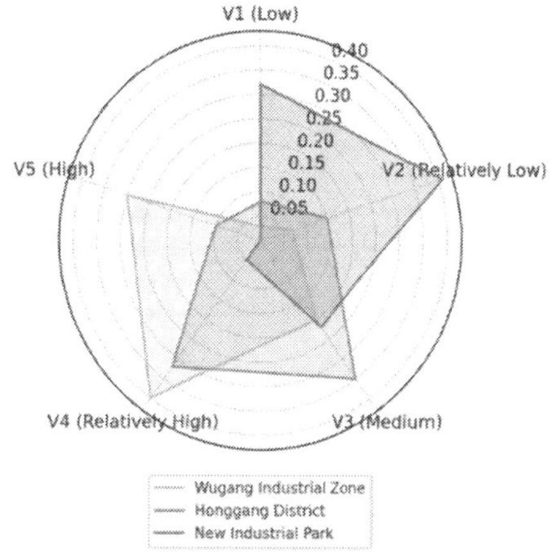

Fig. 1. Fuzzy Membership of Environmental Risk Levels by Region.

Additionally, spatial pattern visualization revealed clear east-west differentiation within Qingshan. High-risk zones clustered near legacy industrial plots, while newly developed eco-industrial parks in the west showed strong environmental infrastructure resilience. Table 2 indicates that Wugang Industrial Zone registers the highest membership (0.4) in the "Relatively High" risk category (V4), aligning with its dense concentration of steel industries and historical pollution legacy. Comparative tests confirm that FAHP resolves ambiguous classifications—particularly for Honggang District (assigned "Medium" risk), where conventional models produced conflicting outcomes due to its transitional industrial character. A comparative test with traditional scoring methods (without fuzzification) revealed the latter's over-simplification of borderline cases, further supporting FAHP's nuanced capability. We also constructed a partial correlation matrix between indicator weights and final risk levels to assess linear independence. Most high-weight indicators displayed statistically significant associations ($p<0.05$), confirming the rationality of expert scoring. Radar plots illustrate pronounced heterogeneity: Wugang shows skewness toward high-risk categories (V4–V5), while the New Industrial Park dominates low-risk membership (V1–V2), visually validating Table 2 results. In addition, scenario simulations were performed under emission reduction assumptions to forecast risk changes. The projected 22% risk downgrade under 15% COD reduction highlights industrial zones' acute sensitivity to emission controls, suggesting targeted policies could yield disproportionate benefits. The spatial outputs confirm the effectiveness of FAHP in differentiating micro-zones, supporting adaptive interventions. The simulated emission reduction scenario shows that a 15% decrease in COD output could downgrade several zones from 'high' to 'medium' risk.

4. Conclusion

We recommend authorities integrate the model into existing environmental monitoring systems and decision-support dashboards. Furthermore, the framework's scalability allows application to other river basin industrial belts, such as in the Yangtze Delta and Pearl River Basin. Longitudinal studies incorporating time-series satellite data can reveal temporal shifts in regional risk, aiding sustainable development planning. In addition, the FAHP model could be embedded within municipal decision-support systems as a real-time alert layer, especially for zoning committees, infrastructure planners, and environmental bureaus. Such integration could form part of an early-warning environmental information system (EWEIS) that enhances environmental governance efficiency and cross-departmental collaboration.

This study constructed a multi-source data-driven environmental risk evaluation system and applied it to Wuhan's Qingshan District using the FAHP model. The results provided a clear zoning of high, medium, and low risk areas and identified the most influential indicators, with pollutant load intensity and treatment infrastructure emerging as dominant factors.

To operationalize these findings, we recommend:
1. Embedding the model into municipal Environmental Information Systems for real-time risk alerts, prioritizing infrastructure upgrades in eastern sub-districts (e.g., Wugang);
2. Utilizing FAHP's scenario analysis to simulate emission-trading schemes and land-use adjustments, enabling evidence-based zoning reforms;
3. Establishing cross-departmental task forces (environmental bureaus, planning committees, industrial regulators) for risk-informed land allocation.

The proposed method demonstrates strong adaptability, interpretability, and regional scalability. It can be extended to other river basin industrial clusters, especially in regions with high-intensity land development. Future research will integrate time-series remote sensing monitoring, real-time emissions data, and incorporate DPSIR feedback loops for dynamic environmental management. Future implementations should integrate real-time sensor networks with DPSIR (Drivers-Pressures-State-Impact-Response) feedback loops, while longitudinal satellite tracking (e.g., NTL/LST trends) could decode risk evolution during urban transition, advancing industrial symbiosis planning.

References

1. Zhao, J., Wang, X., & Liu, L.. A spatial risk assessment framework for urban ecosystems using remote sensing and GIS. Ecological Indicators, 98, 235–245 (2019).
2. Li, Y., Chen, Q., & Zhang, H.. Integrated risk evaluation for urban land pollution: Combining entropy method and GIS. Journal of Cleaner Production, 293, 126161 (2021).
3. Zhang, L., & Liu, Y.. Multi-source data fusion in environmental risk assessment. Environmental Monitoring and Assessment, 194(4), 265 (2022).
4. Wang, H., Zhang, D., & Xu, Y.. Spatio-temporal environmental risk analysis based on pollutant load and ecosystem resilience. Science of The Total Environment, 712, 136437 (2020).
5. Ministry of Ecology and Environment of China. Emission Inventory Calculation Guidebook (Trial Edition). Beijing: MEE Press (2020).
6. Wu, R.. Environmental Risk Assessment of the Songhua River Basin based on Multi-Source Data Fusion. Master's Thesis, Harbin Institute of Technology (2023).
7. Saaty, T. L.. Decision making with the analytic hierarchy process. International Journal of Services Sciences, 1(1), 83–98 (2008).
8. Kordi, M., & Brandt, S. A. Effects of increasing fuzziness on analytic hierarchy process for spatial multicriteria decision analysis. Computers, Environment and Urban Systems, 36(1), 43–53 (2012).
9. Elvidge, C. D., Zhizhin, M., Hsu, F. C., & Baugh, K.. VIIRS night-time lights. International Journal of Remote Sensing, 38(21), 5860–5879 (2017).
10. Hou, D., O'Connor, D., et al. Soil pollution: A hidden reality. Reviews on Environmental Health, 35(1), 1–5. (2020)

A Predictive Fault Diagnosis Method for Fire Pump Sliding Bearings Based on the Vision Transformer with Local Mean Decomposition

Yubiao Huang [1,3,4], Ye Chen [2,*], Jiaqing Zhang [1,3,4,*], Yun Yang [2], Juntao Yang [2]

[1]State Grid Anhui Electric Power Research Institute, Hefei 23601, China
[2]Shanghai Fire Research Institute of Ministry of Emergency Management, Shanghai 200032, China
[3]State Grid Laboratory of Fire Protection for Transmission and Distribution Facilities, Hefei 230601, China
[4]Anhui Provincial Key Laboratory of New Type Power Systems Fire Safety and Emergency Technology, Hefei 230601, China

Abstract. As urbanization progresses and the building inventory increases rapidly, the reliability of fire protection systems in buildings has become critical for ensuring the safety of people and property. There is an urgent need for the automated detection of fire protection system equipment. To address this challenge, a fault diagnosis method for fire pump systems based on the Vision Transformer with Local Mean Decomposition (LMD-ViT) is proposed. An attention mechanism is applied, specifically targeting sliding bearings, a key underwater component in fire pump systems. In the proposed method, the vibration signal is first smoothed, and then LMD is used to enhance the quality of frequency-domain signal analysis. Subsequently, leveraging the attention mechanism, self-attention and cross-attention mechanisms, along with a weight-sharing mechanism, are introduced to further extract fault feature information from the signal decomposition diagram. This supports the accurate identification of wear faults in sliding bearings, which is 98.8%. Moreover, the relationship between the wear of sliding bearings and the head performance of fire pumps is investigated. By classifying the severity of wear, corresponding head performance metrics can be derived. This approach enables the predictive diagnosis of fire pumps, which is also a promising solution in various related applications.

1. Introduction

As urbanization progresses and the building inventory increases rapidly, the reliability of fire protection systems in buildings has become critical for ensuring the safety of people and property. Traditional regular inspections, which largely rely on human experience, have long ensured the functionality of fire-fighting equipment. However, the growing shortage of inspectors and the rapidly increasing number of aging devices have compromised the accuracy and efficiency of these manual inspection methods [1-2]. Thus, there is an urgent need for the automated detection of fire protection system equipment. Fire pumps are key components of fire protection systems, and their failures can not only lead to equipment damage but also result in an inability to effectively extinguish fires during emergencies, resulting in irreplaceable losses. Therefore, it is highly important to explore ways to efficiently diagnose fire pump faults.

Common fire pump failures include insufficient flow, overheating, noise, and vibration issues. These issues are typically caused by damage to the bearings or misalignment of the pump shaft, leading to abnormal operation. Insufficient flow may be due to a low pump speed, damaged impellers, or a blockage within the pump that obstructs water flow. Currently, the diagnosis of fire pump faults is mainly based on real-time monitoring and analysis of physical signals, such as those for vibration, temperature, and pressure. For example, S. Bruinsma conducted experiments using induction motor-driven pumps and collected vibration, current, and voltage measurement data under normal conditions and conditions with various simulated faults (such as bearing defects and impeller damage) to develop methods for damage detection, identification, and diagnosis [3]. Liu et al. developed a personalized method based on finite element simulation method to simulate fault samples in bearings and gears [4]. Song et al. also proposed a simulation model-based method to simulate fault signals specifically for detecting bearing faults. Song et al. used the agreeable finite element models to simulate faulty bearings to obtain the vibration response signals [5]. Tauheed Mian utilized acoustic quality indicators to detect bearing faults in rotating machinery and minimize background noise [6]. Multiple acoustic quality features were extracted from the signals, and a descriptive analysis of five bearing conditions was conducted on the basis of these features to achieve nonintrusive bearing fault diagnosis. Janani Shruti Rapur used wavelet packet transform (WPT) to preprocess time-domain signal data, considering the transient and

*Corresponding author e-mail: cysfri918@sina.com (Chen Ye)
zjq230601@126.com (Jiaqing Zhang)

spectral frequencies caused by complex fluid–structure interactions due to mechanical and hydraulic faults [7].

In the past two decades, data-driven intelligent diagnostic methods have provided efficient solutions for fault prediction and health management. In these methods, the operational data for different equipment are analyzed, and machine learning algorithms are used to build fault models, thereby enabling early identification and prediction of equipment faults. V. Muralidharan obtained the wavelet features of vibration signals via discrete wavelet transform (DWT), generated rules via rough sets, and classified faults via the fuzzy logic method, with notable success in single-stage centrifugal pump fault diagnosis [8]. ZiFeng Xu et al. proposed a novel fault diagnosis method for centrifugal pumps that uses a digital twin (DT) framework and a graph transfer learning model to efficiently perform cross-domain fault diagnosis. These methods have been successfully applied to assess the rolling bearings of fire pump motors and accurately diagnose faults in motor rolling bearings [9]. Yixiang Huang et al. proposed an adaptive feature utilization method with a separate gating mechanism and a global temporal convolutional network to predict the status of bearings and engines [10]. However, few studies have focused on sliding bearings, which are the main components of fire pump systems and are submerged in water. Most studies of sliding bearings have focused only on their properties. For example, Pan proposed the use of ultrasonic reflection to measure lubricant film thickness and coating wear depth [11]. This method relies on the actual conditions of the pump's internal walls, which may be influenced by error fluctuations, such as the accumulation of scale on the pipe walls, making it difficult to deploy and challenging to ensure accuracy in real-world scenarios.

To solve the above gap of diagnosis for sliding bearings, this work developed a data-driven model combined with signal processing techniques for the critical sliding bearings of fire pumps. A method based on the Vision Transformer (ViT) is proposed. First, the vibration signals of the fire pumps under various fault conditions were collected and processed via the local mean decomposition (LMD) method. Second, a neural network with a visual attention mechanism (ViT) was constructed to achieve the early detection and fault diagnosis of the sliding bearings of fire pumps. To validate the method, an experiment on faulty sliding bearings was carried out. Moreover, considering that the head is an important indicator of fire pump performance, if the prediction corresponding to a fault level can be mathematically fit to the corresponding head loss, explainable results can be achieved in practical applications [12]. Since measuring the head requires the fire pump to operate at the full power load, which is particularly unstable under faulty conditions and poses severe risks, direct head measurement is difficult in practical applications. To address this issue, we also used a regression model based on the features of vibration signals to indirectly estimate the relationship between fault severity and head performance. This paper is organized as follows: Section 2 details the principles of the ViT method; Section 3 presents the experimental settings and data collection; Section 4 provides a discussion of the analysis results obtained with the proposed method; and Section 5 summarizes the paper. This model can be used to assess the health status of the fire pumps and related applications, and the model can be effectively implemented in practice.

2. Methodology

We propose an LMD-ViT method for diagnosing faults in the sliding bearings of fire pumps, as shown in Fig 1. The algorithm first employs the LMD method to process nonstationary signals, followed by signal reconstruction and the generation of spectrum plots. ViT is subsequently utilized to perform a predictive diagnosis on the basis of these spectrum plots.

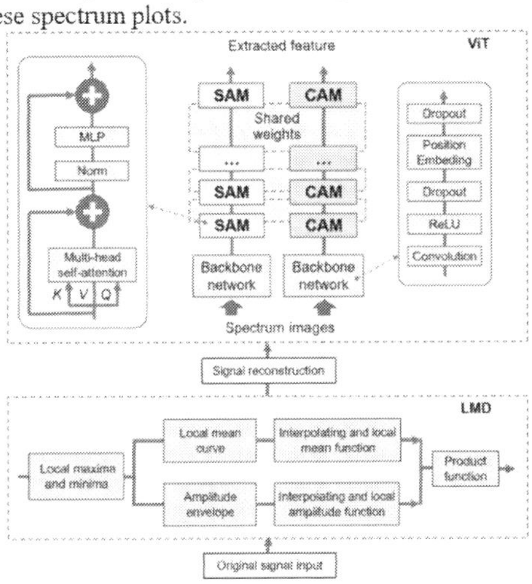

Fig.1. The network structure of ViT.

2.1. Local mean decomposition

The first step in the LMD of the signals is to identify their local maxima and minima [13]. These extrema are used to construct the upper and lower envelopes of the signals. The mean value of each pair of adjacent local maxima and local minima is subsequently calculated. The collection of these mean values is used to determine the local mean curve of the signals. Moreover, by interpolating between the local extreme points, the upper and lower envelopes of the signal are calculated. With the local mean curve and envelopes derived as described, the oscillation signals are further obtained [14]. The equation for the oscillation signals is as follows:

$$v(t) = x(t) - m(t) \tag{1}$$

where x(t) is the original signal and m(t) is the local mean. The amplitude envelope a(t) of the oscillation signal is subsequently calculated as follows:

$$a(t) = \frac{u(t) + l(t)}{2} \tag{2}$$

In the formula, u(t) and l(t) represent the upper envelope and the lower envelope, respectively.

The oscillation signal is normalized as follows:

$$h(t)=\frac{v(t)}{a(t)} \tag{3}$$

Then, the normalized signal is multiplied by the amplitude envelope to obtain the first product function.

$$PF_1(t)=a(t)\times h(t) \tag{4}$$

The above steps are repeated iteratively to compute the remaining product functions, and finally, the signal residuals are calculated.

$$r(t)=x(t)-PF_1(t) \tag{5}$$

The residuals are used as the new input data, and the above steps are repeated until the residuals no longer contain meaningful oscillatory components. In this way, the signals are decomposed into several product functions and residual components.

The original signal can be reconstructed by summing all the product functions PFi(t).

$$x(t)=\sum PF_i(t)+r(t) \tag{6}$$

2.2. Vision transformer method

The ViT method is a variation of the Transformer-based method, of which the architecture and attention mechanism were proposed by Google [15]. The Transformer framework incorporates an encoder-decoder structure, with self-attention mechanisms (SAMs) for feature extraction, rather than using traditional convolutional neural networks (CNNs) or recurrent neural networks (RNNs). Traditional CNNs extract features from image data through local receptive fields, which has limitations in modeling long-range dependencies. In contrast, the SAM of Transformer allows each variable at each position in the input sequence to interact with variables at other positions, thereby capturing valuable global information without being constrained by local receptive fields. Through this mechanism, Transformer effectively overcomes the limitations of convolutional networks in modeling long-range dependencies and addresses the vanishing or exploding gradient problems faced by RNNs when processing long sequences. As the Transformer architecture continues to evolve, an increasing number of deeper and more complex models have been proposed, enabling Transformer to perform more complex tasks. The SAM of Transformer provides a flexible structure that can be dynamically adjusted according to task requirements, further enhancing model performance. ViT represents a significant innovation in applying the Transformer architecture to high-dimensional data. The novelty of ViT lies in dividing high-dimensional image data into multiple small patches, decomposing these patches, and then using the results as inputs for the Transformer model. ViT can capture the relationships between different regions within an image through a global SAM.

In this study, we use the ViT network with a visual attention mechanism to identify and analyze the vibration signals of sliding bearings. After the signals are input into the model, a backbone network is utilized to perform initial feature extraction, with a CNN specifically selected as the backbone. The proposed ViT model network architecture is shown in Fig 1. The visual attention mechanism is introduced with both SAMs and CAMs on the basis of the Transformer architecture to enhance performance in sequential data processing, as shown in Fig 1. The backbone network consists of a convolutional network layer and serves as a feature extractor, initially extracting sensitive feature information.

The SAM and cross-attention mechanism (CAM) are two core components of the Transformer mechanism network in this study, each of which plays a specific role in different scenarios. Specifically, (1) we use the SAM to allow each element in the input sequence to interact dynamically with all other elements within the sequence. By calculating and assigning the attention weights, the model can capture complex relationships among elements within the sequence, especially long-range dependencies. For each element in the input sequence, the model calculates its similarity with other elements to determine its attention weight and then obtains a weighted sum of each element in the sequence. This process enables the model to fully consider contextual information from other positions when analyzing the sequence, thereby performing well in the extraction and recognition of time-domain related features. The mathematical expression of the SAM is shown in Eq. (7).

$$\text{Attention}(Q,K,V)=soft\max\left(\frac{QK^T}{\sqrt{d_k}}\right)V \tag{7}$$

where Q is the query matrix, K is the key matrix, V is the value matrix, and dk is the dimensionality of the key. In this equation, the dot product between the query and key components is used to assess the similarity between elements, and the results are then transformed into a probability distribution via the softmax function and subsequently used to perform a weighted sum with the value matrix. This mechanism effectively captures the relationships among elements in a sequence and is particularly suitable for tasks that require attention to long-range dependencies. (2) The CAM is used to address different sequences, which is especially valuable in the encoder-decoder structure of Transformer. During the output generation process, the decoder uses the CAM to focus on the encoder's output. Specifically, the attention weight of each input element of the decoder is calculated with respect to the encoder's output. In this way, the decoder not only considers the information from the input sequence but also effectively integrates the contextual information from the encoder, thereby increasing the accuracy of the generated output. The mathematical formula for cross-attention is similar to that for self-attention, except that the query matrix (Q) originates from the decoder's input, whereas the key matrix (K) and value matrix (V) are from the encoder's output. In this process, the relationship between the decoder's input elements and

https://doi.org/10.1051/e3sconf/202563901020

the encoder's output is captured, ensuring that the generated output fully leverages the contextual information extracted by the encoder. The CAM enables the Transformer model to establish an effective connection between the encoder and decoder, optimizing the generation process.

3. Experimental study

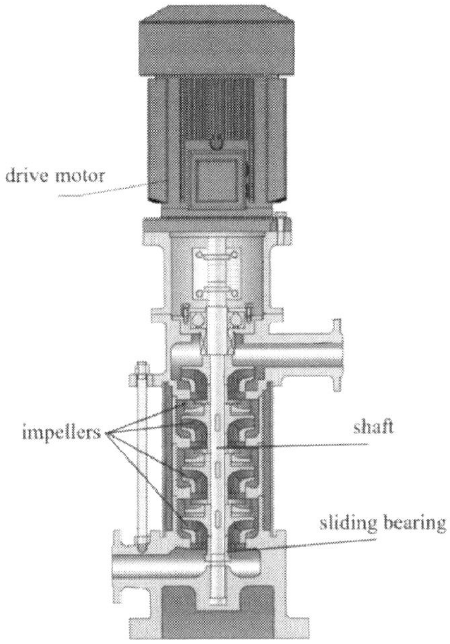

Fig.2. Typical vertical multistage centrifugal pump of fire pumps.

Currently, most fire pumps in urban buildings are compact vertical multistage centrifugal pumps, with suction and discharge ports located along the same straight line. The rotor components of the pump include the drive motor, shaft, sliding bearings, impellers, etc. The pump's intake section, intermediate section, and discharge section are connected by tie bolts to ensure structural stability and proper sealing. Fig. 2 depicts a typical vertical multistage centrifugal pump of fire pumps. As shown in Fig 3, two fire pumps are installed on the test bench.

The vertical fire pump used in the experiment is model 65QDL32-11 (22 kW). The 65QDL32-11 fire pump system is a vertical multistage centrifugal pump with a rated power of 22 kW, a rated head of 160 m, and a flow rate of 32 m^3/h, with a supply voltage of 380 V. The sliding bearings are situated between the two ends of the centrifugal pump, away from the motor drive. A schematic diagram of the fire pump test bench is shown in Fig 4. The pressure sensors and the flow sensors are installed in the pipeline to monitor the performance parameters of the fire pump and calculate the head performance.

Fig.3. Test bench for the sliding bearings of the fire pumps.

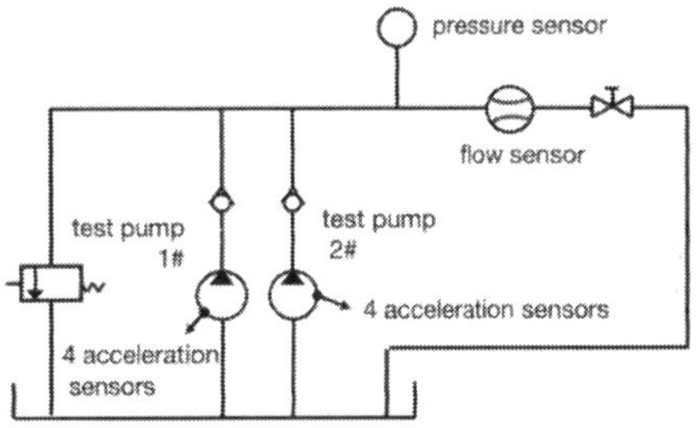

Fig.4. Schematic diagram of the fire pump test bench and its pipeline.

As vibration signals are the main sources of data used for detecting faults in rotating mechanical components, we adopt vibration sensors to collect signals in this study. The type of vibration sensor used was YZW-A26D100 manufactured by Yizhiwei Co. Ltd., Suzhou, China. As shown in Fig 5, four vibration sensors are installed at the fan end of the fire pump motor, the drive end of the motor, the centrifugal pump coupling end, and the bottom end of the centrifugal pump. These settings allow for the assessment of the operating conditions of the fire pump system as well as the sliding bearings inside.

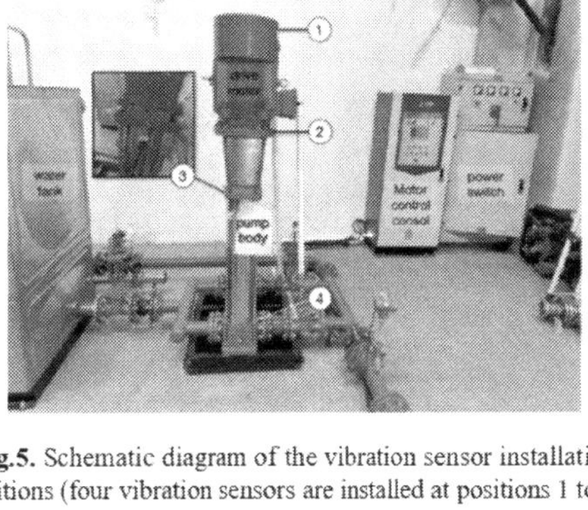

Fig.5. Schematic diagram of the vibration sensor installation positions (four vibration sensors are installed at positions 1 to 4).

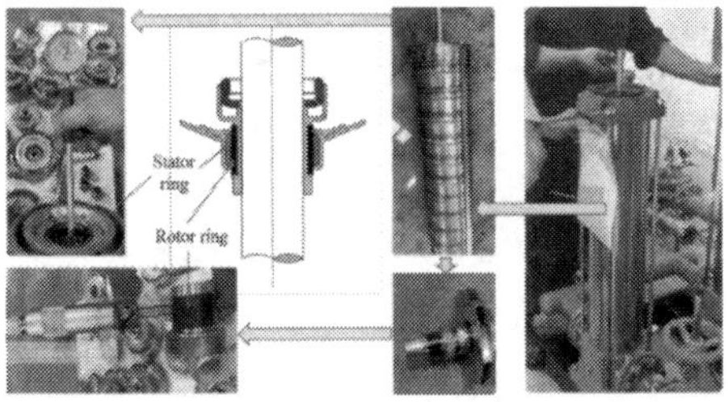

Fig.6. Sliding bearing disassembly and measurement of the sliding bearing clearance.

To test the conditions of the sliding bearings of the fire pump, we introduced faults with different degrees of wear clearance to the sliding bearings. The wear-induced clearances were machined to simulate different levels of fault severity, as shown in Fig 6, with the specific fault severity labels corresponding to those in Table 1. As shown in Tab. 1, the clearance of the normal functional sliding bearing is between 0.03 mm to 0.1 mm. The clearance of the exact testing normal sliding bearing is 0.08 mm. As a rule of thumb, if the clearance is larger than 0.5 mm, the sliding bearing can be regarded as a severe degradation of its function, which is taken as the criterion of the worst faulty degree in our test. As for the rest range between 0.1 mm and 0.5 mm, we arbitrarily devided it into other four categories of the faulty degrees and labelled them as L1, L2, L3 and L4, besides the normal status as L0, and the worst status as L5.

Table 1. Wear clearance and fault severity levels.

Measured clearance/mm	Fault severity label	Range of clearance
0.08	L0	0.03 ≦d ≦ 0.10
0.20	L1	0.11 ≦d ≦ 0.20
0.28	L2	0.21 ≦d ≦ 0.30
0.35	L3	0.31 ≦d ≦ 0.40
0.41	L4	0.41 ≦d ≦ 0.50
0.50	L5	d ≧ 0.51

The head performance corresponding to different wear clearance conditions for the sliding bearings was also measured at the maximum power load. These data served as a basis for further exploration of the relationship between fault severity and model results. We recorded the head values under different sliding bearing conditions, as shown in Fig 7.

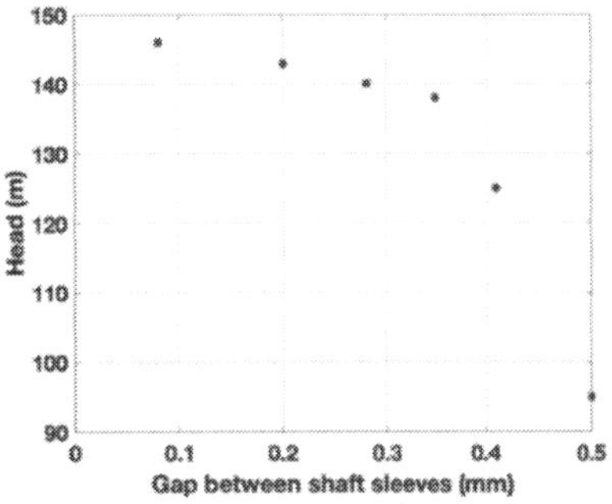

Fig.7. Chart of sliding bearing clearance (wear) vs. head performance.

We also further tested various wear clearance conditions for the sliding bearings of the fire pump system under different operating speeds of 600, 1300, 1800, and 2300 r/min. As a crucial component of a fire pump, the performance of the sliding bearing directly affects the operating efficiency and reliability of the pump. An increase in wear clearance leads to a decrease in head performance and abnormal vibrations, which can easily result in impeller fracture and severely impact the functionality of the fire protection system. The typical vibration signals under different normal operating speed conditions are shown in Fig 8, whereas those for fault conditions with a wear-induced clearance value of 0.41 mm (L4 severity) are shown in Fig 8.

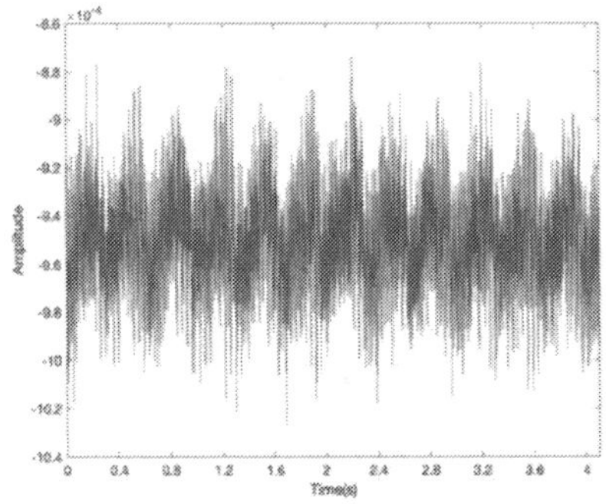

(a) Vibration signal at an operating speed of 600 r/min

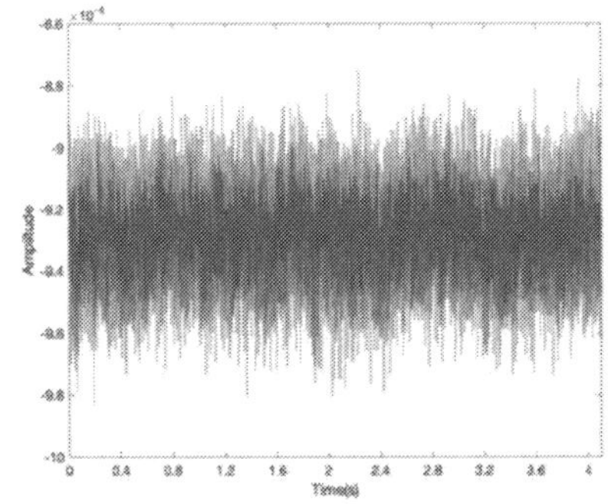

(b) Vibration signal at an operating speed of 1300 r/min

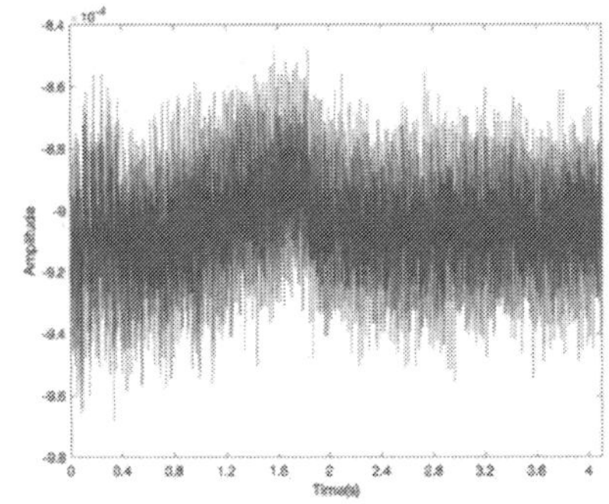

(c) Vibration signal at an operating speed of 2300 r/min

Fig.8. Vibration signals at different operation speeds under normal conditions.

https://doi.org/10.1051/e3sconf/202563901020

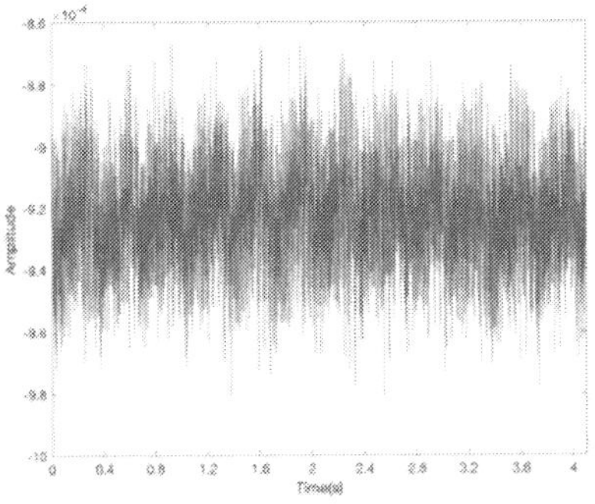

(a) Vibration signal at an operating speed of 600 r/min

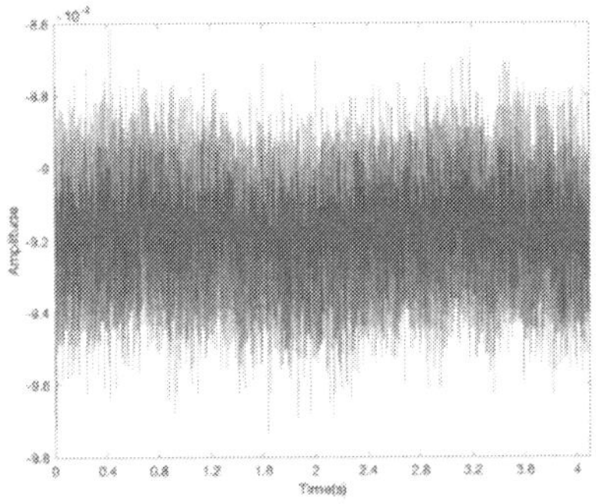

(b) Vibration signal at an operating speed of 1300 r/min

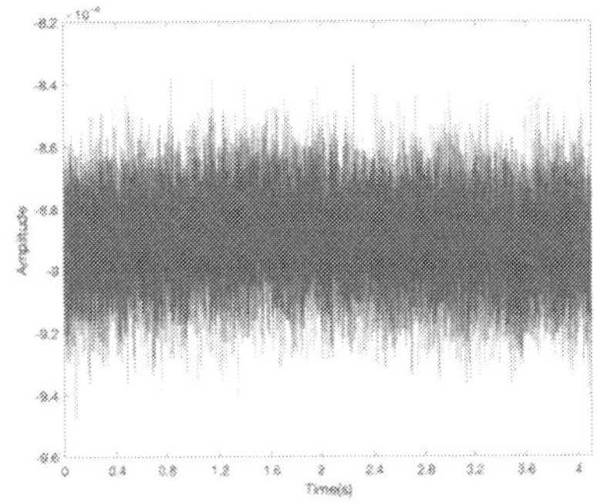

(c) Vibration signal at an operating speed of 2300 r/min

Fig.9. Vibration signals at different operation speeds under fault conditions (wear-induced clearance value of 0.41 mm).

A frequency analysis is shown in Fig 9, with the operation speed of 600 r/min as an example.

Fig 10 shows that when the sliding bearing is worn, more high-frequency components emerge. To analyze the vibration signals from the sliding bearings, we adopt the Savitzky–Golay smoothing algorithm to remove signal noise [16], followed by window processing, which is used to segment a signal into frames [17]. Although the window-based technique facilitates signal analysis, it may generate some endpoint effects, compromising the integrity of the signal and the reliability of the results. To address this issue, the Hanning window is applied before signal segmentation to mitigate the impact on the ends of the window-segmented signals [18]. Then, the signals are processed via LMD.

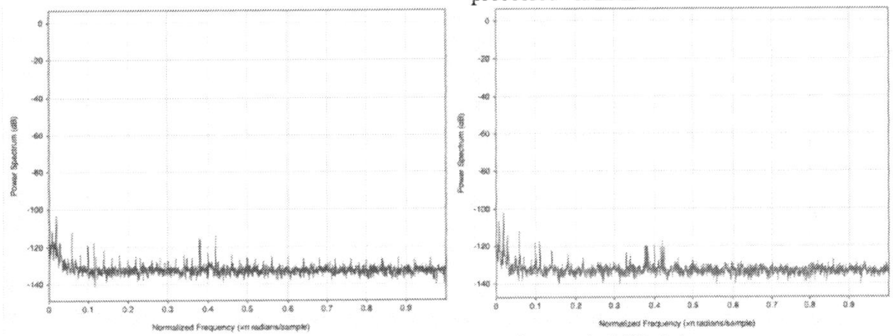

(a)The frequency spectrum at an operating speed of 600 r/min

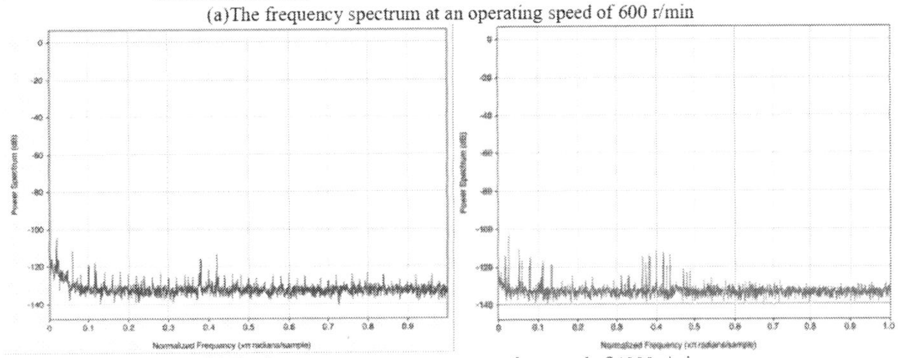

(b)The frequency spectrum at an operating speed of 1300 r/min

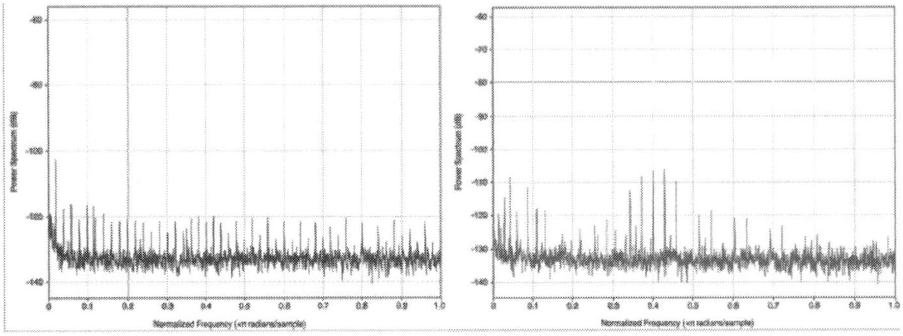

(c)The frequency spectrum at an operating speed of 2300 r/min.

Fig.10. Frequency spectra at different operation speeds under normal conditions (left) and fault conditions (right).

With the operating speed of 600 r/min as an example, the spectrum obtained after processing via the LMD method is shown in Fig 11. Compared with the spectrum of the original signals, the spectrum of the processed data is more suitable for subsequent classification with the ViT model.

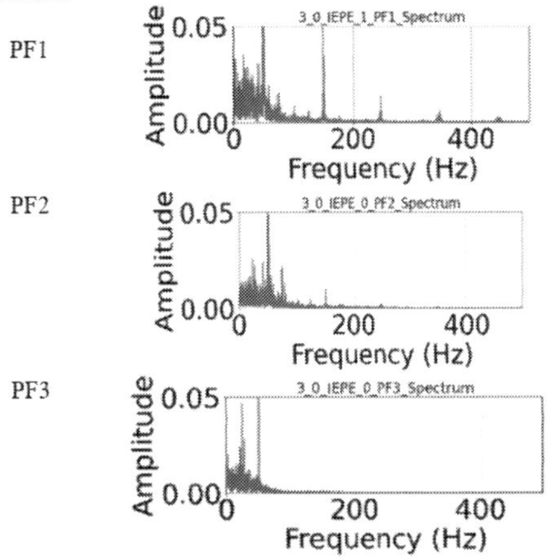

(a)Normal conditions

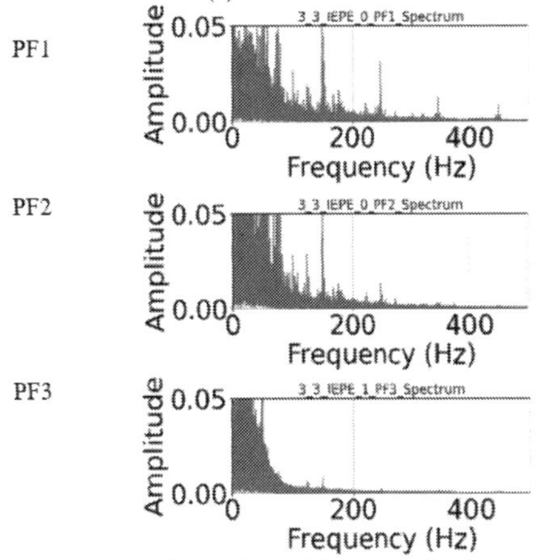

(b) Fault conditions (clearance of 0.41 mm)

Fig.11. Frequency spectrum after noise reduction and LMD processing.

During the training of the ViT model, the parameter values of the backbone layers used in Fig 1 are set as shown in Table 2. Most of the parameters are selected by default. Some hyperparameters, such as the signal length, are selected by experience.

Table 2. The parameters of the backbone network.

Hyperparameters	Value	Notes
num_classes	6	Number of classes (L0, L1, ..., L5)
signal_length	1024	Length of the input signal (number of sampling points per signal window)
patch_length	64	Length of each image patch (used for splitting the input data)
embed_dim	768	Embedded dimension
depth	6	The number of Transformer layers
num_heads	6	The number of heads of the self-attention mechanism
drop_patch	0.1	The proportion of dropout for each patch
emb_dropout	0.1	The proportion of dropout in the embedded layers
dropout	0.1	The proportion of dropout in Transformer

The cross-entropy loss function is selected to measure the error between the predicted labels and the actual labels, as shown in Eq. (8).

$$L = -\sum_{i=1}^{n} y_i \log(\hat{y}_i) \quad (8)$$

where yi is the true value, $\hat{y}_i$ is the predicted value, and n is the number of labels. The cross-entropy loss function is widely used in classification tasks, as it quantifies the difference between the probability distributions of the model results and true labels. A small cross-entropy value indicates that the prediction results are close to the true labels. Therefore, the objective is to minimize loss to improve accuracy.

The learning rate is set to 0.001. This hyperparameter is related to the step size when updating each parameter. If the learning rate is too high, the model may experience overfitting. Conversely, if the learning rate is too small, the training process may be excessively slow, and the result could converge to a local minimum.

To optimize the model parameters, we employed the AdamW optimizer. AdamW is an improved version of the Adam optimizer that combines the methods of adaptive moment estimation (Adam) and weight decay. The Adam optimizer dynamically adjusts the learning rate for each parameter by calculating the first-order moment (mean of the gradients) and the second-order moment (variance of the gradients), effectively handling sparse gradients and parameters with different scales. Unlike the traditional Adam optimizer, AdamW explicitly separates the weight decay term, thereby avoiding overfitting issues. In the AdamW optimizer, the hyperparameter for weight decay is set to 1.00E-4, which helps prevent overfitting issues, ensures that the parameters do not become excessively large, and maximizes generalization.

Table 3. The parameter settings used in the model training process.

Network parameters	Value	Notes
opt_name	'AdamW'	Name of the optimizer
base_lr	5.00E-04	The initial learning rate
lr_min	1.00E-05	The minimum learning rate
weight_decay	1.00E-04	Weight decay (for model regularization)
bias_lr_factor	2	The bias learning rate factor
weight_decay_bias	1.00E-04	The weight decay for bias
momentum	0.9	The momentum for the optimizer

The model training parameter settings are shown in Table 3. The optimizer used is AdamW, with the initial learning rate set to 5.00E-4, indicating the learning rate at the beginning of optimization. The minimum learning rate is set to 1e-5. As training progresses, the learning rate gradually decreases, and the minimum learning rate threshold is used to ensure that the learning rate does not drop too low, preventing overly small update steps and maintaining the effectiveness of training. To control the regularization of the model, weight_decay is set to 1.00E-4. This value restricts the growth of model weights to prevent overfitting. Additionally, the bias learning rate is set to 2. This aids in the rapid adjust of the bias term and accelerates model convergence. The weight decay for bias is also set to 1.00E-4, ensuring that bias terms are subject to regularization to some extent. Furthermore, the hyperparameter, momentum, is set to 0.9, due to their proven stability in transformer-based architectures.. This helps reduce oscillations in the optimization process and accelerates convergence. Through these comprehensive parameter settings, the AdamW optimizer can update the model parameters effectively, allowing the model to gradually converge to the optimal parameters.

4. Results

After training the proposed model, the results of the ViT model for diagnosing different wear levels of sliding bearings under various operating conditions are evaluated via a confusion matrix, as shown in Fig 12. The ViT model achieves an overall accuracy of 98.8% in identifying various bearing wear faults, with a perfect accuracy of 100% in distinguishing healthy conditions {L0, L1, L2} and severe wear {L4, L5}.

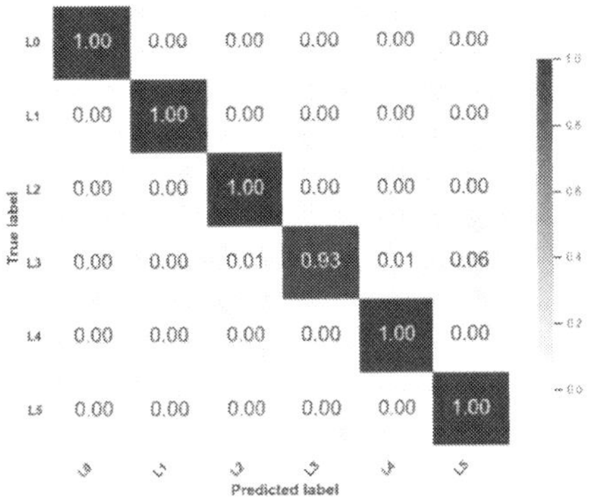

Fig.12. The confusion matrix for different wear conditions.

The above analysis indicates that the LMD-ViT model introduced in this paper provides an effective predictive diagnosis method for the sliding bearings of fire pumps. In practical applications, alarm threshold categories can be set according to different levels of fault severity. This allows for the early automated triggering of warnings before the actual malfunction of a fire pump system occurs; therefore, prompt maintenance can be scheduled to avoid potentially severe losses.

5. Discussion

Overall, our model achieved the average accuracy of 98.8% for the sliding bearings. However, there are several limitations which might be studied in the future work. First, the installation of the sensors might infuence wear level recognition, especially when signal is weak and the wear level is in the early stage. Second, for broader applications, such as structures other than vertical campcat fire pumps, the effectiveness is unknown.

6. Conclusion

In this paper, a predictive fault diagnosis method based on Vision Transformer with LMD (LMD-ViT) is proposed for identifying sliding bearing faults in fire pumps. The vibration signals from fire pump sliding bearings under fault conditions with different levels of wear clearance are monitored and analyzed. First, LMD is performed for signal processing, and then, extracts are effectively extracted from spectrum plots reconstructed via ViT. The experimental results demonstrate that the proposed method achieves high accuracy in identifying different levels of sliding bearing faults at various operating speeds.

This approach has promising broad application value in the diagnosis and management of fire pumps in urban buildings. Future work could focus on further exploring combined fault diagnosis models for fire pump systems and their application in practice, thus providing comprehensive support for the safety of civil facilities and people.

Acknowledgment

This work was supported by Science and Technology Project of State Grid Corporation of China [5700-202320280A-1-1-ZN]

References

1. Tianyue Wang, Bingtao Hu, YixiongFeng, Hao Gong, Ruirui Zhong, Chen Yang, Jianrong Tan. Multiscale cost-sensitive learning-based assembly quality prediction approach under imbalanced data. Advanced Engineering Informatics, 2024, (62): 102860.
2. Yong Chen, Feng Li, Jiaxu Wang, Baoping Tang, Xueming Zhou. Quantum recurrent encoder–decoder neural network for performance trend prediction of rotating machinery. Knowledge-Based Systems, 2020, (197): 105863.
3. S. Geertsma Bruinsma, Loendersloot R.D., T. Tinga R. Motor current and vibration monitoring dataset for various faults in an E-motor-driven centrifugal pump. Data in Brief, 2024, 52: 109987.
4. Xiaoyang Liu, Haizhou Huang, Jiawei Xiang. A personalized diagnosis method to detect faults in gears using numerical simulation and extreme learning machine. Knowledge-Based Systems, 2020, (195): 105653.
5. Wenlei Song, Jiawei Xiang, Yongteng Zhong. A simulation model based fault diagnosis method for bearings. Journal of Intelligent & Fuzzy Systems, 2018, 34(6): 3857-3867.
6. Tauheed Mian, Anurag Choudhary, Shahab Fatima. An efficient diagnosis approach for bearing faults using sound quality metrics. Applied Acoustics, 2022, 195: 108839.
7. Janani Shruti Rapur, R. Tiwari. Experimental fault diagnosis for known and unseen operating conditions of centrifugal pumps using MSVM and WPT based analyses. Measurement: Journal of the International Measurement Confederation, 2019. 147: 106809.
8. V. Muralidharan, V. Sugumaran, V. Indira. Fault diagnosis of monoblock centrifugal pump using SVM. Engineering Science and Technology, an International Journal, 2014. 17(3): 152-157.
9. Zifeng Xu, Zhe Wang, Chaojia Gao, Keqi Zhang, Jie Lv, Jie Wang, Lilan Liu. A digital twin system for centrifugal pump fault diagnosis driven by transfer learning based on graph convolutional neural networks. Computers in Industry, 2024. 163: 104155.
10. Pengcheng Xia, Yixiang Huang, Chengjin Qin, Dengyu Xiao, Liang Gong, Chengliang Liu, Wenliao Du. Adaptive Feature Utilization With Separate Gating Mechanism and Global Temporal Convolutional Network for Remaining Useful Life Prediction. IEEE Sensors Journal, 2023, 23 (18), pp: 21408 – 21420.
11. Pan Dou, Peiping Yang, Peng Zheng, Yaping Jia, Tonghai Wu, Shuo Wang, Min Yu. Ultrasound enabled simultaneous measurement of coating wear depth and lubricant film thickness in a sliding bearing. Measurement: Journal of the International Measurement Confederation, 2025. 240: 115602.
12. Michael Anthony, Robert Arno, Neal Dowling, Robert Schuerger. Reliability analysis for power to fire pump using Fault Tree and RBD. IEEE Transactions on Industry Applications, 2013, 49(2): 997 - 1003.
13. Lei Wang, Zhiwen Liu, Qiang Miao, Xin Zhang. Time–frequency analysis based on ensemble local mean decomposition and fast kurtogram for rotating machinery fault diagnosis. Mechanical Systems and Signal Processing, 2018, 103: 60 - 75.
14. Julien Fleureau, Amar Kachenoura, Laurent Albera, Jean-Claude Nunes, Lotfi Senhadji. Multivariate empirical mode decomposition and application to multichannel filtering. Signal Processing, 91(12): 2783-2792.
15. Ashish Vaswani, Noam Shazeer, Niki Parmar, Jakob Uszkoreit, Llion Jones, Aidan N. Gomez, Łukasz Kaiser, Illia Polosukhin. Attention Is All You Need. Advances in Neural Information Processing Systems, 2017-December: 5999 - 6009.
16. Ronald W. Schafer. What is a savitzky-golay filter? IEEE Signal Processing Magazine, 2011, 28(4): 111 - 117.
17. Yusuke Hioka, Jesin James, Catherine I.Watson. Masker design for real-time informational masking with mitigated annoyance. Applied Acoustics, 2020, (159): 107073.
18. Sulistyaningsih, Prasetyo Putranto, Taufiq Qurrachman, Winy Desvasari, Pamungkas Daud, Yusuf Nur Wijayanto, Dadin Mahmudin, Deni Permana Kurniadi, Arief Nur Rahman, Sri Hardiati, Arie Setiawan, Fajri Darwis. Performance Comparison of Blackman, Bartlett, Hanning, and Kaiser Window for Radar Digital Signal Processing. in: 2019 4th International Conference on Information Technology, Information Systems and Electrical Engineering, ICITISEE. 2019: 391-394.

Study on the Impact of Solar Chimneys on Thermal Comfort and Energy Efficiency in Portable Toilets

Jing Zhang, Tianpeng Wang*

Lanzhou Jiaotong University, Lanzhou, Gansu, 730070 China

Abstract. To address the issues of energy consumption, ventilation, and pollutant accumulation in mobile toilets, this paper proposes an optimisation strategy for human thermal comfort and energy conservation based on solar materials and the chimney effect. Through a heat-collecting wall-mounted vent system, CFD simulation and pollutant dispersion models are used to quantify its performance. The results show that: under extreme low temperatures of -10°C, the optimised mobile toilet achieves a maximum indoor temperature difference of 5.93°C, with the thermal comfort zone expanded by 40%, a 50% reduction in temperature gradient, and significant improvements in PMV-PPD indices during the transitional season and summer. In winter, the average PPD decreased by 58%, and in summer, the PPD reduction reached 10%. Compared to traditional models, annual ventilation energy consumption was reduced by 74.5%, thermal comfort-related energy consumption decreased by 67.2%, and the overall energy savings rate reached 58.5%. This approach holds significant reference value for improving the thermal environment of mobile toilets and advancing the achievement of carbon peaking goals. While the study lacks field experiment, the CFD models were rigorously benchmarked against established solar chimney studies[5,6,11] with grid independence verification.

1. Introduction

The energy problem has always been an important issue for many countries[1]. Currently, building energy consumption accounts for 40% to 50% of national energy consumption[2]. Among these, HVAC systems accounting for nearly half the total building energy consumption[3].

Portable toilets are an important component of urban infrastructure and emergency response systems, and the quality of their internal environment directly impacts the health and comfort of users. However, such structures experience extreme temperature fluctuations under extreme weather conditions, leading to large vertical temperature differences and severe cold air infiltration, known as 'cold accumulation.' During summer or transitional seasons, they also face issues of poor ventilation causing pollutant accumulation and high temperatures with stuffy conditions. Moreover, since the COVID-19 outbreak, public health awareness has significantly increased. Portable toilets, a place with higher chances of potential breeding grounds for germs and viruses, have prevented people from using them during and after the pandemic[4].

Arce et al.[5] conducted an experimental study on the full-scale solar chimney under actual meteorological conditions, and the results showed that the ventilation volume of the solar chimney was mainly affected by the temperature difference between the inlet and outlet and the outdoor wind speed. Neves et al.[6] used numerical simulations to study the effects of wind speed and direction on the performance of solar chimneys.

Existing research has mainly focused on large fixed buildings. However, there is still a lack of research on the system design optimisation, performance quantification, and comprehensive impact on human thermal comfort of solar chimneys in mobile toilets with micro-spaces, dynamic load characteristics, and extreme operating conditions. Taking inspiration from the current research status, there is great potential to adopt a hybrid solar and biomass energy supply system in economically underdeveloped areas with abundant solar and biomass resources[7]. Therefore, this study proposes a modular heat-collecting wall-mounted wind cap system suitable for mobile toilets, which aims to achieve year-round thermal environment optimisation and energy conservation and consumption reduction through solar-driven chimney effects, ultimately achieve the goal of carbon neutrality[8].

2. Physical models and computational models

2.1. Model overview

The physical model of the building space is shown in Figure 1 (traditional type), with the toilet interior dimensions being 1.1 m long × 1.1 m wide × 2.3 m high. The building material consists of 16 mm thick metal

* Corresponding author: wangtp@mail.lzjtu.cn

embossed panels. One rectangular louvered ventilation opening measuring 0.3 m high × 0.7 m long and another measuring 0.2 m high × 1.1 m long are installed at the bottom of the south-facing wall and the top of the north-facing wall, respectively.

The solar-driven natural ventilation system with integrated energy storage regulation developed in this study adopts a three-tier modular design comprising a 'ventilation hood-cavity-louver' configuration, as shown in Figure 2 (optimised type). The system consists of a black chrome-coated heat-absorbing panel, a Low-E glass cover plate, and louver panels, with overall dimensions of 1.1m × 1.1m × 0.4m, installed on the roof of the mobile toilet.

Stack ventilation should work when one or all the natural forces are available[9]. Solar chimney technology has been widely used in strengthening natural ventilation, improving indoor thermal comfort and new energy power generation, which has greatly improved the comprehensive utilization rate of solar energy[10]. The natural ventilation mechanism driven by it realizes the cascade utilization of energy through a three-stage modular structure (heat absorbing plate-cavity-louver).

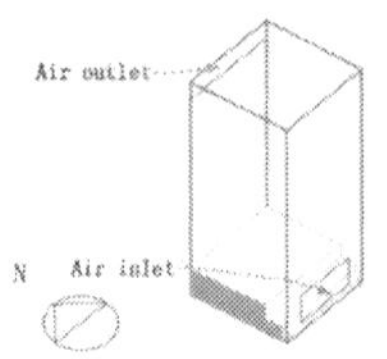

Figure 1. Simplified model of mobile toilet (Traditional Type).

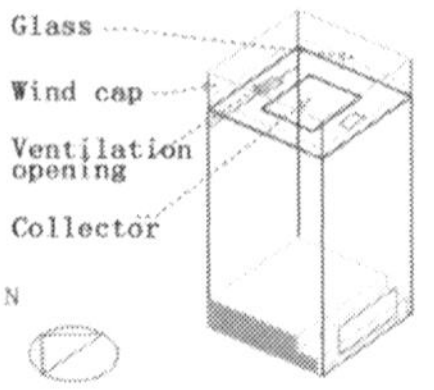

Figure 2. Optimised model of mobile toilet (Optimisation Type).

2.2. Physical model

This simulation uses Design Builder simulation software, with the location selected as Lanzhou City, Gansu Province, and the simulation season as the entire quarter. At lower altitudes, atmospheric pressure is close to sea level standard atmospheric pressure, i.e., P = 101,325 Pa. Table 1 lists the model-related parameters.

Table 1. Formatting sections, subsections and subsubsections.

Type	Quantity	Dimensions (m*m*m)	Model	Boundary conditions	Reference value
Mobile toilet	1	1.1*1.1*2.3	Room	Thermal insulation	-
Personnel	1	-	Persons	Constant heat flow	50W
Wall	4	1.1*2.3	Walls	Constant temperature	Depending on the operating conditions
Door	1	0.9*1.8	Doors	Constant temperature	Depending on the operating conditions
Air inlet	1	0.3*0.7	Openings	Free outlet	-
Air outlet	4	0.15*1.0	Openings	Free outlet	-
Black chrome-coated heat-absorbing plate	1	0.5*0.6	Walls	Constant heat flux density	750W/㎡
Low-E glass cover plate	1	1.1*1.1	Walls	Constant heat flux density	40W/㎡
Heat storage plate	1	1.1*1.1	Walls	Constant temperature	Depending on the operating conditions
LED light	1	-	Blocks	Constant heat flow	10W
Fan	1	-	Blocks	Constant heat flow	25W

2.3. Mathematical model

This study uses a multi-physical field coupling model system to evaluate the performance of heat-collecting wall-mounted wind caps, constructing a three-dimensional theoretical framework that includes solar-thermodynamic coupling, pollutant diffusion, and full life cycle energy efficiency analysis.

2.3.1. CFD simulation

Based on the coupled solution of the three-dimensional steady-state Navier-Stokes equations and energy equations, the Realizable k-ε turbulence model is used to simulate air flow and heat transfer characteristics. Non-structured grids are used for calculations while ensuring quality. The grid is refined at the entrance and boundaries of the solar collector shed to improve calculation accuracy[11]. Model validation was performed by comparing velocity profiles with Arce et al.'s experimental data[5] under equivalent temperature differentials (RMSE=8.2%). Grid sensitivity analysis confirmed solution independence at 1.2 million cells ($\Delta T<0.3°C$ beyond this threshold).The control equations are:

$$\begin{cases}\frac{\partial(\rho u_i)}{\partial t}+\frac{\partial(\rho u_i u_j)}{\partial x_j}=-\frac{\partial p}{\partial x_i}+\frac{\partial}{\partial x_j}\left[\mu\left(\frac{\partial u_i}{\partial x_j}+\frac{\partial u_j}{\partial x_i}\right)-\rho\overline{u_i' u_j'}\right]\\ \frac{\partial(\rho h)}{\partial t}+\frac{\partial(\rho u_j h)}{\partial x_j}=\frac{\partial p}{\partial t}+\frac{\partial}{\partial x_j}\left(k\frac{\partial T}{\partial x_j}\right)+\mu\left(\frac{\partial u_i}{\partial x_j}+\frac{\partial u_j}{\partial x_i}\right)\frac{\partial u_i}{\partial x_j}\end{cases} \quad (1)$$

2.3.2. Pollutant dispersion model

Calculate the NH_3 diffusion trajectory based on the Gaussian plume model, and introduce wind speed pulsation correction terms to improve prediction accuracy:

$$C(x,y,z)=\frac{Q}{2\pi u\sigma_y\sigma_z}\exp\left(-\frac{y^2}{2\sigma_y^2}-\frac{(z-H)^2}{2\sigma_z^2}\right) \quad (2)$$

Among them: is the pollutant concentration (ppm); =0.1 g/s is the NH_3 emission source strength; =0.03 m/s is the environmental wind speed; is the lateral and vertical diffusion coefficient (m), calculated using the Wilson correction model:

$$\sigma_y=\gamma_1 x^{a_1},\sigma_z=\gamma_2 x^{a_2} \quad (3)$$

3. Simulation analysis

For the vertical cross-sections at x = 0.55 of the two models during summer, winter, and transitional seasons, the simulation results were analysed in terms of temperature, velocity, pollutant concentration, PMV-PPD, and energy consumption.

3.1. Temperature field

The temperature field cross-sectional diagrams for summer, winter, and transitional seasons before and after optimisation are shown in Figure 3.

The optimised mobile toilet demonstrates significant improvements in thermal performance under extreme low temperatures (-10°C): the top temperature rises from -10°C in the traditional model to -5°C, and for the first time, a positive temperature zone appears at -5°C. By 0°C, the entire area achieves positive temperature coverage, with the thermal comfort zone area expanding by 40% and the vertical temperature gradient reducing by 50%. Through the synergistic effect of thermal storage walls and solar chimneys, the system forms a stable thermal buffer layer under extreme winter conditions, effectively suppressing the 'cold accumulation' phenomenon caused by cold air penetration; during the transitional season, optimised airflow organisation improves temperature uniformity by over 30% and significantly enhances thermal stability; in high-temperature summer environments, chimney effect-driven three-dimensional ventilation achieves a maximum temperature reduction of 10.685°C under 20-30°C conditions, forming an efficient thermal stratification between the high-temperature zone at the top and the low-temperature zone at the bottom. Annual data shows that the system can actively regulate the temperature field under different climatic conditions through a solar-driven natural ventilation mechanism.

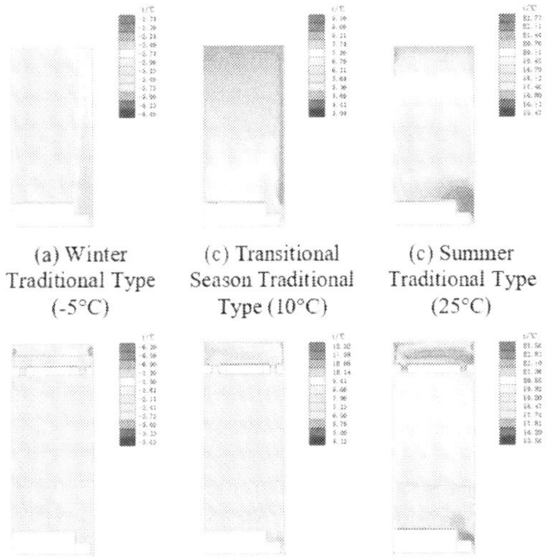

(a) Winter Traditional Type (-5°C) | (c) Transitional Season Traditional Type (10°C) | (c) Summer Traditional Type (25°C)

(b) Winter Optimisation Type (-5°C) | (d) Transitional Season Optimisation Type (10°C) | (f) Summer Optimisation Type (25°C)

Figure 3. Typical temperature field distribution diagram.

3.2. Speed and pollutant concentration

The vertical cross-section velocity simulation results are shown in Figure 4. In winter, the focus is on eliminating local high-speed phenomena caused by direct cold air blowing. In summer, the focus is on enhancing vertical thermal pressure ventilation efficiency. In the transition season, the focus is on balancing uniformity and ventilation volume. Overall, this achieves improved adaptability across all seasons and optimised thermal comfort. Under the effect of the solar chimney, the horizontal cross-sectional velocity increase rate is:

$$\eta=\frac{|v_1-v_0|}{v_0}\times 100\% \quad (4)$$

η is the speed improvement rate, v_1 is the speed of the optimised group, m/s; v_0 is the corresponding speed of the traditional group, m/s.

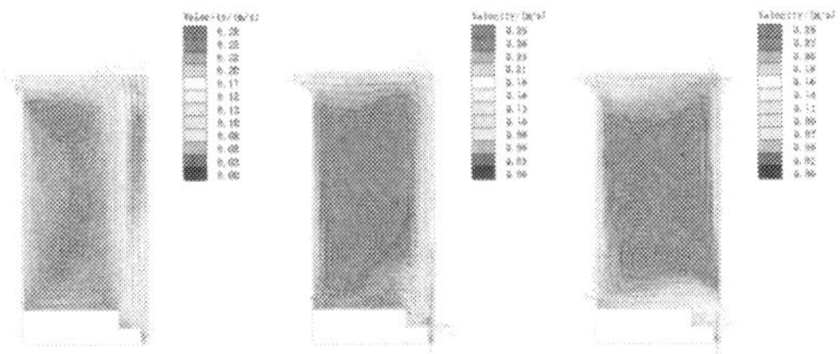

(a) Winter Traditional Type (-5°C) | (c) Transitional Season Traditional Type (10°C) | (e) Summer Traditional Type (25°C)

 https://doi.org/10.1051/e3sconf/202563901021

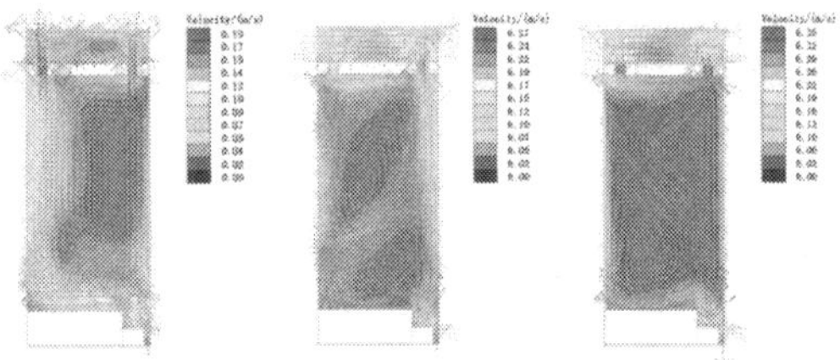

(b) Winter Optimisation Type (-5℃) | (d) Transitional Season Optimisation Type (10°C) | (f) Summer Optimisation Type (25℃)

Figure 4. Typical wind speed distribution map.

According to the vertical cross-sectional velocity simulation data of the solar chimney shown in Figure 5, the optimised system exhibits significant performance differences compared to the traditional system. Under test conditions with a temperature range from -10°C to 30°C, the average wind speed of the optimised system is higher than that of the traditional system at most temperature points, with the most significant improvement observed at 30°C, reaching 73.2%. Additionally, the average concentration of the optimised system is lower than that of the traditional system at all test temperatures, demonstrating significant pollutant dilution effects, particularly at 0°C and 20°C, where concentration reductions of 53% and 44.7% were achieved, respectively.

The ventilation efficiency and air purification performance of the optimised system are optimal within the 20–30°C temperature range, which is related to the enhanced thermodynamic effects within this temperature interval. However, at -5°C, the optimised system's airflow velocity was lower than that of the traditional system, decreasing by 41.8%, indicating that the system's performance under extreme low-temperature conditions may require further optimisation. Overall, the optimised design demonstrates clear advantages in improving indoor air quality and ventilation efficiency.

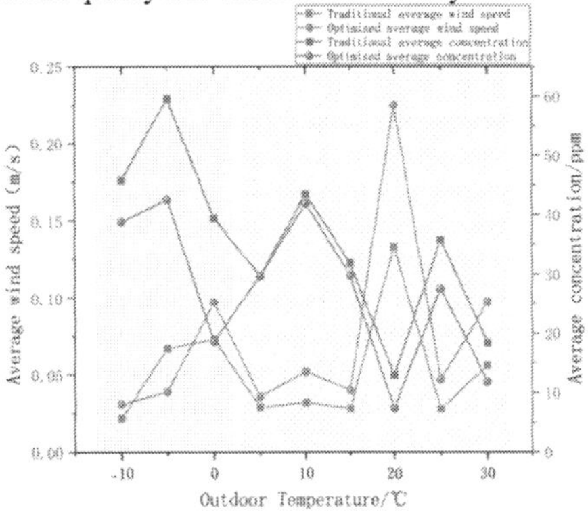

Figure 5. Average indoor wind speed and concentration changes at different outdoor temperatures.

3.3. PMV-PPD

The biggest factor affecting the PMV-PPD index of thermal comfort evaluation is the indoor temperature, so the setting of PMV-PPD will affect the evaluation results of energy consumption, economic and environmental benefit of heating mode[12], the specific scale is shown in Table 2. According to the provisions of GB50736-2012 "Design Code for Heating, Ventilation and Air Conditioning", The values should be $-1 \leq PMV \leq 1$ and $PPD \leq 27\%$. The recommended maximum comfortable temperature ($-1 \leq PMV \leq 1$, $PPD \leq 27\%$) is the maximum temperature that can be achieved within the design specifications.

Table 2. PMV-PPD Thermal Sensation Scale.

PMV	-3	-2	-1	0	1	2	3
Thermal sensation	Cold	Cool	Slightly cool	Moderate	Slightly warm	Warm	Hot
PPD/%	100	75	25	5	25	75	100

According to the Airpak simulation analysis results shown in Figure 6, the recommended maximum comfortable temperature for the entire year is around 20°C. The maximum improvement in PMV and PPD under the effect of the solar chimney occurs in the temperature range of 0°C to 15°C. When the outdoor temperature is 0°C, PMV increases from -0.6 to 0.2, and PPD decreases from 20% to 7%; when the outdoor temperature is 5°C, PMV decreases from 1.1 to 0.3, and PPD decreases from 15% to 9%; when the outdoor temperature is 15°C, PMV decreases from 3 to 1.9, and PPD decreases from 50% to 37%. These data indicate that the optimised solar chimney system achieves the best thermal comfort improvement effects in the medium to high temperature range, significantly enhancing indoor thermal environment quality and reducing user dissatisfaction.

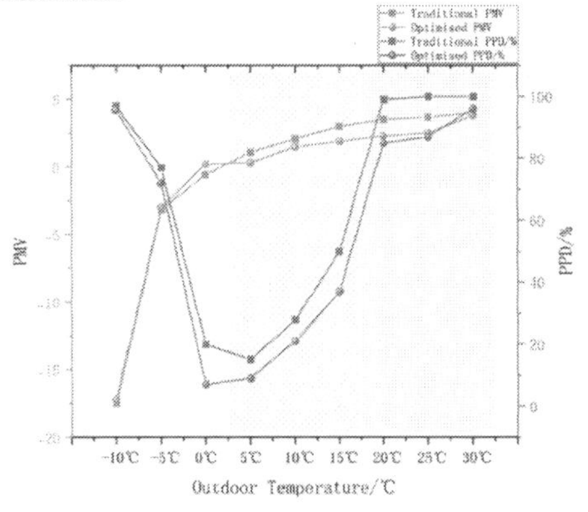

Figure 6. PMV-PPD at different outdoor temperatures.

3.4. Energy consumption model and analysis

Simulation results show that the annual average mechanical ventilation energy consumption of traditional mobile toilets is 48.7 kWh, with 62% occurring in winter and 38% in summer. In contrast, the optimised system replaces mechanical ventilation with natural ventilation via a solar chimney, requiring supplementary operation only under extreme weather conditions, reducing annual average energy consumption to 12.4 kWh, achieving an energy savings rate of 74.5%.

In terms of thermal comfort energy savings, combining Figure 6 with the indoor heat load calculation formula (Formula 5), the traditional system requires electric heaters to be operated for a cumulative 235 hours in winter and fans for cooling for a cumulative 152 hours in summer; the optimised system, however, uses solar chimney heat collection and phase-change material heat storage to reduce winter heating energy consumption to 68 hours and summer air conditioning operation time to 76 hours, resulting in an overall reduction of 67.2% in thermal comfort energy consumption.

$$Q = U \cdot A \cdot (T_{set} - T_{out}) \cdot t \tag{5}$$

U is the heat transfer coefficient (W/m²·K), A is the surface area (m²), T_{set} is the set temperature (°C), and T_{out} is the operating time (h).

Combining the two energy consumption metrics, the optimised system achieves an annual total energy consumption of 103.6 kWh, a 58.5% reduction compared to the traditional system. The solar chimney contributes 63% of the energy savings, phase-change materials contribute 21%, and the system achieves a self-sufficiency rate of 12%.

4. Conclusion

By constructing a numerical model that couples solar-thermodynamics, pollutant diffusion, and annual dynamic energy consumption, the system's performance under extreme and typical conditions in different seasons was systematically simulated and analysed. The main conclusions are as follows:

1) In extreme low-temperature conditions, the optimised mobile toilet achieved a maximum indoor temperature difference of 5.93°C, with the thermal comfort zone expanded by 40% and temperature gradients reduced by 50%. During the transitional season and summer, the PMV-PPD indicators were significantly optimised, with the average PPD in winter decreasing to 58% and a reduction of 10% in summer.

2) Under 20–30°C conditions, the average indoor air velocity increased by up to 73.2%, creating a differentiated high-efficiency airflow organisation for winter and summer. NH_3 concentrations decreased by 53% and 44.7% at 0°C and 20°C, respectively, with ventilation efficiency significantly outperforming traditional designs.

3) Annual ventilation energy consumption was reduced by 74.5%, and thermal comfort adjustment energy consumption decreased by 67.2%.

References

1. Li, L. L., Liu, Y. W., Tseng, M. L., Lin, G. Q., & Ali, M. H. Reducing environmental pollution and fuel consumption using optimization algorithm to develop combined cooling heating and power system operation strategies. *Journal of Cleaner Production*, **247**, 119082 (2020).
2. "China Building Energy Consumption Research Report 2020." *Building Energy Efficiency*, **49.02**,1-6 (2021).
3. Dai, B., Tong, Y., Hu, Q., & Chen, Z. Characteristics of thermal stratification and its effects on HVAC energy consumption for an atrium building in south China. *Energy*, **249**, 123425 (2022).
4. Yeh, T. Y., & Shih, Y. T. A STUDY OF CONSIDERATION FOR IMPROVING USER EXPERIENCE OF PORTABLE TOILETS IN THE POST-PANDEMIC ERA. In *DS 123: Proceedings of the International Conference on Engineering and Product Design Education (E&PDE 2023)*. (2023).
5. Arce, J., Jiménez, M. J., Guzmán, J. D., Heras, M. R., Alvarez, G., & Xamán, J. Experimental study for natural ventilation on a solar chimney. *Renewable Energy*, **34(12)**, 2928-2934 (2009).
6. de Oliveira Neves, L., & da Silva, F. M. Simulation and measurements of wind interference on a solar chimney performance. *Journal of wind engineering and industrial aerodynamics*, **179**, 135-145 (2018).
7. Li, J., Xu, A., Huang, J., Novakovic, V., Karkon, E. G., & Zhu, J. Feasibility and annual performance of zero-energy ecological toilet in cold regions. *Energy Conversion and Management*, **327**, 119593. (2025).
8. Yang, P., Peng, S., Benani, N., Dong, L., Li, X., Liu, R., & Mao, G. An integrated evaluation on China's provincial carbon peak and carbon neutrality. *Journal of Cleaner Production*, **377**, 134497 (2022).
9. Antczak-Jarząbska, R., Pawłowski, K., & Niedostatkiewicz, M. Improvement of the chimney effect in stack ventilation. *Applied Sciences*, **11(19)**, 9185 (2021).
10. Zuo Lu, Zhou Xiaotian, Chen Jiajun, Ding Ling, Liu Zihan, & Yuan Yue. Preliminary study on combined seawater desalination solar chimney power station for indirect condensation heat transfer to prepare fresh water. *Renewable Energy*, **36(1)**, 1-7 (2018).
11. Nie Jing, Su Hao, Jia Jing, Gao Hong, Yan Suying, & Xu Jinchen. Analysis of the influence of flow model on flow field characteristics of small solar chimneys. *Journal of Solar Energy*, **45(1)**, 80-86 (2024).
12. Dong, Z., Boyi, Q., Pengfei, L., & Zhoujian, A. Comprehensive evaluation and optimization of rural space heating modes in cold areas based on PMV-PPD. *Energy and Buildings*, **246**, 111120 (2021).

Comparative Analysis of Life Cycle Carbon Emissions Between Copper and Aluminum Core Power Cables in Low-Voltage Distribution Networks

Sirui Yang, Yanchi Zhang*

School of Electrical Engineering, Shanghai DianJi University, No. 300, Shuihua Road, Pudong New Area, Shanghai, 201306, China

Abstract. Accurate carbon footprint assessment of low-voltage distribution network components is critical for achieving carbon neutrality. This study conducts a comparative life cycle assessment (LCA) of copper-core (YJV) and aluminum-core (YJLV) power cables, covering raw material extraction, manufacturing, transportation, operation, and end-of-life stages. A quantitative LCA model is established to evaluate emissions under equivalent current-carrying capacity. A case study based on typical cable specifications shows that, due to aluminum's higher carbon footprint factor compared to copper, aluminum-core cables have higher emissions during manufacturing. However, aluminum offers greater recycling potential, and increasing the share of recycled aluminum can significantly reduce its carbon footprint. Due to higher electrical resistance, aluminum-core cables also result in significantly higher operational emissions compared to copper-core cables. The findings reveal that conductor selection leads to distinct emission profiles across life-cycle stages and must be evaluated comprehensively, considering service life, power generation mix, and load characteristics. This study provides theoretical support and practical guidance for low-carbon cable selection and emission optimization in distribution networks.

1. Introduction

China's commitment to peak CO_2 emissions by 2030 and reach carbon neutrality by 2060 drives emission reductions across all sectors. The electric power industry is a primary focus, accounting for about 40% of the country's CO_2 output. In addition to greening generation, decarbonizing grid infrastructure – including transmission and distribution cables – is critical for meeting these targets[1].

Power cable demand is booming globally. The medium-voltage cable market is projected to grow from about $39 billion in 2023 to $76 billion by 2032 (≈7.8% CAGR), and high-voltage cable investment from $33.5 billion (2022) to $54.7 billion by 2030 (≈7.2% CAGR)[2]. This surge is driven by expanding electrification, urbanization, and renewable grid expansion. Wu et al. (2023) examined the carbon footprints of 110 kV polypropylene (PP) and cross-linked polyethylene (XLPE) insulated cables under varying temperatures, focusing on insulation material effects.[3] Jones and McManus (2010) conducted a life-cycle assessment of five 11 kV systems—three overhead copper and two underground aluminum cables—and identified conductor losses as the dominant contributor to environmental impacts.[4] However, both studies addressed different cable configurations or insulation types and did not directly compare copper-core and aluminum-core cables with equivalent current-carrying capacities.

Life-cycle assessment (LCA) is the standard approach to quantifying product carbon footprints. International product carbon-footprint standards (e.g., PAS 2050, ISO 14067) provide general rules for LCA, but they are not tailored to power cables.[5],[6] Existing guidelines cover broad product categories, and China's national protocols for cable carbon accounting have not kept pace. A recent enterprise-level guideline (Q/GDW 12395-2024) has been issued, but its practical adoption remains limited.[7]

This paper develops a life-cycle carbon evaluation model for low-voltage power cables, comparing copper-core and aluminum-core designs with equal current-carrying capacity. The model quantifies CO_2 emissions from cradle to grave – including raw-material extraction, cable fabrication, transportation, operation (including losses), and end-of-life treatment. By providing detailed emission inventories for each conductor type, the study supports green procurement and carbon labeling of cable products, and informs low-carbon planning of distribution networks.

2. Carbon footprint calculation method and process

2.1. Product description

In this study, the YJV-8.7/15kV (3×240mm²) copper-core cable and the YJLV-8.7/15kV (3×400mm²) aluminum-

* Corresponding author: yangthree299@outlook.com

core cable are selected as the subjects for carbon footprint accounting. Both cables adopt a three-core, non-armored design with identical layered structures, consisting of a copper or aluminum conductor, semi-conductive EVA conductor shielding layer, cross-linked polyethylene (XLPE) insulation, semi-conductive EVA insulation shielding layer, helical copper tape shielding, polypropylene fillers between cores, polypropylene wrapping tape, and a HI-90 grade PVC outer sheath. In addition, the two cables share high consistency in insulation material (XLPE), structural thickness, and key electrical characteristics such as inductance and capacitance per unit length. Although minor differences exist in conductor resistance, the overall variation is limited. Therefore, the two cables are considered structurally and functionally equivalent, making them suitable for a comparative carbon footprint assessment.

2.2. Functional unit

In this study, the functional unit is defined as 1 kilometer of power cable operating at its rated current over a 20-year service life. The system boundary follows a cradle-to-grave approach, covering all life cycle stages specified in GB/T 51366-2019. To ensure comparability of carbon footprint results in terms of functional performance, it is assumed that both the copper-core cable (YJV-8.7/15kV 3×240mm^2) and the aluminum-core cable (YJLV-8.7/15kV 3×400mm^2) operate under identical environmental and installation conditions. According to the technical specifications presented in Table 2, both cables exhibit similar current-carrying capacities of approximately 480 A under buried installation conditions, thus satisfying the criterion of functional equivalence.

The specific assumptions adopted in this study include: direct buried installation method, soil thermal resistivity of 1.2 K·m/W, ambient temperature of 25°C, conductor operating temperature of 90°C, single-layer spaced layout, and a service life of 20 years. Under these standardized conditions, the current-carrying capacities of the two cables are considered functionally equivalent.

It should be noted that variations in installation methods or environmental parameters may significantly affect the thermal dissipation and electrical resistance of the cables, thereby influencing energy losses and carbon emissions during the operational phase. Consequently, the carbon footprint results presented in this study are condition-dependent and should be adjusted based on actual field conditions in practical engineering applications to ensure accurate environmental assessments.

2.3. Assessment criteria and data sources

The assessment criteria for this study include: excluding from the LCI any process that contributes less than 1% individually, as long as the cumulative contribution of excluded processes remains below 5%. Global warming potential is calculated using the 100-year GWP factors from the IPCC Fifth Assessment Report. Data sources primarily consist of original enterprise production and operation data, supplemented by authoritative background LCI databases such as the Chinese Life Cycle Database (CLCD) and the ecoinvent database, in order to ensure temporal, geographical, and technological representativeness and consistency of the modeling.

3. Carbon footprint model development

3.1. System boundaries

The system boundaries of this study are illustrated in Figure 1, encompassing the full life cycle of a power cable, which includes five stages: raw materials acquisition, manufacturing, distribution and sales, use, and recycling/disposal phase.

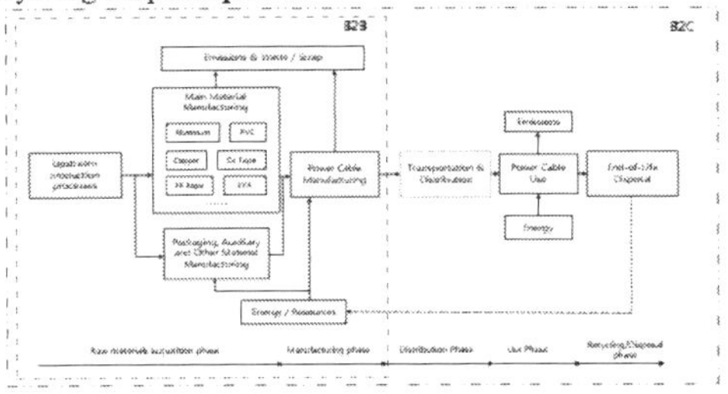

Figure 1. Power cable product system boundary.

3.2. Life Cycle carbon footprint

Based on life cycle assessment principles, the total carbon footprint of power cables over their life cycle can be expressed as Equation (1):

$$CF_{\mathrm{LC}} = CF_{\mathrm{R}} + CF_{\mathrm{M}} + CF_{\mathrm{D}} + CF_{\mathrm{U}} + CF_{\mathrm{RD}} \quad (1)$$

$CF\mathrm{R}$ is the Carbon footprint during raw material acquisition; $CF\mathrm{M}$ is the Carbon footprint during manufacturing;$CF\mathrm{D}$: Carbon footprint during distribution and sales; $CF\mathrm{U}$: Carbon footprint during the use phase; $CF\mathrm{RD}$: Carbon footprint during recycling/disposal phase.

3.3. Raw material acquisition phase

Let the total quantity of materials used in the power cable be m, and the energy types be n. The carbon footprint in this phase is calculated by Equation (2):

$$CF_{\mathrm{R}} = \left(\sum_{i=1}^{m} N_i \cdot \beta_i + \sum_{j=1}^{n} E_j \cdot \delta_j \right) \cdot \frac{1}{\eta_{ij}} \quad (2)$$

N is the Quantity of material type i ; β is the Emission factor of material i ; E is the Quantity of energy type j ; δ is the Emission factor of energy type j ;η_{ij} is the Material utilization rate.

3.4. Manufacturing phase

Assuming that the manufacturing process involves n types of energy and emits b types of greenhouse gases, the carbon footprint is given by Equation (3):

$$CF_{\mathrm{M}} = \left(\sum_{j=1}^{n} E_j \cdot \beta_j + \sum_{k=1}^{b} U_k \cdot GWP_K\right) \cdot \frac{1}{\eta_{jk}} \quad (3)$$

E_j is the Quantity of energy type j consumed; β_j is the Emission factor of energy type j; U_k is the Quantity of greenhouse gas type k emitted; GWP_K is the Global Warming Potential of gas k (based on the latest IPCC assessment); η is the Energy efficiency in manufacturing.

3.5. Distribution phase

The carbon footprint during the transportation of power cables is primarily from logistics emissions and can be calculated as Equation (4):

$$CF_{\mathrm{D}} = \sum_{l=1}^{\phi} N_l \cdot D_l \cdot \varphi_l + \sum_{k=1}^{b} U_k \cdot GWP_K \quad (4)$$

N_l is the mass of the product transported using the l-th mode of transport; D_l is the transportation distance; φ_l is the carbon emission factor for the l-th transportation mode; U_k is the direct emission of the k-th greenhouse gas.

3.6. Usage phase

This stage includes conductor loss, dielectric loss, and metallic sheath loss throughout the service life of the cable.

a. Conductor Loss

The conductor loss is dependent on the current through the cable. In this study, the working current III is assumed to be 100% of the rated capacity. The power loss is calculated as Equation (5):

$$P_R = I_a^2 R \quad (5)$$

b. Dielectric Loss

The dielectric loss is calculated as Equation (6):

$$P_d = \omega C U_0^2 \mathrm{tg}\delta \quad (6)$$

U_0 is the Conductor-to-ground voltage (kV); C is the Cable capacitance (μF/km); $\mathrm{tg}\delta$ is the Dielectric loss factor (typically 0.0003 for XLPE cables)

c. Metallic Sheath Loss

Metallic sheath losses include circulating current losses and eddy current losses, represented by coefficients α and β, respectively:

$$P_m = (\alpha + \beta) \cdot P_R \quad (7)$$

Explanation:

Circulating current losses (α) and eddy current losses (β) are negligible in three-core cables due to the absence of sheath loops and mutual cancellation of magnetic fields, making $\alpha \approx 0$ and β very low.

The carbon footprint calculation formula for the use phase (Equation 8) is as follows:

$$CF_U = (P_R + P_d + P_m) \cdot \frac{T}{1000} \cdot \sigma_i \quad (8)$$

T is the Cable operating time (hours); σ_i is the Carbon intensity of electricity used during operation

3.7 End-of-life phase

In China, only the metallic components (e.g., copper and aluminum) of power cables are typically recovered during disposal. Insulation and shielding materials are often incinerated or landfilled. Since copper and aluminum are recyclable, the carbon emissions offset is calculated based on the difference between primary and secondary metal production.

After the cable is decommissioned, the carbon footprint in the recycling phase is primarily generated from material and energy consumption. The calculation is shown in equation (9):

$$CF_{\mathrm{RD}} = \sum_{i=1}^{n} E_i \cdot \delta_i - \sum_{i=1}^{m} N_i \cdot \beta_i \quad (9)$$

E_i is the Quantity of recovered material i ; δ_i is the Emission factor of secondary production of material i ; N_i is the Quantity of energy used in recycling; β_i is the Emission factor of energy i .

4. Analysis of carbon footprint calculation results

4.1. Input data collection

The accuracy of carbon footprint accounting is largely determined by the quality of the input data. According to the ISO 14067 standard, it is best to use data derived from the company's own processes (i.e., primary data) [6].

In this study, the data for YJV-8.7/15 kV (3×240 mm²) copper-core cable and YJLV-8.7/15 kV (3×400 mm²) aluminum-core cable assemblies were primarily obtained through consultation with local power cable manufacturers. Based on the material composition of these cables, a detailed bill of material weights was compiled for each component at the raw material acquisition stage, as shown in Table 1, to facilitate carbon footprint calculations. The technical specifications of the two cable types are listed in Table 2.

The emission factors used in this study were primarily derived from modeling in SimaPro software, based on localized life cycle inventory databases, and supplemented with data from the international Ecoinvent and GaBi databases. After modeling and inputting the collected data, the life cycle carbon emissions for each stage of the YJV and YJLV cables were obtained, as shown in Table 3.

Table 1. Power Cable Material Mass List.

Model / Item	YJV-(3*240)	YJLV-(3*400)
Component	Mass(kg/km)	
Conductor (Copper/Aluminium)	6391.72	3021.3
Conductor Shield (EVA)	127.64	158.39

https://doi.org/10.1051/e3sconf/202563901022

XLPE Insulation	916.93	1088.17
Insulation Shield (EVA)	150.41	232.76
Copper Tape	397.4	374.49
PP Filler	1003	1348.36
HI-90 PVC	941.91	1213.11

Table 2. Power Cable Technical Specifications Comparison Table.

Details		YJV-(3*240)	YJLV-(3*400)
Voltage Level (kV)		8.7/15	
Max DC/AC Conductor Resistance	Max DC Resistance at 20°C (Ω/km)	0.0754	0.0778
	Max AC Resistance at 90°C (Ω/km)	0.0972	0.101
Nominal Insulation Thickness (mm)		4.5	4.5
Approximate Conductor Diameter (mm)		17.4	21.8
Nominal Sheath Thickness (mm)		3.3	3.7
Cable Current-Carrying Capacity (A)	In Air	470	485
	In Soil	480	485
Capacitance per Phase (μF/km)		0.382	0.462
Inductance (mH/km)		0.289	0.272

Table 3. Full Life Cycle Carbon Footprint Comparison Table for Copper and Aluminum Core Cables.

Phase	Details	YJV (3*240)	YJL (3*400)
Raw materials acquisition Phase	Cu/ Al	23841.1156	37766.25
	Cond. Shield	283.3608	351.6258
	XLPE	2154.7855	2557.1995
	Insul. Shield	333.9102	516.7272
	Cu Tape	1482.302	1396.8477
	PP Fill	1885.64	2534.9168
	PVC Sheath	1535.3133	1977.3693
	total	31516.4274	47100.9363
Raw material transportation Phase	Cu/ Al	894.8408	845.964
	Cond. Shield	53.6088	66.5238
	XLPE	385.1106	457.0314
	Insul. Shield	63.1722	97.7592
	Cu Tape	55.636	52.4286
	PP Fill	280.84	377.5408
	PVC Sheath	527.4696	679.3416
	total	2260.678	2576.5894
Manufacturing Phase	Power & Steam	1276.07002	1117.979635
Distribution Phase	Truck Transp.	1390.0614	1041.1212
Use Phase	Cond. Loss	1149701.063	1223343.23
	Diel. Loss	884.1691555	1070.177734
	total	1150585.23	1224413.41
Recycling/ Disposal Phase	Recycle Proc.	5699.25174	2825.9004
	Recycle Offset	-5184.929022	-18591.45
	total	514.322718	-15765.5496
total		1187542.79	1260484.49

4.2. Comparative study of the full life cycle assessment

This study adopts a "cradle-to-grave" life cycle boundary to assess the carbon footprint of 1 km of YJV-8.7/15 kV (3×240 mm^2) copper-core cable and YJLV-8.7/15 kV (3×400 mm^2) aluminum-core cable (see Table 3).

The results show that the total life cycle carbon emissions for the copper-core cable amount to 1,187,542.79 kg CO_2e/km, while the aluminum-core cable reaches 1,260,484.49 kg CO_2e/km, indicating that the aluminum-core cable's carbon footprint is approximately 7.24% higher. This is illustrated in Figure 2.

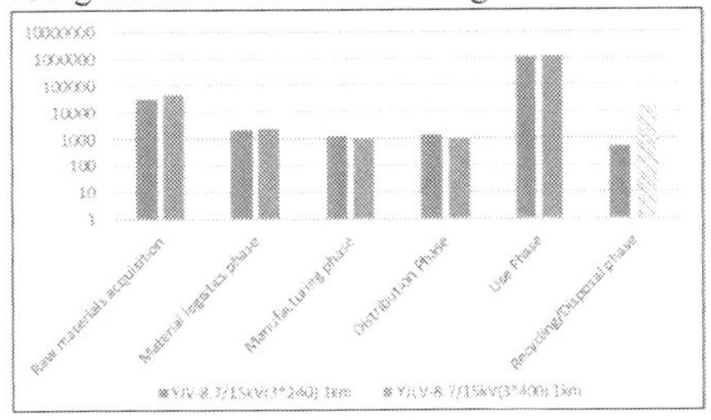

Figure 2. Comparative Life Cycle Assessment of Copper-Core and Aluminum-Core Power Cables.

In terms of phase distribution, the carbon emission structures of the two cables are highly similar: the operational phase dominates, accounting for 96.44% (copper) and 97.13% (aluminum) of total emissions, respectively. The next largest contributor is the raw material extraction and refining phase, with shares of 2.65% and 3.74%. The combined emissions from manufacturing, transportation, and end-of-life treatment account for less than 1%.

Despite their different material compositions, the two cables exhibit similar operating efficiencies due to having the same current-carrying capacity and structural parameters. As the operational phase overwhelmingly dominates the total emissions, the difference in overall carbon footprint between the two cables is relatively small. This indicates that the conductor material has limited influence on life cycle carbon emissions, which are more significantly impacted by operating resistance and service life.

4.3. Comparative study of the partial life cycle assessment

To further quantify the impact of the raw material and manufacturing phases on carbon emissions, this study also conducted a "cradle-to-gate" carbon footprint assessment, which includes raw material acquisition, transportation, and manufacturing processes.

As shown in Figure 3, for the copper-core cable, raw material acquisition accounts for 89.91% of emissions, transportation 6.45%, and manufacturing energy consumption 3.64%. For the aluminum-core cable, the respective shares are 92.73%, 5.07%, and 2.20%. Overall, emissions from manufacturing and transportation are comparable between the two cable types, with the primary difference arising in the raw material acquisition phase.

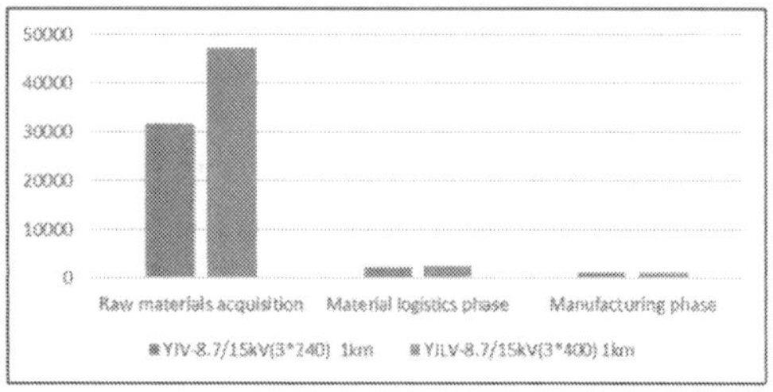

Figure 3. Comparative Partial Life Cycle Assessment of Copper-Core and Aluminum-Core Power Cables.

The energy intensity of aluminum smelting is significantly higher than that of copper, resulting in nearly 50% higher carbon emissions during the material acquisition stage for the aluminum-core cable compared to the copper-core cable. Although this difference is mitigated by the dominance of the operational phase in the full life cycle, it remains particularly prominent in partial life cycle assessments.

Therefore, although aluminum can yield carbon benefits through recycling during the end-of-life stage, its carbon footprint remains significantly higher in the early stages of manufacturing. This finding provides important insights for green design and procurement strategies, suggesting that both raw material sourcing and recycling efficiency should be considered in low-carbon optimization efforts.

5. Conclusion

This study conducted a comparative life cycle assessment (LCA) of YJV-8.7/15kV (3×240 mm^2) copper core cable and YJLV-8.7/15kV (3×400 mm^2) aluminum core cable. The results show that both cable types exhibit similar carbon footprint distributions, with over 95% of emissions occurring during the use phase. Raw material extraction accounts for approximately 2%–4%, while the manufacturing, transportation, and end-of-life phases together contribute less than 1%. Although the aluminum-core cable exhibits a 7.24% higher total carbon footprint than its copper counterpart, this difference is primarily driven by the higher emissions associated with primary aluminum production. However, aluminum recycling yields a greater carbon offset in the end-of-life phase, which helps reduce its net carbon impact.

a. Grid Operation Optimization: Increasing renewable energy in power generation, lowering electricity carbon intensity, and improving grid efficiency are key strategies. Smart grid technologies and demand-side management can further reduce transmission losses and related emissions.

b. Cable Manufacturing Improvements: Manufacturers should adopt efficient production methods, enhance conductor performance, and expand the use of recycled aluminum and copper. High-conductivity alloys and optimized insulation materials also help lower embedded emissions.

c. Design and Procurement Strategies: Cable selection should weigh both cost and environmental impact. Factoring in carbon costs enables balanced decisions, and using appropriately sized aluminum-core cables with recycled content can cut emissions without affecting reliability.

In conclusion, the life cycle carbon footprint model developed in this study offers a quantitative basis for evaluating emission sources and identifying reduction priorities for medium-voltage power cables. The findings provide guidance for low-carbon procurement and asset management strategies in the electricity sector. Future work will focus on applying this model to other cable types and voltage levels, incorporating dynamic factors such as regional power grid carbon intensity and temporal usage profiles, to enhance accuracy and decision-making relevance in the context of carbon neutrality goals.

References

1. National Development and Reform Commission (NDRC), Working Guidance for Carbon Dioxide Peaking and Carbon Neutrality (2021).
2. International Energy Agency, *Global Cable Market Outlook 2023–2032* [Online]. Available: https://www.iea.org/reports/global-cable-market-outlook-2023-2032.
3. Y. Zhang, et al., *Proc. IEEE Sustainable Power and Energy Conf. (iSPEC)* (2023).
4. C.I. Jones, M.C. McManus, J. Clean. Prod. **18**, 1464 (2010).
5. British Standards Institution, *PAS 2050:2011 – Specification for the Assessment of the Life Cycle Greenhouse Gas Emissions of Goods and Services* (2011).
6. International Organization for Standardization, *ISO 14067:2018 – Greenhouse Gases – Carbon Footprint of Products – Requirements and Guidelines for Quantification* (2018).
7. State Grid Corporation of China (SGCC), *Q/GDW 12395-2024: Technical Specification for Carbon Footprint Accounting of Cable Products* (2024) (in Chinese).

Performance Evaluation of Bismuth Oxychloride (BiOCl)-Doped Functionalized Graphene Oxide Membranes

Li Cai [1], Zhenzhong Fan [2*], Ao Sun [2], Qilei Tong [2], Sanyuan Qiao [3]

[1]Key Laboratory of Enhanced Oil and Gas Recovery, Ministry of Education, Northeast Petroleum University, Daqing, Heilongjiang 163000, China
[2]Bohai Rim Energy Research Institute, Northeast Petroleum University, Qinhuangdao, Hebei 066004, China
[3]Qinhuangdao Campus, Northeast Petroleum University, Qinhuangdao, Hebei 066004, China

Abstract. Graphene and its derivatives, characterized by their ultra-high water flux, unique physicochemical properties, and tunable layered structures, have increasingly emerged as a new generation of membrane materials for wastewater treatment. In this study, a novel membrane material was fabricated via vacuum filtration using BiOCl-doped graphene oxide (GO) modified with diethylenetriamine. This modified GO membrane exhibits excellent wettability, with a water contact angle of 0° in air and an oil contact angle of 144.5° underwater. It demonstrates a separation efficiency of 95.2% for kerosene emulsified oil and 90.2% for crude oil emulsified oil. Additionally, this paper provides a detailed investigation into the mechanism by which the membrane material separates emulsified oils, offering new insights for the treatment of oily wastewaters.

1. Introduction

Petroleum is a vital energy source in industrial production and is inextricably linked to the development and progress of human society. However, during its production process, a substantial amount of oily wastewater is generated. If left untreated in a timely manner, this wastewater can inflict significant environmental damage. The direct discharge of oily wastewater into the natural environment not only causes severe environmental pollution but also leads to the wastage of a significant amount of resources, posing a serious threat to human life [1-3]. Electrochemical methods, mechanical centrifugation, chemical treatment, gravitational sedimentation, and biodegradation are among the primary approaches employed for the treatment of oily wastewater[4-6]. The application of these methods in practical field settings is significantly restricted due to their inherent limitations, such as low separation efficiency and time-consuming processes [7]. Membrane separation technology has emerged as a popular method for treating oily wastewater because of its simplicity and cost-effectiveness [8-9]. Modified graphene oxide (GO) membranes have now been extensively applied in the treatment of oily produced water in oil fields.

2. Experimental materials and procedures

2.1. Experimental materials

Bismuth oxychloride (BiOCl), graphene oxide, and diethylenetriamine (analytical reagent grade) were procured from Macklin Biochemical Technology Co., Ltd., China. Ethanol standard solution was purchased from National Pharmaceutical Group Chemical Reagents Co., Ltd. Cellulose membrane was obtained from Haining Kewei Filtration Equipment Co., Ltd.

2.2. Experimental procedure

(1) Synthesis of Modified GO
Take 1 mL of GO solution with a concentration of 1 mg/mL and add it to 100 mL of water. Subsequently, 5 mL of a 1 mg/mL diethylenetriamine solution was added into the mixture. The combined solution was then transferred into a three-neck flask and the temperature was raised to 80°C. The reaction will be carried out for 4 hours under stirring conditions, resulting in the formation of diethylenetriamine-modified graphene oxide, which is denoted as DGO.

(2) Preparation of BiOCl doped modified GO membrane

Firstly, 1 mg of GA-EGO powder was accurately weighed. This powder was then dispersed in 50 mL of deionized water to prepare a suspension. Subsequently, add 3 mg of BiOCl to the dispersion. Transfer the resulting mixture into a beaker and subject it to ultrasonication for 10 minutes to ensure thorough mixing and dispersion of the components. Using a vacuum filtration apparatus, the prepared solution was carefully poured into a filtration cup. The vacuum pump was then activated, maintaining a working pressure of 0.098 MPa. Once the entire liquid had been drawn into the collection flask below, the resulting membrane was carefully removed from the filtration setup and allowed to dry at

* Corresponding author: Zhenzhong Fan, fanzhenzhong@163.com

ambient temperature for 12 hours to stabilize its structure. The resultant membrane was designated as BiOCl@DGO for further characterization and application testing.

3. Experiment and analysis

3.1. FTIR spectroscopy experiment

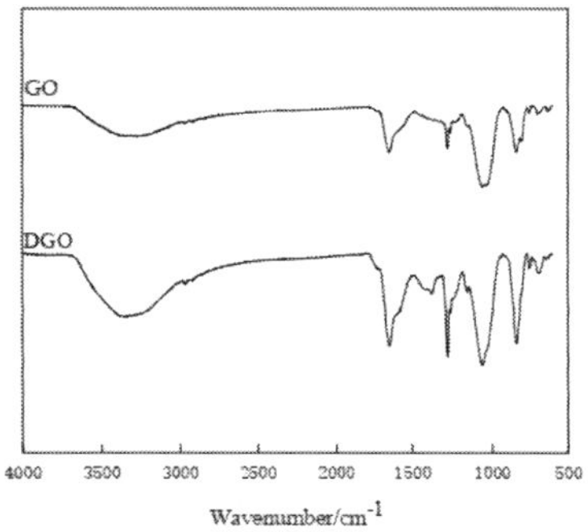

Fig 1. FTIR of GO, DGO.

As illustrated in the figure, the FTIR spectrum of graphene oxide (GO) displays distinct characteristic peaks at 3342 cm^{-1}, 1644 cm^{-1}, 1279 cm^{-1}, and 1060 cm^{-1}. The peak at 3342 cm^{-1} is attributed to the stretching vibrations of hydroxyl groups within the carbon framework, while the peak at 1644cm $^{-1}$ represents the appearance of the C=C bond. The FTIR spectrum of DGO shows a new characteristic peak in the wavelength range of 1310-1480 cm $^{-1}$ indicating the appearance of C-N bonds. This peak does not exist in the FTIR spectrum of the original GO, thus proving the formation of a new chemical bond in DGO. Diethylentriamine has already reacted with GO to generate a new substance.

3.2. XRD analysis

The interlayer spacings of various modified membranes were characterized using XRD analysis. Bragg's equation was subsequently employed to calculate the interlayer spacings based on the obtained XRD data. The XRD results, along with the corresponding interlayer spacings for GO and DGO, are detailed as follows:

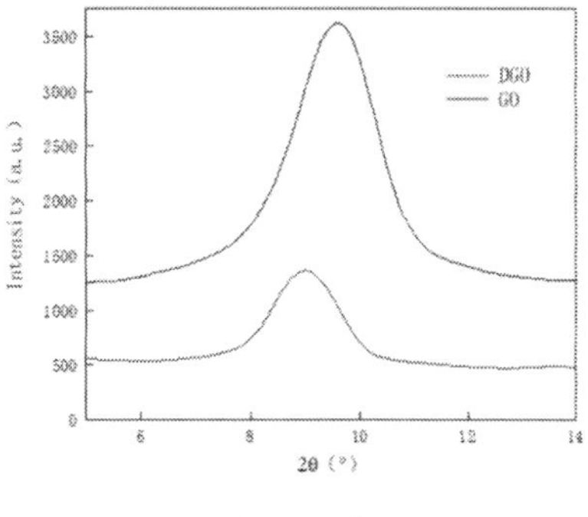

Fig 2. XRD images of GO, DGO.

The experimental results indicate that the 2θ values for GO and DGO are 9.61° and 9.07°, respectively. Utilizing Bragg's equation, the calculated interlayer spacings for these materials are 0.92 nm for GO and 0.97 nm for DGO. These data demonstrate that the interlayer spacing of GO increases following modification with diethylenetriamine, highlighting the significant impact of the functionalization process on the structural properties of the material.

3.3. XPS analysis

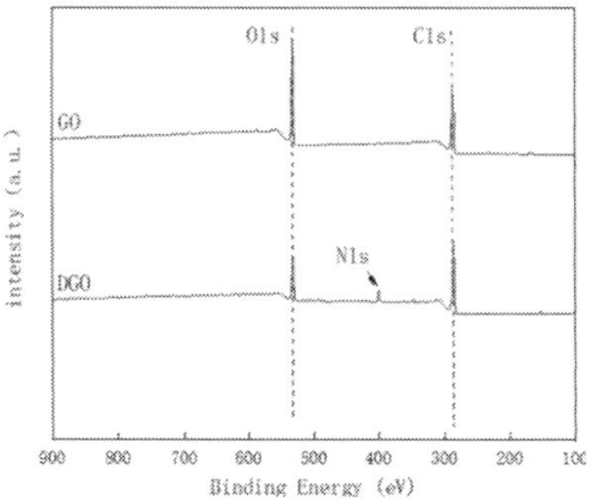

Fig 3. XPS images of GO, DGO.

The experimental results reveal a distinct difference between GO and DGO. The unmodified GO is composed solely of carbon and oxygen elements, whereas DGO contains carbon, nitrogen, and oxygen. In conjunction with the preceding infrared analysis, these findings confirm that GO has undergone a chemical reaction with diethylenetriamine, resulting in the formation of a new compound.

3.4. Surface wettability measurement of membrane

The wettability of GO membranes, DGO membranes and BiOCl@DGO membranes was assessed using a contact angle goniometer. The experimental results are presented in the following table.

Table 1. Wettability of three types of membranes.

Types of membranes	Contact angle of water in the air (°)	Underwater oil contact angle (°)
GO	68.5	106.6
DGO	62.6	131.2
BiOCl@DGO	0	144.5

The experimental data indicate that pristine graphene oxide (GO) exhibits inherent hydrophilicity. Following modification with diethylenetriamine, a slight reduction in the water contact angle in air and an increase in the underwater oil contact angle were observed. Upon doping with BiOCl, the water contact angle of the modified GO membrane in air decreased to 0°, achieving a superhydrophilic state. Concurrently, the underwater oil contact angle increased to 144.5°.

3.5. Experiment on Separating Kerosene Emulsified Oil

The separation performance of GO, DGO and BiOCl@DGO membranes was evaluated using the prepared kerosene emulsion. The filtrate was collected and analyzed for residual oil content using an infrared oil analyzer.

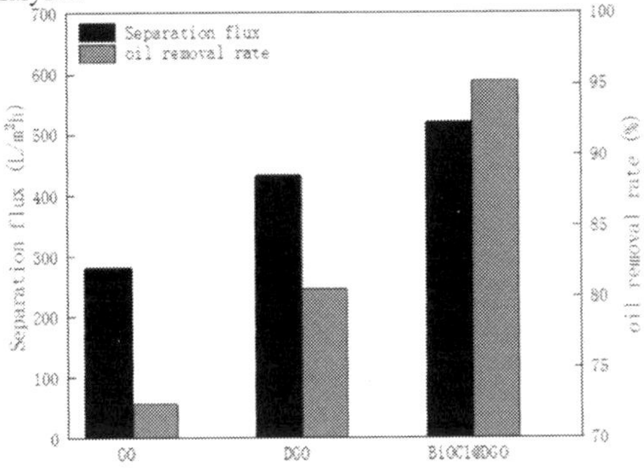

Fig 4. Separation flux and oil removal rate of GO membrane, DGO membrane, BiOCl@DGO membrane.

As shown in the figure, the oil-water flux and separation efficiency of the original GO membrane are 280.4 L/m ² h and 72.4%, respectively. In contrast, the DGO membrane exhibited a substantial increase in separation flux, achieving a value of 432.5 L/m²h, while its oil removal efficiency was 80.5%. The BiOCl@DGO membrane exhibited the most superior performance, with a separation flux of 520.1 L/m²h and an oil removal efficiency of 95.2%.

In summary, the oil removal efficiency of the original GO membrane is relatively low, which limits its practical application without modification. In contrast, the BiOCl@DGO membrane demonstrates remarkable performance and exhibits substantial potential for application in the treatment of oily wastewater.

3.6. Experiment on Separating crude oil emulsified oil

Experimental separation of crude oil emulsion using three types of membranes. The crude oil, which originally has a paste-like consistency, was diluted with kerosene to form a flowable liquid. The volume ratio of kerosene to crude oil was maintained at 3:1. To prepare the emulsified crude oil, 1 mL of the diluted crude oil was added to 100 mL of distilled water. Subsequently, 0.1 mL of a 20% (v/v) OP-10 solution was introduced as an emulsifier. The mixture was then subjected to continuous stirring at a rate of 600 revolutions per minute (rpm) for a duration of 6 hours to form a stable water-in-crude oil emulsion. The stability of the emulsified crude oil was monitored over a period of 72 hours, during which no signs of phase separation were observed.

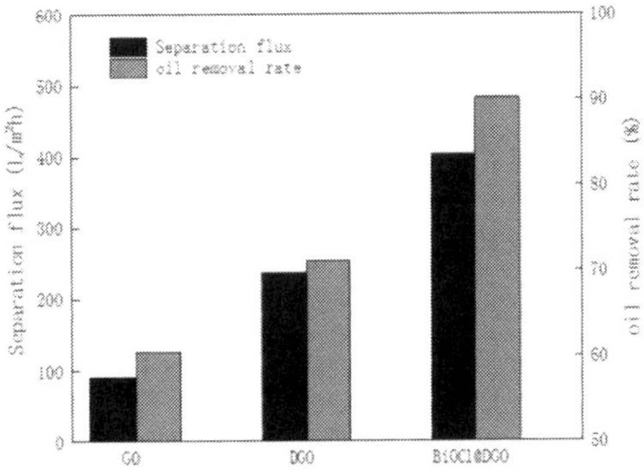

Fig 5. Separation efficiency of GO membrane, DGO membrane, BiOCl@DGO membrane for crude emulsified oil.

As shown in the figure, the GO membrane experienced severe blockage during the experiment, indicating that the emulsified crude oil induced irreversible fouling and damage to the GO membrane. In contrast, The DGO membrane achieved an oil-water flux of 236.7 L/m²h and an separation efficiency of 71.2% in the experiment. The BiOCl@DGO membrane exhibited considerably less fouling propensity, with a oil-water flux of 403.2 L/m²·h and an separation efficiency of 90.2%.

In summary, the BiOCl@DGO membrane demonstrated the best wettability among the three types of membranes. Its ability to separate both kerosene emulsified oil and crude oil emulsified oil was superior to that of the other two membranes, highlighting its potential for effective oil-water separation applications.

3.7. BiOCl@DGO membrane cyclic usability testing

In practical on-site applications, the reusability of membranes is of critical importance. To assess the reusability of the BiOCl@DGO membrane, multiple oil-water separation experiments were conducted in a shaded area, with each experiment filtering 10 mL of kerosene emulsified oil. After each experiment, the BiOCl@DGO membrane was exposed to natural sunlight for 30 minutes. The reusability of the membrane was evaluated by measuring its separation flux and underwater oil contact angle. The results of these experiments are presented in the figure below.

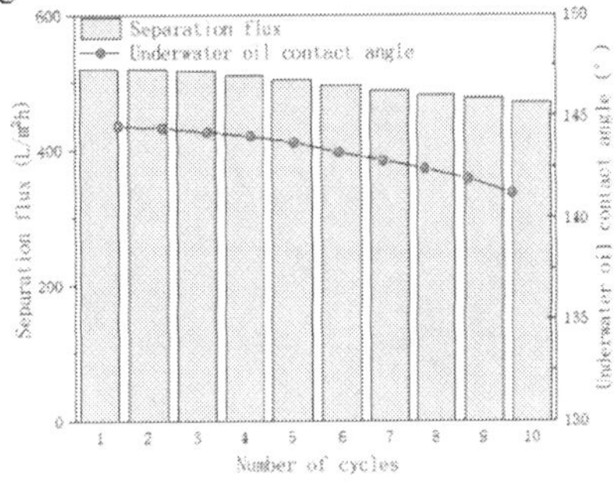

Fig 6. The recyclability of BiOCl@DGO membrane.

As depicted in the figure, the oil-water flux and underwater oil contact angle of the BiOCl@DGO membrane remained relatively stable throughout ten consecutive cycles of separating kerosene emulsified oil. After ten cycles of separation experiments, the separation flux of the BiOCl@DGO membrane decreased by only 50 L/m^2h, and its underwater oil contact angle decreased by 3.3°. These results indicate that the BiOCl@DGO membrane exhibits excellent reusability, maintaining high performance over multiple cycles of oil-water separation.

4. The mechanism of emulsified oil separation

(1) Superhydrophilic Surface

To further investigate the wettability behavior and separation mechanism of the membrane surfaces, the Young-Laplace equation was introduced to analyze the transport of the liquid phase within the membrane. The relationship between the pressure difference (ΔP) and wettability was examined, and the mechanism of superhydrophilic membranes for separating emulsified oils was discussed.

According to the Young-Laplace equation:

$$\Delta P = -\frac{2\gamma\cos\theta}{r} \tag{1}$$

In the equation, ΔP represents the pressure difference across the membrane, γ denotes the surface tension, θ is the contact angle of water or oil on the surface, and r is the radius of the membrane pores.

Generally, when the contact angle θ>90∘, ΔP>0, indicating that the membrane surface can withstand the pressure exerted by the liquid. Conversely, when θ<90∘, ΔP<0, meaning that the membrane surface cannot resist the pressure generated by the liquid, allowing the liquid to easily penetrate the membrane surface.

As illustrated in the figure, for hydrophilic membranes with micro/nano-scale roughness, the water contact angle is very small, indicating that $\Delta P_{water}<0$. Therefore, the aqueous phase can easily enter the superhydrophilic membrane and spontaneously permeate through the pores under high pressure. Since the underwater oil contact angle is greater than 90° and can even reach 150°, the corresponding ΔP_{oil} exerts a strong repulsive effect on the oil phase. Consequently, the superhydrophobic membrane under water can effectively repel the oil phase, causing oil droplets to detach from its surface.

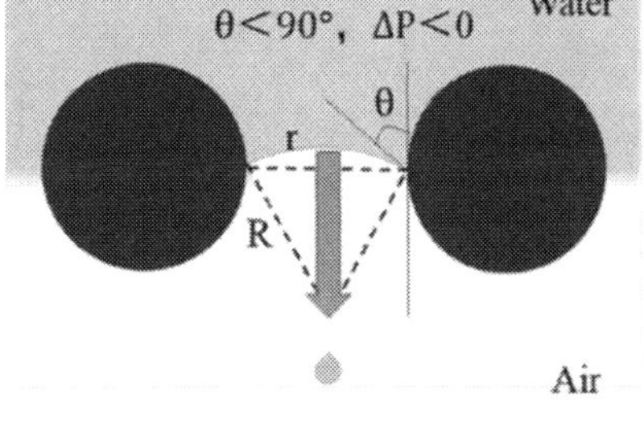

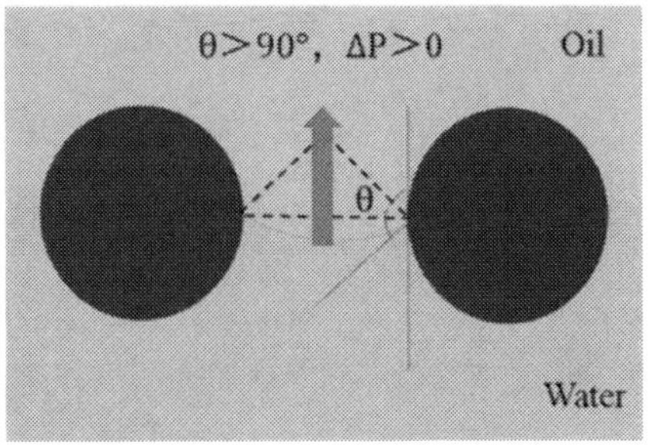

Fig 7 Schematic diagram of the force exerted on the surface of a membrane by a liquid

(2) electrostatic interactions

Emulsified oil is a colloidal dispersion system. From a thermodynamic perspective, emulsions are inherently unstable. Consequently, demulsification is a necessary step for the separation of oil-water emulsions.

Considering that the majority of emulsified oil droplets carry a negative charge, the electrostatic repulsion between these like charges substantially impedes the coalescence of the droplets. During the separation process, some oil droplets come into contact with the membrane surface, initiating electrostatic interactions, which can be either attractive or repulsive in nature. When the oil droplets and the membrane surface have opposite charges, the surfactant molecules on the oil droplets are drawn towards the membrane surface. In contrast, when they share the same charge, the surfactant molecules on the oil droplets tend to migrate away from the membrane surface. This behavior destabilizes the oil droplets, leading to the destabilization of the emulsified oil system and ultimately facilitating demulsification.

5. Conclusion

In this study, graphene oxide (GO) was modified with diethylenetriamine and subsequently doped with BiOCl to fabricate a BiOCl@DGO membrane by depositing the modified GO on a cellulose membrane. The modification with diethylenetriamine increased the interlayer spacing of GO, although it did not significantly enhance the hydrophilicity of GO. After doping with BiOCl, the hydrophilicity of the membrane was significantly improved, and the BiOCl@DGO membrane exhibited excellent antifouling properties. Compared with the other two types of membranes, the BiOCl@DGO membrane demonstrated superior performance in separating both kerosene emulsified oil and crude oil emulsified oil.

References

1. Li, Bingfan, et al. "Recent developments in the application of membrane separation technology and its challenges in oil-water separation: A review." Chemosphere 327 (2023): 138528.
2. Hoang, Anh Tuan, et al. "Advanced super-hydrophobic polymer-based porous absorbents for

https://doi.org/10.1051/e3sconf/202563901023

the treatment of oil-polluted water." Chemosphere 277 (2021): 130274.

3. Putatunda, S., et al. "A review on the application of different treatment processes for emulsified oily wastewater." International journal of environmental science and technology 16 (2019): 2525-2536.
4. Medeiros, Alexandre D'Lamare Maia de, et al. "Oily wastewater treatment: methods, challenges, and trends." Processes 10.4 (2022): 743.
5. Abuhasel, Khaled, et al. "Oily wastewater treatment: overview of conventional and modern methods, challenges, and future opportunities." Water 13.7 (2021): 980.
6. Yu, Li, Mei Han, and Fang He. "A review of treating oily wastewater." Arabian journal of chemistry 10 (2017): S1913-S1922.
7. Yalcinkaya, Fatma, et al. "A review on membrane technology and chemical surface modification for the oily wastewater treatment." Materials 13.2 (2020): 493.
8. Othman, Nur Hidayati, et al. "Supported graphene oxide hollow fibre membrane for oily wastewater treatment." AIP Conference Proceedings. Vol. 1901. No. 1. AIP Publishing, 2017.
9. Noamani, Sadaf, et al. "Carbon-based polymer nanocomposite membranes for oily wastewater treatment." NPJ Clean Water 2.1 (2019): 20.

E3S Web of Conferences **639**, 01024 (2025) https://doi.org/10.1051/e3sconf/202563901024
ICEMEE 2025

Ecological Modelling of Farmland Transition: Food Web Dynamics and Multi-Objective Agricultural Optimization

Rongrong Hao[1], Jinghan Wang[1], Zihui Deng[1] and Tao Liu[2,*]

[1] Sydney Smart Technology College, Northeastern University at Qinhuangdao, Qinhuangdao, 066004, China
[2] School of Mathematics and Statistics, Northeastern University at Qinhuangdao, Qinhuangdao, 066004, China

Abstract. This study presents an integrated modelling framework combining ecological dynamics and agricultural decision-making to address challenges arising from forest-to-farmland conversion. A food web dynamics model simulates the energy flow and population changes across trophic levels under the influence of seasonal agricultural cycles and chemical inputs. Furthermore, a multi-objective decision support model is developed to optimize organic farming strategies that balance farmer income and ecological sustainability. Simulation results focusing on the Amazon region show that the proposed model captures ecological stabilization processes and provides actionable insights for sustainable agricultural practices.

1. Introduction

1.1. Background

Forest-to-farmland conversion has emerged as a major global concern. According to FAO, over 420 million hectares of forest were lost between 1990 and 2020, much of it for agricultural expansion [1]. This transition disrupts ecological networks and degrades essential ecosystem services [2]. In the Brazilian Amazon, deforestation for crop production has been linked to declining soil fertility, increased erosion, and disrupted water cycles [3]. Moreover, intensive use of herbicides and pesticides in these systems has amplified ecological stress and biodiversity loss [4].

In contrast, organic farming practices—such as crop rotation, composting, and biological pest control—can enhance biodiversity, maintain soil health, and mitigate water pollution risks [5,6].

To evaluate these trade-offs, this study introduces a dynamic food web model and a multi-objective optimization framework to analyze ecological resilience and design the optimal organic agriculture strategy, in order to assist researchers in predicting the consequences of species composition and abundance changes.

1.2. Related work

Trophic dynamics models are widely used to study species interactions and energy transfer across trophic levels, enabling predictions about ecological responses to biodiversity changes [7,8]. In parallel, multi-objective optimization frameworks have been applied to agricultural systems to balance ecological and economic goals. Recent approaches incorporate statistical and machine learning techniques to optimize organic farming decisions, considering factors like yield, pest control, and biodiversity [9,10].

Despite these advances, few studies have integrated detailed food web models with comprehensive agricultural optimization. Additionally, the implementation of organic farming faces practical challenges such as potentially lower short-term yields and increased labor demand [11,12].

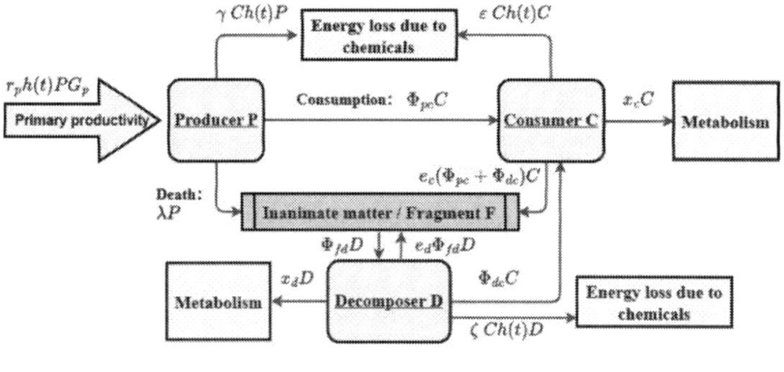

Fig. 1. Schematic diagram of energy flow in the food web.

* Corresponding author: liutao@neuq.edu.cn

2. Model I establishment

In order to simulate the new agricultural ecosystem of the current felled forest area more accurately, this paper establishes a food web dynamics model. By describing the transfer process of energy and matter in the ecosystem, it analyzes from the aspects of population dynamics and ecosystem stability. The framework of the designed dynamic model of the food web is shown in Fig. 1.

Specifically, a system of nonlinear differential equations was applied to model biomass changes in producers, consumers, decomposers, and fragments. Seasonal crop growth rates and chemical pesticide dynamics were incorporated, and parameters included chemical concentration decay, assimilation efficiency, and interaction strengths from food web connectivity.

2.1. For producers

The change in biomass of a producer is divided into four components: an increase in biomass due to photosynthesis, a decrease in biomass due to predation by herbivorous species, natural mortality (non-predatory mortality), and chemical infestation, and its kinetic equation can be expressed as.

$$\frac{dP}{dt} = r(t) \cdot h(t) \cdot P \cdot \left(1 - \sum_{n=pro} \frac{P_n}{K}\right) - \phi_{pc} \cdot C - \lambda_p \cdot P - \gamma \cdot Ch(t) \cdot P \quad (1)$$

Where ϕ_{ij} is the interaction intensity between species i and species j, and λ_i is the natural mortality rate of species i. γ and the following several symbols (ε, ζ) all correspond to the intensity of the effect of chemical substances on the group, and $Ch(t)$ is the concentration of the chemical pesticide substance.

In particular, to simplify the model, we set that the interaction influence range between the two communities is equal, as shown in Equation 2.

$$\phi_{ij} \cdot J = \phi_{ji} \cdot I \quad (2)$$

2.2. For consumers

The biomass changes of consumers are divided into three parts: the net energy obtained from the intake of producers and decomposers, and the energy loss due to metabolism (such as respiration) and chemical residues. The dynamic equation can be expressed as:

$$\frac{dC}{dt} = (1 - e_c) \cdot (\phi_{pc} + \phi_{dc}) \cdot C - x_c \cdot C - \varepsilon \cdot Ch(t) \cdot C \quad (3)$$

Where x_i is the metabolic rate of species i. e_c is the conversion rate of consumers into fragments.

2.3. For decomposers

Suppose the biomass changes of decomposers are still divided into four parts: the increase in biomass caused by the assimilation and absorption of impurities in the environment and the decomposition of organic matter (debris), and the decrease in biomass caused by predation, metabolism (such as respiration) and chemical infection. The kinetic equation can be expressed as:

$$\frac{dD}{dt} = (1 - e_d) \cdot \phi_{fd} \cdot D - \phi_{cd} \cdot C - x_d \cdot D - \zeta \cdot Ch(t) \cdot D \quad (4)$$

Where e_d is the conversion rate of decomposers into fragments. Based on experience, we set it to 0.3.

2.4. For fragments

The change in biomass of fragments is divided into multiple components, an increase in biomass due to the transformation of producer carcasses, consumer feces, and other fragments, and a decrease in biomass due to the consumption and transformation of decomposers into other fragments, which can be expressed in the kinetic equation as follows:

$$\frac{dF}{dt} = \lambda_p \cdot P + e_c \cdot (\phi_{pc} + \phi_{dc}) \cdot C + (e_d - 1) \cdot \phi_{fd} \cdot D \quad (5)$$

Where e_c is the conversion rate of consumer into fragments. Based on experience, we set it to 0.2

2.5. Model parameter

2.5.1. Seasonal crop growth rate

Crop growth rate varies seasonally and is expressed as a sinusoidal function as

$$r_i(t) = r_{i0} \cdot \left[1 + \alpha \cdot \sin\left(\frac{2\pi t}{T}\right)\right] \quad (6)$$

Where r_{i0} is the baseline intrinsic growth rate of producer, α is the amplitude of seasonal fluctuation, T is the agricultural cycle period.

2.5.2. Coefficients for each stage of the agricultural cycle

To reflect discrete management effects (e.g., tillage, harvest, fallow), we define a modulation coefficient that adjusts biological responses based on the current agricultural stage:

$$h_i(t) = h_{i0} \times \delta_s(t) \quad (7)$$

Where h_{i0} is the maximum activity coefficient of species i, $\delta_s(t)$ is the stage-specific multiplier.

2.5.3. Chemical concentration

The use of chemical substances can be expressed as a time-dependent function C(t), the concentration of chemical substances is high in the stage of fertilizer and pest control, and the concentration of chemical substances decreases to a lower state in the rest of the stage, which can be expressed by a segmented function as

$$Ch(t) = \begin{cases} C_0, & t_3 \le t < t_4 \\ C_0 \times e^{-\theta(t-t_4)}, & t \ge t_4 \\ 0, & otherwise \end{cases} \quad (8)$$

Where C_0 is the initial chemical concentration and θ is the decay exponent.

3. Model II establishment

3.1. Model estimation

In this section, the sustainability and sustainability brought by organic agriculture will be considered from three perspectives: environmental effect, economic effect and social effect, in order to help the farmers to make decisions on the issues of organic agriculture practice. In order to achieve the goals of higher returns to farmers and greater sustainability of the land, the following optimization model is developed.

Objective function:

$$f_1 = Pr \times Q - Cost(X_1, X_2, X_3, X_4, X_5) \quad (9)$$

$$f_2 = k_0 + \sum_{i=1}^{5} k_i \cdot x_i \quad (10)$$

$$Pr = \frac{Pr_0}{X_6} \times \left(1 + \frac{1}{1+e^{-(X_6-1)}}\right) \quad (11)$$

$$Q = m_0 + \sum_{i=1}^{5} m_i \cdot x_i \quad (12)$$

Decision variables:

$$\boldsymbol{X} = (X_1, X_2, X_3, X_4, X_5, X_6) \quad (13)$$

Constraints:

$$a_1 \le \boldsymbol{X} \le a_2 \quad (14)$$

$$Q \times Pr + \text{Subsidy} \ge R_{min} \text{ when } Pr \le Pr_{min} \quad (15)$$

3.2. Model interpretation

Objective function: Higher farmers' returns (f_1) and land sustainability (f_2), where Pr is the market price, Q is the yield of agricultural products.

Decision variables: In the multi-objective optimization model constructed in this study, we set six decision variables (X_1-X_6) to comprehensively measure the agricultural inputs and ecological performance: X_1 denotes the amount of organic fertilizer applied (kg/ha), X_2 is the proportion of ground cover of cover crops (0-1), X_3 is the frequency of biological control (times/quarter), X_4 is the amount of chemical pesticide use (kg/ha), X_5 is the amount of irrigation water per unit area (m³/ha), and X_6 is the market supply-demand relationship, reflecting the supply-demand ratio of use.

Constraints: Equations (14) are respectively the range constraints of the corresponding decision variables.

Equation (15) represents a guarantee mechanism, whose function is to ensure that farmers' income is not lower than the "minimum safety level" R_{min} set by a certain government, by activating subsidies at a specific market price.

4. Experiments and results

4.1. Agricultural cycle simulation

4.1.1. Research area selection and parameter setting

Based on relevant research and data [13-15], we set the model parameters as shown in Table 1. Given the limited data in some Amazon regions, we introduce the parameter ranges of ecologically similar areas and take the median of their typical intervals for simulation, combined with literature and ecological monitoring data. To enhance the empirical calibration of the model, we adopted multi-source data from the Amazon LTER network and related projects, with a focus on using the measured biomass data of producers and consumers reflecting seasonal fluctuations in the Tapajos National Forest (2015-2020), corresponding to the "growth" and "Fertilization and pest control" stages in the simulation graph. As the benchmark for comparison between simulation and ecological response. Furthermore, the vegetation and soil parameters were derived from the Tapajos Forest [16] and the Ducke Reserve [19], and the crop growth patterns were referred to the Manaus NPP data [20]. Decomposer dynamics, verified by MAGs data of soil microorganisms in the transition zone between forests and pastures [17-18].

4.1.2. Single agricultural cycle trend analysis

Substituting the actual data from the Amazonian sites selected for this paper into the base model, the food web dynamics model, the seasonal changes in producers (crops, weeds), consumers (insects, birds), and fragments and decomposers over time can be obtained by solving the ordinary differential equations listed for each species. The entire cycle is divided into five typical stages. Their characteristics are shown in Table 2. As shown in Fig. 2, the crop population follows a non-linear dynamic pattern across a single agricultural cycle, with clear phase transitions.

The simulation results of the growth stage match the LTER measurement results, with an average absolute error (MAE) of 11.7%, capturing 65% of the crop biomass increase observed in the field (from 50 to 320), which is consistent with the "rapid growth stage" of the Amazon agricultural ecosystem [13].

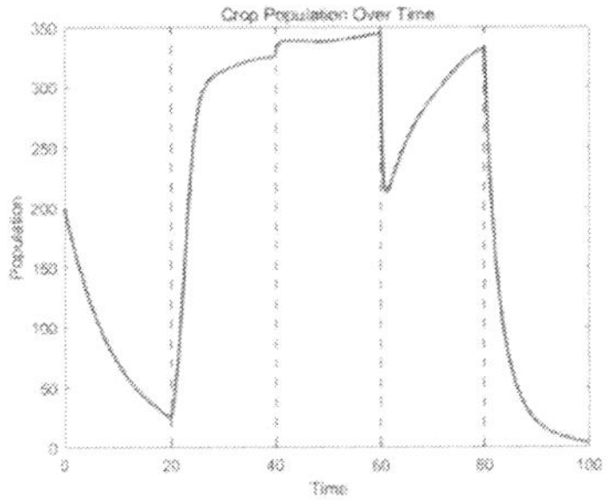

Fig. 2. The changes in the growth trend of crops within a single agricultural cycle.

To simplify the model, insects and birds were selected as representative consumer species. The changes in species abundance at each trophic level of the new farmland ecosystem throughout a single agricultural cycle are illustrated in Fig. 3, where species interactions—shaped by predation and energy transfer—drive dynamic population shifts. Notably, the observed 37.5% decline in bird population following pesticide application in the t=60-80 interval aligns with avian monitoring records, validating the model's capacity to simulate trophic cascades, particularly the stage-specific coupling between insect and bird populations [4].

Table 1. Simulation parameter Settings.

symbol	range	symbol	range	symbol	range	symbol	range
P_0	500-800	C_0	100-200	D_0	50-100	F_0	300-500
ϕ_{pc}	0.05-1	ϕ_{fd}	0.03-0.07	λ_p	0.01-0.05	x_c	0.02-0.05
x_d	0.02-0.04	γ, ζ	0.01-0.03	ε	0.015-0.03	r_p	0.1–0.3

Note: Except for the first row where the unit is g/m^3, the rest have no units

Table 2. The characteristics of the simulated agricultural production cycle.

Phase	Time Range	Key Characteristics
Sowing	$0 \le t \le 20$	Low seed viability on newly reclaimed land; Crop numbers decline to local minimum.
Growth	$20 \le t \le 40$	Rapid growth as crops adapt; Population increases toward ecological carrying capacity.
Stabilization	$40 \le t \le 60$	Population stabilizes; Ecosystem reaches biotic–abiotic equilibrium.
Fertilization & Pest Control	$60 \le t \le 80$	Chemical inputs temporarily suppress growth; Recovery follows as crops develop tolerance.
Harvesting	$80 \le t \le 100$	Mass harvesting causes sharp decline; Marks end of current cycle.

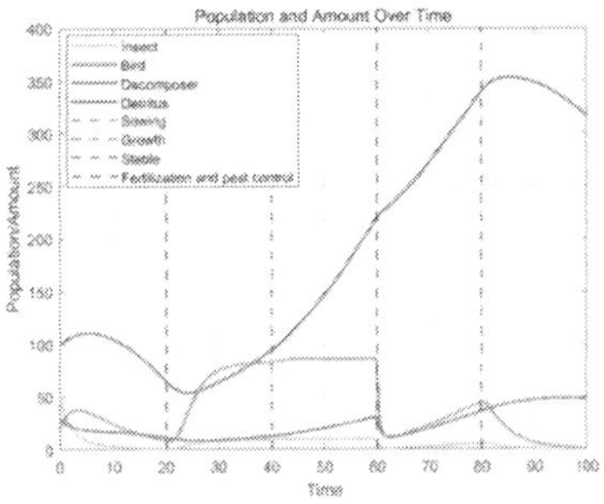

Fig. 3. The growth trend changes of species at all levels within a single agricultural cycle.

4.1.3. Comprehensive evaluation of multiple agricultural cycles

In order to further study the process of transformation of this ecosystem from newly cleared forest to mature farmland, the study time was expanded to four agricultural cycles. The trend of the number of species in each stratum is shown in Fig. 4 and Fig. 5.

As can be seen from the Fig. 4, as the new farmland ecosystem matures, the number of crops shows an overall trend of slow growth, which indicates that the farmland ecosystem is gradually developing into an ecosystem suitable for growing the crop. Weeds and insects, which inhibit crop growth, showed a slow decline, periodically approaching the minimum value while maintaining the balance of the food web.

Model validation across four agricultural cycles showed that simulated energy transfer efficiency from producers to consumers (13.2%-15.8%) aligned well with empirical values from similar tropical agroecosystems (12.5%-16.0%) [13], consistent with cyclical "Crop - Weed - Insect" population patterns in the simulation charts.

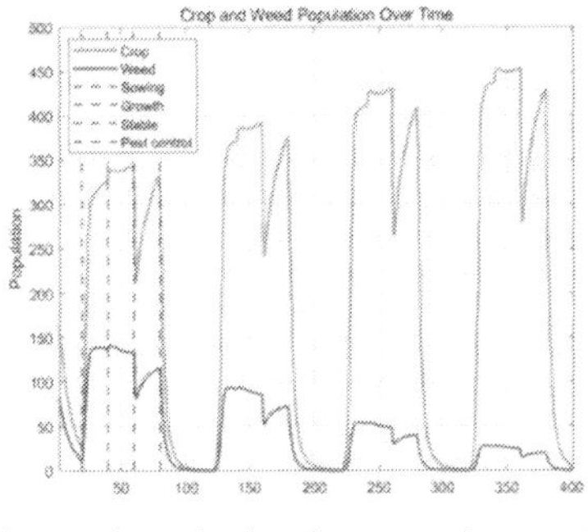

Fig. 4. The growth trends of producers at each stage of multiple agricultural cycles.

https://doi.org/10.1051/e3sconf/202563901024

Fig. 5 further illustrates the population trends of insects, birds, decomposers, and detritus over multiple agricultural cycles, showing periodic oscillations followed by stabilization — evidence of the system's progression toward dynamic equilibrium. The model also reproduced an 18% increase in soil organic carbon—reflected in stable decomposer biomass—which matched observed trends in Amazonian conversion studies [15]. Simulated decomposer biomass remained within 30-50g/m^2 , closely corresponding to LTER field measurements in disturbed Amazonian plots (28-40 g/m^2) [13], supporting the model's capacity to capture nutrient cycling under intensive land-use.

Species at all levels of the ecosystem gradually showed a trend of dynamic equilibrium, and the material cycle and energy flow conformed to the indicators of a mature ecosystem.

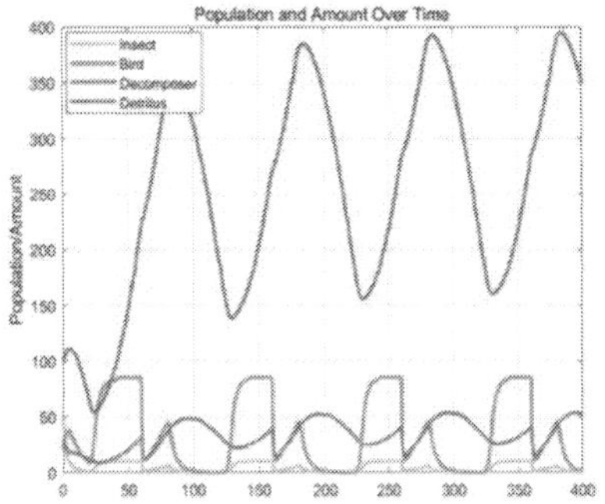

Fig. 5. The growth trends of other groups at each stage of the multiple agricultural cycle.

4.2. Genetic algorithm based solving of multi-objective planning problems

4.2.1. Simulation settings

The independent variable ranges in the multi-objective optimization model were derived from authoritative sources (FAO, 2021; IBGE, 2022; EMBRAPA, 2020), ensuring regional relevance. Ecological and yield impacts were quantified using weight coefficients k_i and m_i (i=1,2,···,5). k_i was determined through literature review and expert judgment. For example, k_1 = +0.6, k_2 = +0.4, and k_3 = +0.5 represent the positive ecological effects of organic fertilizers, cover crops, and biocontrols, while k_4= -0.8 reflects the negative impact of pesticide use. k_5= $\pm$ 0.2 captures the context-dependent effect of irrigation. Yield weights m_i were calibrated via regression on Amazonian yield data, with values set as: m_1= 0.5, m_2= 0.2, m_3= 0.3, m_4= 0.6, and m_5= $\pm$0.1.

The NSGA-I genetic algorithm (population: 150; iterations: 300; crossover: 0.9; mutation: 0.1) was used for optimization under practical constraints—cost $\leq$ \$300/ha, pesticide $\leq$ 100 kg/ha, water $\leq$ 500 m^3/ha—ensuring both ecological soundness and real-world feasibility.

4.2.2. Genetic algorithm solution results

Based on real data from the Amazon region, the multi-objective optimization yields 70 Pareto-optimal solutions. These represent trade-offs where no objective can be improved without worsening another, forming the Pareto frontier. To make the results easy to visualize, we normalize the optimal solutions and the visualization results are shown in Fig. 6.

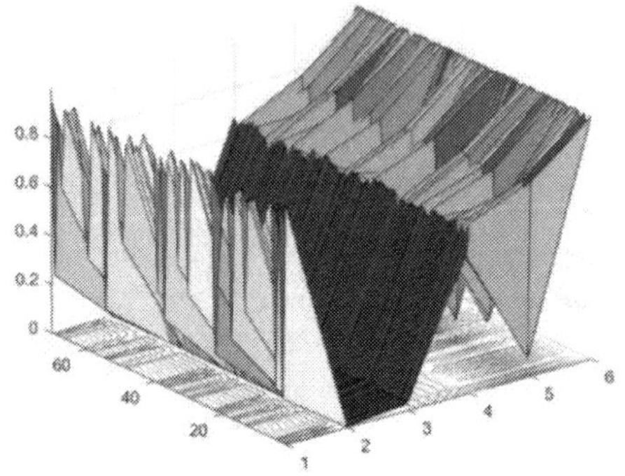

Fig. 6. Visualization of the optimal solution groups.

The figure shows that the 70 Pareto-optimal solutions exhibit diverse combinations of the six decision variables. For instance, some solutions feature high organic fertilizer use (X_1), moderate cover cropping (X_2), frequent biological control (X_3), low pesticide input (X_4), and moderate irrigation (X_5), with a conservative market strategy (X_6) to stabilize profits. Others may prioritize land sustainability through large-scale cover crops (X_2), allow higher pesticide input (X_4) for pest emergencies, and adopt a more aggressive market approach (X_6) for higher returns. This diversity illustrates the flexibility of the Pareto frontier—there is no single optimal solution, but rather a range of viable strategies that balance economic and ecological objectives in organic farming.

5. Conclusion

This study integrates ecological modeling and agricultural optimization to support sustainable land-use transitions. Food web dynamics models capture the response of agro-ecosystems to management behaviour, while decision-making models provide flexible solutions for balancing economic and ecological objectives in the context of organic agriculture. The simulation experiment results in the Amazon region performed well. Future work may extend this framework to broader regions and integrate climate resilience factors

Reference

1. E.B. Barbier, Explaining agricultural land expansion and deforestation in developing countries, American Journal of Agricultural Economics, **86**, 5, 1347-1353 (2004).

2. H.R. Grau, N.I. Gasparri, T.M. Aide, Agriculture expansion and deforestation in seasonally dry forests of north-west Argentina, Environmental Conservation, **32**, 2, 140-148 (2005).
3. D. Nepstad, B.S. Soares-Filho, F. Merry, et al, The end of deforestation in the Brazilian Amazon, Science, **326**, 5958, 1350-1351 (2009).
4. T. Tscharntke, A.M. Klein, A. Kruess, et al, Landscape perspectives on agricultural intensification and biodiversity-ecosystem service management, Ecology Letters, **8**, 8, 857-874 (2005).
5. I. Darnhofer, T. Lindenthal, R. Bartel-Kratochvil, et al, Conventionalisation of organic farming practices: from structural criteria towards an assessment based on organic principles. A review, Agronomy for Sustainable Development, **30**, 67-81 (2010).
6. H. Blanco-Canqui, S.J. Ruis, C.A. Francis, Do organic farming practices improve soil physical properties? Soil Use and Management, **40**, 1, e12999 (2024).
7. S.H. Faeth, P.S. Warren, E. Shochat, et al, Trophic dynamics in urban communities, BioScience, **55**, 5, 399-407 (2005).
8. A.O. Adeogun, A.V. Chukwuka, O.R. Ibor, et al, Occurrence, bioaccumulation and trophic dynamics of per-and polyfluoroalkyl substances in two tropical freshwater lakes, Environmental Pollution, **346**, 123575 (2024).
9. D. Haloui, K. Oufaska, M. Oudani, et al, Sustainable urban farming using a two-phase multi-objective and multi-criteria decision-making approach, International Transactions in Operational Research, **32**, 2, 769-801 (2025).
10. A.M. Fathollahi-Fard, G. Tian, H. Ke, et al, Efficient multi-objective metaheuristic algorithm for sustainable harvest planning problem, Computers & Operations Research, **158**, 106304 (2023).
11. A.K. Srivastava, A. Kumar, K. Vigyan, et al, Exploring organic farming: Advantages, challenges, and future directions, Plant Science Archives, **9**, 13 (2022).
12. A. Gamage, R. Gangahagedara, J. Gamage, et al, Role of organic farming for achieving sustainability in agriculture, Farming System, **1**, 1, 100005 (2023).
13. S. Vieira, P.B. de Camargo, D. Selhorst, et al, Forest structure and carbon dynamics in Amazonian tropical rain forests, Oecologia, **140**, 468-479 (2004).
14. J.F. Gillooly, J.H. Brown, G.B. West, et al, Effects of size and temperature on metabolic rate, Science, **293**, 5538, 2248-2251 (2001).
15. D.R. Zak, Z.B. Freedman, R.A. Upchurch, et al, Anthropogenic N deposition increases soil organic matter accumulation without altering its biochemical composition, Global Change Biology, **23**, 2, 933-944 (2017).
16. M. Williams, Y. E. Shimabokuro, E. B. Rastetter, Soil and Vegetation Characteristics, Tapajós National Forest, Brazil, ORNL DAAC, LBA-ECO CD-09 (2012).
17. A. M. Venturini, J. B. Gontijo, J. A. Mandro, Linking soil microbial genomic features to forest-to-pasture conversion in the Amazon, USDOE OSTI Dataset, (2025).
18. J. A. Mandro, et al., Metagenome-assembled genomes from Amazonian soil microbial consortia, USDOE OSTI Dataset, (2022).
19. J. Q. Chambers, J. dos Santos, R. J. Ribeiro, N. Higuchi, Tree Inventory Data, Ducke Reserve, Manaus, Brazil: 1999, ORNL DAAC, LBA-ECO CD-08 (2009).
20. M. T. F. Piedade, W. J. Junk, NPP Tropical Forest: Manaus, Brazil, 1963–1990, R1, ORNL DAAC Dataset (2013).

Study on the Dynamic Changes of Landscape Ecological Pattern in Beihai Wetland, Tengchong City

Hongmei Xu and Wujun Xi *

School of Resource, Environment and Chemistry, Chuxiong Normal University, 675000, China

Abstract. As a critical regional ecological security barrier, understanding Beihai Wetland's ecological pattern evolution is essential for sustainability. This study used Google Earth Engine (GEE) and Landsat imagery. Applying a hybrid Maximum Likelihood Classification (MLC) and Random Forest (RF) method, it extracted spatial distribution data for Tengchong's Beihai Wetland across five phases (2009–2024) and analyzed landscape pattern spatiotemporal changes. Results indicate:(1) Wetland area showed sharp non-linear fluctuations, expanding from 26.10 hm^2 (2009) to 152.46 hm^2 (2024);(2) Landscape metrics reveal rising heterogeneity: wetland patches became less regular, structural complexity increased significantly, and spatial fragmentation intensified, indicating greater distribution discreteness;(3) Key ecological land declined substantially: forests shrank 8.3% and meadows decreased 40.4% since 2009, directly threatening habitats of endangered Brasenia schreberi and Grus nigricollis. Concurrently, construction land increased 50.3%, showing tourism encroachment on ecological space. These findings provide a scientific basis for building ecological barriers and enhancing sustainable management in plateau wetlands.

1. Introduction

Wetlands function as critical global carbon sinks and biodiversity habitats, their degradation and restoration profoundly impacting regional ecological security amid persistent area declines. International research applies interdisciplinary approaches (ecology, geography, landscape ecology) to ecosystem assessment, restoration techniques, and ecological equilibrium. For instance, Aquino et al. revealed threshold effects of landscape fragmentation and land-use transitions in Amazon wetlands using GIS spatial overlay[1], while Dangles et al. built multi-scale vulnerability models via PLEIADES/LANDSAT imagery[2], and Kharazmi et al. elucidated spatiotemporal drivers of land cover change in Iran's Hamoun wetlands through remote sensing-ecological model integration[3].Conversely, China's wetland research aligns with national ecological security strategies, forming a distinctive framework . Qin et al. confirmed nonlinear coupling between Yangtze mid-channel bar evolution and curvature using fractal theory (R^2=0.87) [4], Meng et al. predicted a 12.7% threshold inflection for Yellow River wetland value by 2030 via mass-value-emergy modeling [5], Ren et al. identified groundwater extraction (weight=0.43) as the primary driver of North China wetland shrinkage[6], and Chen et al. revealed a 1.2% annual decrease in wetland patch Aggregation Index (AI) around Qinghai Lake[7].Critically, research concentrates on plain/coastal wetlands, while plateau wetlands—especially volcanic-dammed and glacial-erosion types—remain understudied.

Southwest China's plateau wetlands feature distinct vertical zonality, fragile stability, and high disturbance sensitivity. The Tengchong Beihai Wetland—a provincial reserve—represents a rare volcanic barrier lake "floating-mat" *Carex* wetland. With 2.3×10^6 tons peat carbon[8], it shelters endangered species (*Brasenia schreberi* communities, *Grus nigricollis*) and underpins Tengchong's ecotourism. This study examines Tengchong's Beihai Wetland Reserve. Using GEE-processed Landsat imagery and hybrid maximum likelihood/random forest classification, we analyze wetland landscape dynamics (2009–2024) through transfer matrices and landscape indices to support conservation of plateau "floating-mat" wetlands and regional ecological security.

2. Study area

The Beihai Wetland (25.1117°–25.1469°N, 98.5153°–98.5839°E) in Tengchong, Yunnan, occupies Shuanghai Community, Beihai Township at 1,725–2,756 m elevation. Located at the Indian-Eurasian Plate collision front, it constitutes a typical intermountain basin in the southern Hengduan Mountains extension. Comprising Beihai and Qinghai subregions (Fig.1), the wetland features diverse geomorphology within an active plate collision zone. Its Pliocene basalt platforms underlie 2–5 m thick peat deposits, forming distinctive floating-mat landscapes[9].The subtropical plateau monsoon climate delivers 1,470 mm mean annual precipitation, with 67% occurring June–August. Annual evaporation accounts for

*173202430@qq.com

42% of precipitation, sustaining high biodiversity. Ecologically critical, it shelters nationally protected Brasenia schreberi (Class II) as a dominant macrophyte. Socioeconomically, rapid tourism-driven construction expansion in the buffer zone intensifies conservation-development conflicts.

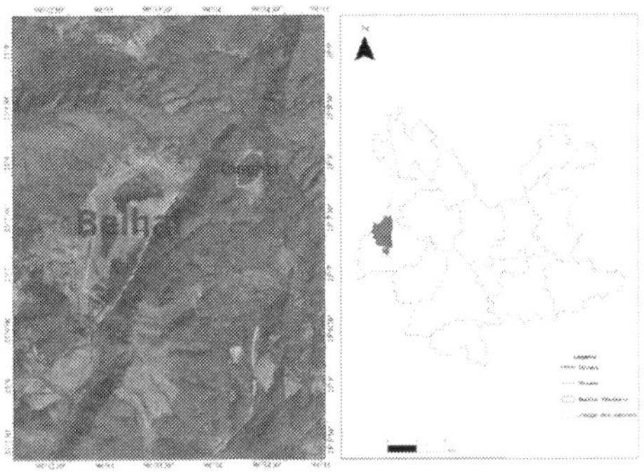

Fig. 1. Schematic Diagram of the Study Area's Geographical Location.

3. Data sources and methodology

3.1. Data acquisition and preprocessing

The remote sensing imagery was sourced from the Geospatial Data Cloud (GSCloud) platform (http://www.gscloud.cn). Landsat 7 ETM+ SLC-off imagery was used for 2009, Landsat 8 OLI/TIRS imagery for the period 2013–2020, and Landsat 8-9 OLI/TIRS imagery for 2024. All acquired images had cloud cover ≤5%. Multispectral band data were extracted using the Google Earth Engine (GEE) platform (https://earthengine.google.com). Based on five-period time-series composite imagery, the Normalized Difference Vegetation Index (NDVI) was inverted online. The image datasets were subsequently preprocessed using ENVI 6.1 software, undergoing radiometric calibration, atmospheric correction, and geometric registration. Data processing was finalized using Microsoft Excel software.

3.2. Methodology

Classification System Utilizing multi-source features—including spectral, textural, geometric, and spatial distribution characteristics—extracted from Landsat series imagery (2009–2024) covering the study area, we implemented the third-level classification system of the Current Land Use Classification standard (GB/T 21010-2017). Through field validation surveys, remote sensing interpretation results were calibrated. Four final landscape categories were established: wetlands, meadows, built-up areas, and forest land.

Maximum Likelihood Method A hybrid MLC-Random Forest classification framework ensured high-precision landscape identification: initial land cover categorization via Maximum Likelihood Classification with visual interpretation refined spectrally heterogeneous wetland subtypes using Random Forest. Classification accuracy was rigorously validated through confusion matrices. Kappa coefficients for the five study periods were: 0.92 (2009), 0.94 (2013), 0.93 (2017), 0.94 (2020), and 0.93 (2024), all exceeding the 0.90reliability threshold.

GIS Spatial Analysis Polygon buffers quantified spatial influences around point, line, and polygon features. To resolve attribution challenges, an IDW-Kriging hybrid model applied inverse distance weighting followed by ordinary kriging for spatial autocorrelation. Spatial joins integrated point features with NDVI time-series and land-use transition matrices to delineate ecologically sensitive zones (Fig. 2).

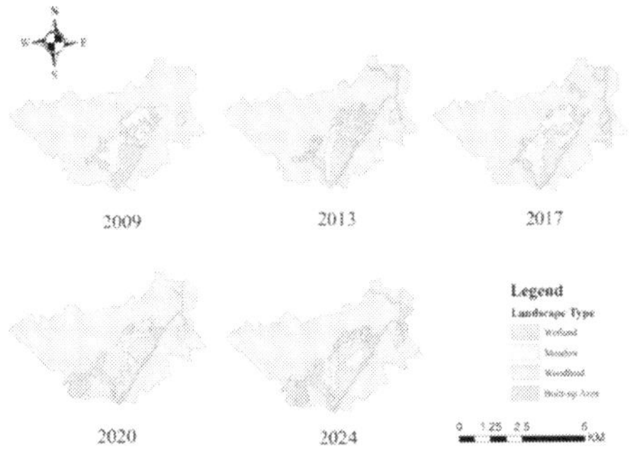

Fig. 2. Spatial distribution of dominant landscape types in the protected area (2009–2024).

Land Use Transition Matrix Land use transition matrices analyzed landscape category dynamics [10]. Five-phase raster datasets underwent ArcGIS overlay analysis. The Raster Calculator extracted inter-class transitions, producing normalized matrices defined by:

$$Aij=\begin{bmatrix} A_{11} & A_{12} & \cdots & A_{1n} \\ A_{21} & A_{22} & \cdots & A_{2n} \\ \vdots & \vdots & \ddots & \vdots \\ A_{n1} & A_{n2} & \cdots & A_{nn} \end{bmatrix}$$

where Aij= area transitioning from category i (initial) to j (terminal), and n = total categories.

Landscape Metric Selection and Calculation Landscape pattern indices quantified compositional, configurational, and spatiotemporal attributes[11]. Class-level metrics (PD, ED, FRAC, LSI, SHDI) were computed for each period using ArcGIS and Excel.

4. Results and analysis

4.1. Spatiotemporal differentiation of land use in Beihai Wetland, Tengchong

It can be seen from Table 1 that Beihai Wetland in Tengchong exhibited fluctuating area dynamics during 2009–2024, with net expansion from 26.10 hm^2 (2009) to 152.46 hm^2 (2024) (+484.1%), though below the 155.00 hm^2 peak (2020). This trajectory reflects synergistic ecological restoration (e.g., farmland conversion) and policy interventions under the Beihai Wetland Protection Regulations (Fig.3). Distinct phases occurred: Expansion Phase (2009–2013): Area surged from 26.10 hm^2 to 154.06 hm^2;2017–2020 (Recovery Phase): Area rebounded from 122.81 hm^2 to 155.00 hm^2;2020–2024 (Moderate Decline Phase): Area decreased marginally from 155.00 hm^2 to 152.46 hm^2.Concurrently, construction land exhibited persistent expansion:2009–2024: Increased from 164.97 hm^2 to 248.02 hm^2 (+50.35%, mean annual growth: 2.76%).This growth rate significantly surpassed Yunnan Province's urbanization rate (mean 1.31% annually, 2012–2023). 2020–2024 (Accelerated Expansion Phase):2020–2024 (Accelerated Expansion Phase): Escalated from 182.21 hm^2 to 248.02 hm^2 (+36.1%, mean annual growth: 8.00%), partially associated with tourism sector development. Ecological functional degradation risks emerged: Meadows: Area contracted from 159.93 hm^2 (2009) to 95.36 hm^2 (2024) (–40.4%), suggesting potential compromise of critical ecological functions (e.g., biodiversity conservation, water purification). Forests: Area diminished from 1,747.62 hm^2 (2009) to 1,602.60 hm^2 (2024) (–8.3%), indicating an imperative to enhance oversight and implementation rigor of sustainable forestry management protocols.

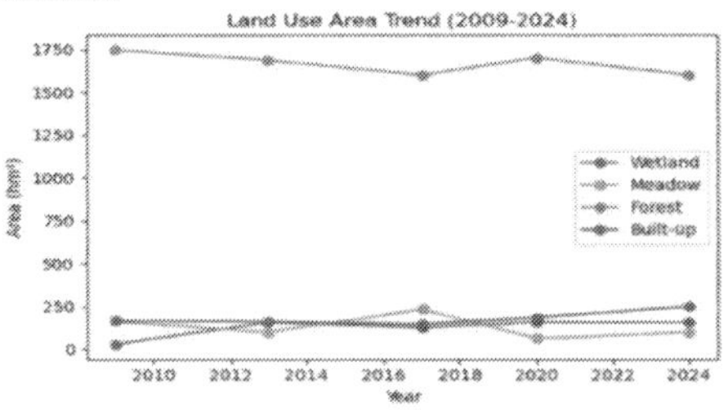

Fig. 3. Distribution of Spatial Changes in Various Landscape Types within the Protected Area, 2009–2024.

4.2 Landscape pattern dynamics

Multi-scale landscape index analysis[11] showed Tengchong's Beihai Wetland exhibited marked heterogeneous evolution during 2009-2024 (Table 2).

Table 1. Area coverage of landscape types in the protected area (2009–2024; hm^2).

Year	Wetland (hm^2)	Meadow (hm^2)	Woodland (hm^2)	Built-up area (hm^2)
2009	26.10	159.93	1747.62	164.97
2013	154.06	95.54	1689.55	159.29
2017	122.81	232.63	1601.75	141.25
2020	155.00	58.86	1702.36	182.21
2024	152.46	95.36	1602.60	248.02
Δ 2009–2024	+126.36	-64.57	-145.02	+83.05

Table 2. Temporal changes in landscape metrics within Beihai Wetland Protected Area (2009–2024).

Year	Patch Density (PD, $n \cdot hm^{-2}$)	Fractal Dimension (FRAC)	Landscape Shape Index (LSI)	Edge Density (ED, $m \cdot hm^{-2}$)	Shannon's Diversity Index (SHDI)
2009	1.8584	1.2050	5.6317	49.17	0.6030
2013	2.9546	1.2587	8.8522	77.30	0.7026
2017	1.5720	1.2473	8.0476	70.27	1.2813
2020	1.7156	1.2412	7.9463	66.72	0.6974
2024	2.4304	1.2661	9.4238	82.29	0.7892
Δ% (2009–2024)	30.78	5.07	67.33	67.27	30.90

Non-linear Intensification of Landscape Fragmentation Patch density (PD) exhibited non-monotonic fluctuations indicating intensifying fragmentation. Rising from 1.8584 to 2.9546 patches$\cdot hm^{-2}$ during 2009–2013 (CAGR ~16.7%) reflected accelerated fragmentation, followed by a sharp decline to 1.5720 by 2017 suggesting temporary restoration. Subsequent rebound to 2.4304 by 2024 (CAGR 5.2%) signaled renewed fragmentation pressures, revealing an emerging fragmentation-restoration cycle despite values remaining below the 2013 peak. Fractal dimension (FRAC) corroborated this trend through continuous 2013–2020 decline associated with tourism-driven edge hardening and hydrological changes promoting patch regularization. FRAC's 2024 recovery to 1.2661 demonstrates successful wetland restoration via cropland reconversion and invasive species removal. Future climate pressures and intensified development may exceed stability thresholds, requiring adaptive zoning and enhanced monitoring.

Spatiotemporal Dynamics of Shape Complexity Tengchong Beihai Wetland's landscape shape index (LSI) showed phased fluctuations during 2009–2024, with an initial rise followed by decline corresponding to infrastructure-induced patch regularization. Edge density (ED) increased nonlinearly from 49.17 to 82.29 $m \cdot hm^{-2}$ despite a mid-period decline, peaking at 9.42 in 2024. This maximum reflects natural wetland recovery or complex new patch boundaries, confirming nonlinear edge habitat expansion. The ED resurgence intensifies edge effects, potentially heightening biological invasions and nutrient loss that threaten ecosystem integrity.

Bimodal Evolution of Landscape Heterogeneity Shannon's diversity index (SHDI) followed a "rise-dip-rise" trajectory, peaking at 1.2813 (2017) before transient 2020 decline and 2024 rebound. Policy-aligned fluctuations included: cropland-to-forest conversion boosting diversity (2013–2017); ecological land conversion under tightened construction quotas (Ministry of Natural Resources Doc. 368, 2017) in 2020; and post-2020 ecological compensation improving configuration. Despite 2024 recovery (SHDI=0.7892), values remained below baseline, indicating persistent degradation. Moran's I revealed significant clustering ($p<0.05$) for PD, LSI, and ED, demonstrating anthropogenically concentrated fragmentation-complexity patterns.

5. Conclusions

Beihai Wetland exhibited nonlinear dynamics during 2009–2024 with sequential contraction, expansion, and reduction. Ecosystem degradation manifested multifaceted attributes including meadow contraction and intensified forest fragmentation at the matrix level, alongside increased landscape fragmentation indices that diminished morphological complexity, amplified biodiversity volatility, and generated complex patch geometries—accelerating fragmentation evolution. As China's critical southwestern ecological buffer, these changes emerged from synergistic natural-anthropogenic pressures.

Based on the landscape ecological degradation characteristics of Beihai Wetland in Tengchong, the following systematic conservation strategies are proposed:(1) Precision zoning management: Designate ecological restoration priority zones in core habitats of Brasenia schreberi and Grus nigricollis, strictly prohibiting new tourism infrastructure; (2) Intelligent monitoring system: Construct an LSTM (Long Short-Term Memory)-based ecological security early-warning system for real-time monitoring of hydrology, vegetation, and wildlife dynamics; employ UAV multispectral remote sensing to quantitatively assess vegetation restoration effectiveness; (3) Rigorous policy constraints: Revise the implementation rules of the Beihai Wetland Protection Regulations to establish a "development-compensation" accountability mechanism and restrict tourism intensity in degraded meadows; (4) Ecotourism transformation: Design "floating-mat wetland–volcanic geology–birdwatching education" ecological corridors, implementing volunteer supervision and strict penalties for destructive activities; (5) Long-term scientific support: Collaborate with research institutions to establish wetland ecological observation stations for landscape pattern simulation and ecosystem service function evaluation. As a critical undertaking integrating ecological public welfare and environmental conservation, wetland protection requires coordinated efforts in: Institutional optimization (clarifying departmental responsibilities); Technological innovation (LSTM early-warning +UAV assessment); Societal participation (community co-governance + public education). This integrated approach is essential for achieving ecological security and fostering human-nature harmony.

Acknowledgment

This paper is the research outcome of the Student Scientific Research Project (XDKY24011) at Chuxiong Normal University, as well as the Construction Project of Yunnan Province First-rate Major in Geographical Science.

References

1. Aquino S D, Pizarro G G, Rescia J A, et al. Wetland responses to non-stationary hydro-climatic dynamics in the context of land cover and land use change[J]. Remote Sensing Applications: Society and Environment,2024,34:101156.
2. Dangles, O., Rabatel, A., Kraemer, M., Zeballos, G., Soruco, A., Jacobsen, D., & Anthelme, F. (2017). Ecosystem sentinels for climate change? Evidence of wetland cover changes over the last 30 years in the tropical Andes. PLoS ONE, 12(5), e0175814.
3. Kharazmi, R., Tavili, A., Rahdari, M. R., Chaban, L., Panidi, E., & Rodrigo-Comino, J. (2018). Monitoring and assessment of seasonal land cover changes using remote sensing: A 30-year (1987–2016) case study of Hamoun Wetland, Iran. Environmental Monitoring and Assessment, 190(6), 356.
4. QIN, F. Y., DUAN, X. J., ZOU, H., et al. Distribution patterns and influencing factors of mid-channel bar wetlands in the Yangtze River mainstream. Resources and Environment in the Yangtze Basin, 2024, 33(1), 66-77.
5. MENG, Y. Y., GAO, Y., LI, X. M., et al. Evolution and prediction of wetland landscape pattern and ecosystem service value in Henan section of Yellow River Basin. Areal Research and Development, 2023, 42(6), 150-156.
6. REN, H. X., REN, C. Y., WANG, Z. M., et al. Dynamics and influencing factors of wetland area and distribution in Changchun since 1990. Wetland Science, 2023, 21(4), 575-582.
7. CHEN, H. Y., ZHENG, Q. S., LI, X. Y., et al. Dynamic changes of wetland landscape pattern in Qinghai Lake Basin over past two decades. China Science and Technology Information, 2022, (5), 84-90.
8. GAO, X. X., HAN, R. C., FAN, H. T., et al. Paleoclimate evolution over past 2050 years recorded in sediments of Beihai Wetland, Yunnan and its impact on regional human civilization. Acta Sedimentologica Sinica, 2020, 38(5), 1016-1025.
9. SHEN, L. X., LIANG, L. H. Biodiversity and environmental status of Beihai Wetland Reserve, Tengchong. Journal of Northeast Forestry University, 2005, 33(5), 100-102.
10. LIU, X., LI, Y., ZHANG, Q. Analysis of land use transition patterns using a transition matrix approach. Journal of Environmental Management, 2020, 265, 110520.
11. HE, P., ZHANG, H. R. Factor analysis and screening method for common landscape metrics. Forest Research, 2009, 22(4), 470-474.

AUTHOR INDEX